SQUEEZED AND CORRELATED STATES OF QUANTUM SYSTEMS

Proceedings of the Lebedev Physics Institute
Academy of Sciences of Russia

Volume 165	**Solitons and Instantons, Operator Quantization,** Edited by V.L. Ginzburg
Volume 166	**The Nonlinear Optics of Semiconductor Lasers,** Edited by N.G. Basov
Volume 167	**Group Theory, Gravitation and Elementary Particle Physics,** Edited by A.A. Komar
Volume 168	**Issues in Intense-Field Quantum Electrodynamics,** Edited by V.L. Ginzburg
Volume 169	**The Physical Effects in the Gravitational Field Black Holes,** Edited by M.A. Markov
Volume 170	**The Theory of Target Compression by Longwave Laser Emission,** Edited by G.V. Sklizkov
Volume 171	**Research on Laser Theory,** Edited by A.N. Orayevskiy
Volume 172	**Phase Conjugation of Laser Emission,** Edited by N.G. Basov
Volume 173	**Quantum Mechanics and Statistical Methods,** Edited by M.M. Sushchinskiy
Volume 174	**Nonequilibrium Superconductivity,** Edited by V.L.Ginzburg
Volume 174 Supplemental Volume 1	**Thermodynamics and Electrodynamics of Superconductors,** Edited by V.L. Ginzburg
Volume 174 Supplemental Volume 2	**Electron Liquid Theory of Normal Metals,** Edited by V.P. Silin
Volume 175	**Luminescence Centers of Rare Earth Ions in Crystal Phosphors,** Edited by M.D. Galanin
Volume 175 Supplemental Volume	**Luminescence and Anisotropy of Zinc Sulfide Crystals,** Edited by M.D. Galanin
Volume 176	**Classical and Quantum Effects in Electrodynamics,** Edited by A.A. Komar

Volume 176 Supplemental Volume — **Quantization, Gravitation and Group Methods in Physics,** Edited by A.A. Komar

Volume 177 — **Stoichiometry in Crystal Compounds and Its Influence on Their Physical Properties,** Edited by Yu. M. Popov

Volume 178 — **The Delfin Laser Thermonuclear Installation: Operational Complex and Future Directions,** Edited by G.V. Sklizkov

Volume 179 — **X-Ray Plasma Spectroscopy and the Properties of Multiply-Charged Ions,** Edited by I.I. Sobel'man

Volume 179 Supplemental Volume 1 — **Interaction of Ultrashort Pulses with Matter,** Edited by M.D. Galanin

Volume 179 Supplemental Volume 2 — **Electron-Excited Molecules in Nonequilibrium Plasma,** Edited by N.N. Sobolev

Volume 180 — **Inelastic Light Scattering in Crystals,** Edited by M.M. Sushchinskiy

Volume 181 — **Pulsed Metal and Metal Halogenide Vapor Lasers,** Edited by G.G. Petrash

Volume 182 — **Luminescence of Wideband Semiconductors,** Edited by M.D. Galanin

Volume 183 — **Invariants and the Evolution of Nonstationary Quantum Systems,** Edited by M.A. Markov

Volume 184 — **Electron Processes in MIS-Structure Memories,** Edited by A.F. Plotnikov

Volume 185 — **Injection Lasers in Optical Communication and Information Processing Systems,** Edited by Yu. M. Popov

Volume 186 — **Photoproduction of Pions on Nucleons and Nuclei,** Edited by A.A. Komar

Volume 187 — **Quantum Field Theory, Quantum Mechanics and Quantum Optics: Pt. 1. Symmetries and Algebraic Structures in Physics,** Edited by V.V. Dodonov and V.I. Man'ko

Volume 188 — **Integral Systems, Solid State Physics and Theory of Phase Transitions. Pt. 2. Symmetries and Algebraic Structures in Physics,** Edited by V.V. Dodonov and V.I. Man'ko

Volume 189	**Clusters of Galaxies and Extragalactic Radio Sources,** Edited by A.D. Kuz'min
Volume 190	**Metal-Optics and Superconductivity,** Edited by A.I. Golovashkin
Volume 190 Supplemental Volume	**Superconductors with A15 Lattice and Bridge Contacts Based Upon Them,** Edited by M. M. Sushchinskiy
Volume 191	**Polarization of the Vacuum and a Quantum Relativistic Gas in an External Field,** A.Ye. Shabad
Volume 192	**Atomic and Ionic Spectra and Elementary Processes in Plasma,** Edited by I.I. Sobelman
Volume 193	**Research on Chemical Lasers,** Edited by A.N. Orayevskiy
Volume 194	**Biophysical Approach to Complex Biological Phenomena,** Edited by E. Volkov
Volume 195	**Physical Characteristics and Critical Temperature of High Temperature Superconductors,** Edited by M.M. Sushchinskiy
Volume 196	**Pulsars,** Edited by A.D. Kuzmin
Volume 197	**Reliability Problems of Semiconductor Lasers,** P.G. Eliseev
Volume 198	**Theory of Nonstationary Quantum Oscillators,** Edited by M.A. Markov
Volume 201	**Generation of Nonlinear Waves and Quasistationary Currents in Plasmas,** Edited by L.M. Kovrizhnykh
Volume 202	**Gas Dynamics,** Edited by Yu.I. Koptev
Volume 203	**Fiber Optics: Research and Development,** Edited by Ye.M. Dianov
Volume 204	**Sakharov Memorial Lectures in Physics,** Edited by L.V. Keldysh and V.Ya. Fainberg
Volume 205	**Squeezed and Correlated States of Quantum Systems** Edited by M.A. Markov
Volume 206	**Mathematical Physics, Applied Mathematics and Informatics,** Edited by Yu.I. Koptev
Volume 207	**Generalized Mayer Series in Classical Statistical Mechanics** Igor J. Ivanchik

Volume 208 **Physical Properties of High Temperature Superconductors,** Edited by Yu.I. Koptev

Volume 210 **Theory and Technology of High Temperature Superconductors,** Edited by Yu.I. Koptev

Volume 216 **Semiconductors and Insulators: Optical and Spectroscopic Research,** Edited by Yu.I. Koptev

Proceedings of the Lebedev Physics Institute
Russian Academy of Sciences

Series Editor N.G. Basov

Volume 205

SQUEEZED AND CORRELATED STATES OF QUANTUM SYSTEMS

Edited by M.A. Markov

Translators: V.V. Dodonov, O.V. Man'ko and V.I. Man'ko

NOVA SCIENCE PUBLISHERS, INC.

Deputy Series Editor: M.A. Man'ko

Art Director: Christopher Concannon
Graphics: Elenor Kallberg and Maria Ester Hawrys
Book Production: Michael Lyons, June Martino,
Tammy Sauter, and Michelle Lalo

Library of Congress Cataloging-in-Publication Data
available upon request

ISBN 1-56072-117-0

6080 Jericho Turnpike, Suite 207
Commack, New York 11725
Tele. 516-499-3103 Fax 516-499-3146
E Mail Novasci1@aol.com

Printed in the United States of America

CONTENTS

Group Quantum Systems in Coherent and Correlated States Representation
N.A. Gromov, V.I. Man'ko

Physical Effects in Correlated Quantum States
V.V. Dodonov, A.B. Klimov, V.I. Man'ko

Squeezed and Correlated Light Sources
S.M. Chumakov, M. Koserovsky, A.A. Mamedov, V.I. Man'ko

Correlated States of String and Gravitational Waveguide
V.V. Dodonov, V.I. Man'ko, O.V. Man'ko

GROUP QUANTUM SYSTEMS IN COHERENT AND CORRELATED STATES REPRESENTATION

N.A. Gromov, V.I. Man'ko

Readership: Quantum Physics, Theoretical Physics

Group quantum systems with stationary Hamiltonians which are linear functions of group generators realized by means of creation and annihilation operators of both bosonic and fermionic types are investigated. Orthogonal Cayley-Klein groups which can be obtained from the special orthogonal n-dimensional group by means of multidimensional contractions and analytic continuations in the space of group parameters are considered. The propagators (matrix elements of secondly quantized operators of finite group transformations) of systems under discussion are found for groups with dimensions less than five in the coherent and correlated states representations.

GROUP QUANTUM SYSTEMS IN CORRELATED AND COHERENT STATE REPRESENTATIONS

1. INTRODUCTION

Multidimensional quantum systems with Hamiltonians which are inhomogeneous quadratic forms (with time-dependent coefficients in general case) in Bose or Fermi creation and annihilation operators describe, at least in first approximation, the large variety of real phenomena in different branches of physics. The general quadratic systems include the so-called group quantum systems with Hamiltonians which are linear forms in Lie group generators expressed in terms of Bose or Fermi creation and annihilation operators. Schwinger [1] constructed the representations of the rotation group namely using this method of creation and annihilation operators for Bose-oscillators.

The effective method to describe such systems is the method of time-dependent quantum integrals of motion and coherent states [2] connected with integrals which was suggested and elaborated in detail in studies [3-7] (see also review article [8]). In addition to coherent states, the new types of states of quantum systems (squeezed states, correlated states, etc.) which are the generalizations of coherent states have been considered in scientific publications (see studies [9-12] and references therein). These new types of states have specific properties which are useful for solving concrete physical problems, at the same time, the partial cause of so-called continuous representation (generalized coherent state representation) introduced by Klauder [13-16] (see the monograph [17]).

Let us note that the dynamical symmetries of different quantum systems and of nonlinear equations have been studied in papers [18-21] where the notion of the symmetry of equation has also been discussed. This concept, considered in studies [22,23], was applied to such a system as the hydrogen atom which has the dynamical symmetry group O(4,2) that was found in these papers.

The suggestion to study a set of different physical theories on the basis of the given symmetry groups each connected with the others by the limit transitions was published in study [24]. One of the possible generalizations of the dynamical symmetry notion was proposed [12] for the set of physical systems with the different Hamiltonians. The appropriate method to study a set of different physical systems, at least on the basis of the group theory is the unified description of groups in the Cayley-Klein spaces [25-31], i.e. groups which may be obtained from the classical ones by the Inonu-Wigner contractions [26], as well as by the analytical continuations of group parameters.

Here we consider the stationary quantum systems with the Hamiltonians which are linear forms in the secondary quantized generators of the special orthogonal groups in Cayley-Klein spaces [27-29] (expressed in terms of the creation and annihilation operators). These groups are described in detail in Section 2. In Section 3 we consider the method of the "bosonization" and "fermionization", i.e. the expression of the generator matrix elements of groups under study in terms of the creation and annihilation operators either of the Bose-oscillators or of the Fermi-oscillators. The propagators of the stationary group systems (Green functions) in a coherent state basis are connected by the analytical continuations with the matrix elements of the secondary quantized operators of the finite group transformations in the same basis. Because of this, in Section 4 we describe the matrix elements of the operators of the finite rotations in Cayley-Klein spaces and obtain the explicit expressions for the matrix elements in cases of the groups of low dimensions. In the case of Bose-oscillators representation, we find the matrix elements for these groups in Fock basis and express these elements in terms of Hermite polynomials of several variables at the argument equal to zero. From the group multiplication rule we obtain the adding formulae for the Hermite polynomials.

In Section 5 we construct the transformations which connect the expressions for the linear integrals of motion and the propagators of the multidimensional quadratic systems in the coherent state basis with the linear integrals of motion and propagators of the same systems in the correlated coherent state representation. We demonstrate these transformations using the examples of systems corresponding to the one-parametric rotation group and one-parametric Galilean group.

2. SPECIAL ORTHOGONAL GROUPS IN CAYLEY-KLEIN SPACES

In geometry it is well known [36] that there are 3^n n-dimensional real spaces with a projective metric – the spaces of constant curvatures. The unified axiomatic description of these spaces is given in the study [37]. According to the fundamental idea of F. Klein each of these spaces is connected with the group of transformations – the group of motion of the space. The spaces of the constant curvatures are maximally homogeneous, they admit the group of the motion with the largest dimensionality $n(n+1)/2$, and these spaces are intensively used in physics. It is enough to remind the spaces of Euclide, Galilei, Lobachevski, Minkowski, De-Sitter, anti-De-Sitter, etc. So, in the study [38] it is shown that all the possible kinematics described in the study [39] may be modelled by the 4-dimensional spaces of the constant curvatures. Such a variety of the spaces of the constant curvatures and the corresponding groups of the motion creates some difficulties for concrete description of the spaces. The appropriate method of studying these groups (spaces) is the method of the transitions between the groups developed in papers [25-31] and giving the unified description of these groups. The main idea of the method of the transitions is the procedure of the transformation of the well known algebraic constructions (generators, Casimir operators, etc.) characterizing the orthogonal groups to the corresponding algebraic constructions which characterize the rotation groups in Cayley-Klein spaces.

We define the (n+1)-dimensional real Cayley-Klein space $\Re_{n+1}(j)$ as the (n+1)-dimensional vector space with the metric

$$\mathbf{x}^2(\mathrm{j}) = x_0^2 + \sum_{k=1}^{n} \left(\prod_{m=1}^{k} j_m^2 \right) x_k^2, \tag{2.1}$$

where x_0, x_k are the Cartesian coordinates and the set of parameters $j=(j_1, j_2, \ldots, j_n)$ determines the Cayley-Klein space. Each of the parameters j_k, $k=1,2,\ldots,n$, may take three values $j_k=1, \iota_k$, i. Here ι_k are purely dual units introduced by Clifford, each of them are not equal to zero $\iota_k \neq 0$; the product of the different purely dual units is not equal to zero too, $\iota_k \bullet \iota_m = \iota_m \bullet \iota_k \neq 0$, $k \neq m$, but the product of a dual unit multiplied by itself is equal to zero $\iota_k^2=0$. Division of a complex number by a dual unit is not defined, but the division of a dual unit by itself is equal to the real unit $\iota_k \bullet \iota_k^{-1}=1$ (but not $\iota_k \bullet \iota_m^{-1}$ or $\iota_m \bullet \iota_k^{-1}$, $k \neq m$, these contractions are not defined). These numbers may be considered as the different nonzero roots squared from zero $\iota_k=\sqrt{0}$. These numbers are described in the study [40]. The set of all the parameters j give all 3^n different Cayley-Klein spaces $\Re_{n+1}(j)$. It must be emphasized that usually the spaces which have the metric (2.1) with the identical signatures and its motion groups not distinguished, i.e. the space

$\Re_3(1,i)$ with the metric $x_o^2+x_1^2-x_2^2$ and the space $\Re_3(i,i)$ with the metric $x_o^2-x_1^2+x_2^2$ are the same. We have fixed the Cartesian coordinate axes in $\Re_{n+1}(j)$ labeling the axis by fixed numbers, and for us the spaces $\Re_3(1,i)$ and $\Re_3(i,i)$ are the different spaces.

The rotations of the Cayley-Klein space $\Re_{n+1}(j)$ form the group which we define as $SO_{n+1}(j)$. The spheres $S_n(j)$ of the radius R in the space $\Re_{n+1}(j)$ which are determined by the equation

$$S_n(j)=\left\{x \mid x^2(j)=x_0^2+\sum_{k=1}^{n}\left(\prod_{m=1}^{k} j_m^2\right)x_k^2=R^2\right\}, \tag{2.2}$$

are invariant with respect to the rotations from the group $SO_{n+1}(j)$. It is known [37] that the geometry of the n-dimensional spaces of constant curvature is realized on the spheres (2.2). The groups $SO_{n+1}(j)$ are isomorphic to the motion groups of the n-dimensional spaces of constant curvature.

The Cayley-Klein spaces $\Re_{n+1}(j)$ may be obtained from the Euclidean spaces $\Re_{n+1}$ the map

$$\psi:\ \Re_{n+1}\rightarrow\Re_{n+1}(j),$$
$$\psi x_0=x_0,\qquad \psi x_k=x_k\prod_{m=1}^{k} j_m,\qquad k=1,2,\dots,n. \tag{2.3}$$

The map induces the transformation of the rotation group SO_{n+1} of the Euclidean space (the special orthogonal group) into the group $SO_{n+1}(j)$. The matrix generators $X_{\mu\upsilon}$ of the rotations in two-dimensional planes (x_μ, x_ν), $\mu=0,1,\dots, n-1$, $\nu=1,2,\dots,n$, $\mu<\nu$, of the Euclidean space $\Re_{n+1}$ are transformed into the generators $X_{\mu\nu}$ of the rotations in two-dimensional planes (x_μ, x_ν) of the space $\Re_{n+1}(j)$ according to [25-27]

$$X_{\mu\nu}(j)=\left(\prod_{m=\mu+1}^{\nu} j_m\right)\tilde{X}_{\mu\nu}(\rightarrow), \tag{2.4}$$

where by the symbol $\tilde{X}_{\mu\nu}(\rightarrow)$ we denote the transformed generator $\tilde{X}_{\mu\nu}$ with the nonzero matrix elements $(\tilde{X}_{\mu\nu}(\rightarrow))_{\mu\nu}=(\prod_{m=\mu+1}^{\nu} j_m)(\tilde{X}_{\mu\nu})_{\mu\nu}$, $(\tilde{X}_{\mu\nu}(\rightarrow))_{\nu\mu}=(\prod_{m=\mu+1}^{\nu} j_m^{-1})(\tilde{X}_{\mu\nu})_{\nu\mu}$. Taking into account that $(\tilde{X}_{\mu\nu})_{\mu\nu}=-1$, and $(\tilde{X}_{\mu\nu})_{\nu\mu}=1$ we obtain the following expressions for the nonzero matrix elements of the generator $X_{\mu\nu}(j)$

$$(X_{\mu\nu}(j))_{\mu\nu}=-\prod_{m=\mu+1}^{\nu} j_m^2,\qquad (X_{\mu\nu}(j))_{\nu\mu}=1,\qquad \mu<\nu. \tag{2.5}$$

If one calculates using the formula (2.4), the generators $\tilde{X}_{\mu\nu}$ in terms of the generators $X_{\mu\nu}(j)$, and then applies these generators to the well known commutation relations of the orthogonal group SO_{n+1}, the commutators of the group $SO_{n+1}(j)$ may be found

$$[X_{\mu_1\nu_1},X_{\mu_2\nu_2}]=\begin{cases}\left(\prod_{m=1+\mu_1}^{\nu_1} j_m^2\right)X_{\nu_1\nu_2}, & \mu_1=\mu_2,\ \nu_1<\nu_2,\\ \left(\prod_{m=1+\mu_2}^{\nu_2} j_m^2\right)X_{\mu_1\mu_2}, & \mu_1<\mu_2,\ \nu_1=\nu_2,\\ -X_{\mu_1\nu_2}, & \mu_1<\nu_1=\mu_2<\nu_2.\end{cases} \tag{2.6}$$

The map (2.3) not only induces the transformation (2.4) of the generators of the special orthogonal SO_{n+1} into the generators of the group $SO_{n+1}(j)$ of the rotations in the Cayley-Klein spaces, but it also induces the transformation of all the other algebraic constructions characterizing the group SO_{n+1}. Let us consider the Casimir operators, i.e. the basic operators of the center in the universal enveloping algebra. The Casimir operators of the group SO_{n+1} are well known [41,42]. In the case of the even number n=2m these operators are the operators $\tilde{C}'_{2p}$, p=1,2,..., m, the Casimir operators $\tilde{C}'_{2p}$ being defined as the sum of the main minors of the 2p-th order of the antisymmetric matrix constructed from the generators $\tilde{X}_{\alpha\beta}$, $\alpha,\beta=0,1,2,\ldots, n$, $\alpha\neq\beta$. Instead of the operators $\tilde{C}'_{2p}$ we can consider the set of the operator C_{2p} determined by the expression

$$C_{2p}(1) = \sum_{\alpha_1,\ldots,\alpha_{2p}=0}^{n} \tilde{X}_{\alpha_1\alpha_2}\tilde{X}_{\alpha_2\alpha_3}\ldots\tilde{X}_{\alpha_{2p}\alpha_1}. \tag{2.7}$$

In the case of the odd number n=2m+1 to the operators $\tilde{C}'_{2p}$ or $\tilde{C}_{2p}$ the following operator $\tilde{C}^{1/2}$ must be added:

$$C^{1/2}(1) = \sum_{\alpha_1,\ldots,\alpha_{n+1}=0}^{n} \varepsilon_{\alpha_1\ldots\alpha_{n+1}}\tilde{X}_{\alpha_1\alpha_2}\tilde{X}_{\alpha_3\alpha_4}\ldots\tilde{X}_{\alpha_n\alpha_{n+1}}, \tag{2.8}$$

where $\varepsilon_{\alpha_1\ldots\alpha_{n+1}}$ is the antisymmetric unit tensor. In the studies [27 and 43] it was shown that for any set of the parameters j the rotation group $SO_{n+1}(j)$ has exactly $[(n+1)/2]$ of the independent Casimir operators (here [x] implies the integer part of the number x), and all these operators may be obtained from the Casimir operators of the orthogonal group by the following transformations induced by the map (2.3):

$$C'_{2p}(j) = \prod_{k=1}^{p-1}(j_k j_{n-k+1})^{2k}\prod_{l=p}^{n-p+1} j_l^{2p}\cdot C'_{2p}(\rightarrow), \tag{2.9}$$

$$C_{2p}(j) = \prod_{k=1}^{n} j_k^{2p}\cdot C_{2p}(\rightarrow), \tag{2.10}$$

$$C^{1/2}(j) = j_{(n+1)/2}^{(n+1)/2}\prod_{k=1}^{(n-1)/2} j_k^k j_{n-k+1}^k\cdot C^{1/2}(\rightarrow), \tag{2.11}$$

where the symbol $\tilde{C}(\rightarrow)$ means that the generators $\tilde{X}_{\mu\nu}$ which determine the Casimir operator $\tilde{C}$ of the group SO_{n+1} are replaced by the expressions $(\prod_{m=\mu+1}^{\nu} j_m^{-1})X_{\mu\nu}$ according to the relation (2.4).

The matrix generators $X_{\mu\nu}(j)$ described by the formulae (2.5) form the basis of the Lie algebra $so_{n+1}(j)$ of the rotation group $SO_{n+1}(j)$. To the general element

$$X(\mathbf{r}, j) = \sum_{\lambda=1}^{n(n+1)/2} r_\lambda X_\lambda(j) \tag{2.12}$$

of the algebra $so_{n+1}(j)$ corresponds the finite group transformation (finite rotation) of the group $SO_{n+1}(j)$ in the Cayley-Klein space determined by means of the exponential map

$$\Xi\,(\mathbf{r},\ \mathbf{j}) = \exp X\,(\mathbf{r},\ \mathbf{j}). \tag{2.13}$$

In the formulae (2.12) and (2.13) we have introduced the label λ which is in one-to-one correspondence with the pair of indices μ, ν, $\mu<\nu$ due to the relation

$$\lambda = \nu + \mu\,(n-1) - \mu\,(\mu-1)/2 \tag{2.14}$$

and it takes the values from 1 to $n(n+1)/2$. Due to Cayley-Hamiltonian theorem [44] the matrix Ξ is expressed algebraically by the matrices $X^m(r,j)$, $m=0,1,\ldots,n$. But the explicit expressions for the matrix (2.13) in the cases $n\geq 4$ are complicated. We will restrict consideration by the detailed study of the groups $SO_2(j_1)$, $SO_3(j_1,j_2)$ and $SO_4(j_1, j_2, j_3)$.

In the case of the group $SO_2(j_1)$ the map (2.3)

$$\psi:\ \mathfrak{R}_2 \to \mathfrak{R}_2\,(j_1),$$
$$\psi x_0 = x_0, \qquad \psi x_1 = j_1 x_1, \tag{2.15}$$

transforms the Euclidean plane $\mathfrak{R}_2$ with the metric $x^2=x^2_0+x^2_1$ into the plane $\mathfrak{R}_2(j_1)$ with the metric

$$\mathbf{x}^2\,(j_1) = x_0^2 + j_1^2 x_1^2. \tag{2.16}$$

Due to the transformation (2.4) the matrix generator $X_{01} = \begin{Vmatrix} 0 & -1 \\ 1 & -0 \end{Vmatrix}$ of the one-dimensional rotation group $SO_2\equiv SO_2(1)$ transforms into the generator $X_{01}(j_1)= \begin{Vmatrix} 0 & -j_1 \\ 1 & 0 \end{Vmatrix}^2$ of the group $SO_2(j_1)$. The matrix of the finite rotation may be easily found using the formula (2.13), and it has the form

$$\Xi\,(r_1, j_1) = \begin{Vmatrix} \cos j_1 r_1 & -j_1 \sin j_1 r_1 \\ j_1^{-1} \sin j_1 r_1 & \cos j_1 r_1 \end{Vmatrix} \tag{2.17}$$

For j_1-1 transformations $x'=\Xi(r_1,j_2)x$, $x\in\mathfrak{R}_2(j_1)$ are the usual (trigonometrical) rotations of the Euclidean plane $\wp_2=\mathfrak{R}_2(1)$; for $j_1=i$ they are the hyperbolic rotations (or the one-dimensional) Lorentz transformations, if one considers the axis x_o as the time axis, and x_1 as the space one) of the Minkowski plane $\mathfrak{R}_2(i)$; and, at last, for $j_1=\iota_1$ these transformations are the rotations of the Galilean plane $\mathfrak{R}_2(\iota_1)$, i.e. the one-dimensional Galilean transformations.

The replacement of the real parameter $j_1=1$ in the relation (2.17) by the pure imaginary number $j_1=i$, i.e. the analytical continuation of the group parameter r_1 belonging to the real numbers into the pure imaginary number, transforms the rotation group $SO_2(1)$ into the Lorentz group $\Lambda=SO_2(i)$. The replacement of the real or imaginary parameters j_1 by the pure dual unit $j_1=\iota_1$ transforms the rotation group $SO_2(1)$ or Lorentz group $SO_2(i)$ into the Galilean group $G=SO_2(\iota_1)$ and it corresponds to contraction (squeezing, limit process) of the rotation group or the Lorentz group into the Galilean group, i.e. to the procedure which has been introduced by Wigner and Inonu in study [32]. As a result, the real group parameter $r_1 \in R$ or the imaginary one ir_1 become the pure dual parameter $\iota_1 r_1$. From this point of view the group contraction is the continuation of group parameters into the field of the purely dual numbers.

The rotation in the two dimensional plane (x_μ, x_ν), $\mu<\nu$ of the Cayley-Klein space $\mathfrak{R}_{n+1}(j)$ has the character of the trigonometrical, hyperbolic or parabolic (Galilean transformation) rotation. It corresponds to the real, imaginary or pure dual values of the product of the parameters $\prod_{m=\mu+1}^{\nu} j_m$. For $j_1=1$ the group parameter takes the values from the interval $[0,2\pi]$, for $j_1=i$, ι_1 the group parameter takes all the values from the real axis R. The group parameter r_λ which corresponds to the rotation in the plane (x_μ, x_ν), $\mu<\nu$, where λ is related to the numbers μ and ν by the formula (2.14), takes the values from the interval $[0,2\pi]$ if the product $\prod_{m=\mu+1}^{\nu} j_m$ is the real number, and $r_\lambda \in R$ if this product is imaginary or a pure dual number.

In the case of the group $SO_3(j)$, $j=(j_1, j_2)$ the map (2.3) has the form

$$\psi : \mathfrak{R}_3 \to \mathfrak{R}_3(j),$$
$$\psi x_0 = x_0, \quad \psi x_1 = j_1 x_1, \quad \psi x_2 = j_1 j_2 x_2, \tag{2.18}$$

and it transforms the three-dimensional Euclidean space $\mathfrak{R}_3$ into the Cayley-Klein space $\mathfrak{R}_3(j)$ with the metric

$$\mathbf{x}^2(j) = x_0^2 + j_1^2 x_1^2 + j_1^2 j_2^2 x_2^2. \tag{2.19}$$

It is known that the geometry of the two-dimensional planes with constant curvature may be realized on the spheres $S_2(j)=(x \mid x^2(j)=x_0^2+j_1^2x_1^2+j_1^2j_2^2x_2^2=R^2)$, with the real radius R. The parameters $j_1=1$, ι_1, i and $j_2=1$, ι_2, i give nine possible combinations. All these nine planes of constant curvature are demonstrated in the table where curvature K is also shown. For the name of the group $SO_3(j)$ we shall use the names of the corresponding planes.

Table. Planes of constant curvature

j_2	$j_1=1$	$j_1=\iota_1$	$j_1=\iota$
1	Spherical K>0 $SO_3(1,1)\equiv SO_3$	Euclidean K=0 $SO_3(2,1)\equiv E(2)$	Lobachevski K<0 $SO_3(i,1)\int H(2)$
ι_2	Semispherical K>0 $SO_3(1,\iota_2)$	Galilean K=0 $SO_3(\iota_1, \iota_2)\equiv G(2)$	Semihyperbolic K<0 $SO_3(i,\iota_2)$
i	Anti de Sitter K>0 $SO_3(1,i)\simeq SO(2,1)$	Minkowski K=0 $SO_3(\iota 1,i)\equiv P(2)$	de Sitter K<0 $SO_3(i,i)\simeq SO(2,1)$

As the independent generators of the group $SO_3(\mathbf{j})$, we take the generators of the rotations in the two-dimensional planes (x_0, x_1), (x_0, x_2), (x_1, x_2) of the space $\Re_3(\mathbf{j})$. The matrix representation for the generators may be found from the matrix generators X_{01}, X_{02}, X_{12} of the rotation group SO_3 using the formula (2.4) which in this case gives

$$X_1(\mathbf{j}) = X_{01}(\mathbf{j}) = j_1 \mathcal{X}_{01}(\to), \qquad X_2(\mathbf{j}) = X_{02}(\mathbf{j}) = j_1 j_2 \mathcal{X}_{02}(\to),$$
$$X_3(\mathbf{j}) = X_{12}(\mathbf{j}) = j_2 \mathcal{X}_{12}(\to) \tag{2.20}$$

and in accordance with equation (2.5) they have the form

$$X_1(\mathbf{j}) = \begin{Vmatrix} 0 & -j_1^2 & 0 \\ 1 & 0 & 0 \\ 0 & 0 & 0 \end{Vmatrix}, \quad X_2(\mathbf{j}) = \begin{Vmatrix} 0 & 0 & -j_1^2 j_2^2 \\ 0 & 0 & 0 \\ 1 & 0 & 0 \end{Vmatrix}, \quad X_3(\mathbf{j}) = \begin{Vmatrix} 0 & 0 & 0 \\ 0 & 0 & -j_2^2 \\ 0 & 1 & 0 \end{Vmatrix} \tag{2.21}$$

Here the index λ=1,2,3 is connected with the indices μ, ν by the formula (2.14) for n=2. The commutation relations of the rotation group SO_3

$$[\mathcal{X}_{01}, \mathcal{X}_{02}] = \mathcal{X}_{12}, \qquad [\mathcal{X}_{02}, \mathcal{X}_{12}] = \mathcal{X}_{01}, \qquad [\mathcal{X}_{01}, \mathcal{X}_{12}] = -\mathcal{X}_{02} \tag{2.22}$$

transform into the commutation relations of the group $SO_3(\mathbf{j})$: in fact we find from the formula (2.20) the generators $x_{01} = j_1^{-1} X_1, X_{02} = j_1^{-1} j_2^{-1} X_2, X_{12} = j_2^{-1} X_3$. Then these generators are used in the formulae (2.22). The first commutator in the formulae (2.22) is multiplied by the number $j_1^2 j_2$, the second one by the number $j_1 j_2^2$ and the third one by the number $j_1 j_2$. As a final result we have the commutator relations

$$[X_1, X_2] = j_1^2 X_3, \qquad [X_2, X_3] = j_2^2 X_1, \qquad [X_1, X_3] = -X_2. \tag{2.23}$$

The only Casimir operator of the second order for the rotation group has the form

$$C_2 = \mathcal{X}_{01}^2 + \mathcal{X}_{02}^2 + \mathcal{X}_{12}^2. \tag{2.24}$$

From this expression we derive the Casimir operator $C_2(\mathbf{j})$ using the formulae (2.10)

$$C_2(\mathbf{j}) = j_1^2 j_2^2 (j_1^{-2} X_1^2 + j_1^{-2} j_2^{-2} X_2^2 + j_2^{-2} X_3^2) = j_2^2 X_1^2 + X_2^2 + j_1^2 X_3^2. \tag{2.25}$$

For the special values of the parameters j_1 and j_2 the operator (2.25) coincides with the known Casimir operators of the Galilean group, Poincare group, de Sitter group, etc., (see study [27] and reference therein). According to the formula (2.13) the finite rotation for the group $SO_3(\mathbf{j})$ has the form

$$\Xi(\mathbf{r}, \mathbf{j}) = E \cos r + X(\mathbf{r}, \mathbf{j}) \frac{\sin r}{r} + X'(\mathbf{r}, \mathbf{j}) \frac{1 - \cos r}{r^2}, \tag{2.26}$$

where E is the three-dimensional unity matrix, and

$$X(\mathbf{r}, \mathbf{j}) = \sum_{\lambda=1}^{3} r_\lambda X_\lambda = \begin{Vmatrix} 0 & -j_1^2 r_1 & -j_1^2 j_2^2 r_2 \\ r_1 & 0 & -j_2^2 r_3 \\ r_2 & r_3 & 0 \end{Vmatrix}, \tag{2.27}$$

$$X'(\mathbf{r}, \mathbf{j}) = \begin{Vmatrix} j_2^2 r_3^2 & -j_1^2 j_2^2 r_2 r_3 & j_1^2 j_2^2 r_1 r_3 \\ -j_2^2 r_2 r_3 & j_1^2 j_2^2 r_2^2 & -j_1^2 j_2^2 r_1 r_2 \\ r_1 r_3 & -j_1^2 r_1 r_2 & j_1^2 r_1^2 \end{Vmatrix}, \tag{2.28}$$

$$r^2 = j_1^2 r_1^2 + j_1^2 j_2^2 r_2^2 + j_2^2 r_3^2. \tag{2.29}$$

It is interesting to note that all these nine groups $SO_3(j)$ may be analogously parametrized to the rotation group parametrization by F.I. Fedorov [45] which turned out to be useful for some physical applications [46].

Let us consider the groups $SO_4(j)$, $j=(j_1, j_2, j_3)$ depending on six group parameters r_λ, $\lambda=1,2,\ldots,6$. The map (2.3) has the form

$$\psi: \Re_4 \to \Re_4(j),$$
$$\psi x_0 = x_0, \quad \psi x_1 = j_1 x_1, \quad \psi x_2 = j_1 j_2 x_2, \quad \psi x_3 = j_1 j_2 j_3 x_3. \tag{2.30}$$

It transforms the four-dimensional Euclidean space $\Re_4$ into the Cayley-Klein space $\Re_4(j)$ with metric

$$x^2(j) = x_0^2 + j_1^2 x_1^2 + j_1^2 j_2^2 x_2^2 + j_1^2 j_2^2 j_3^2 x_3^2. \tag{2.31}$$

The geometry of the three-dimensional homogeneous spaces of constant curvatures is realized on the spheres $S_3(j)=(x \mid x^2(j)=R^2)$ of the real radius R. The groups of rotations $SO_4(j)$ are isomorphic to the motion group of the mentioned spaces. The set of groups $SO_4(j)$ contains the groups used in physics which have the special names, for examples, $SO_4(\iota_1, 1,1)$–the motion group of the three-dimensional Euclidean space, $SO_4(i, 1,1)$–the motion group of three-dimensional Lobachewski space, $SO_4(\iota_1, i,1)$–the motion group of three-dimensional Minkowski space (three-dimensional Poincare group), $SO_4(\iota_1, \iota_2, 1)$–the motion group of three-dimensional Galilean space. It also contains other groups labeled only by the parameters j. These groups have no special names.

Six generators $X_{\mu\nu}(j)$ of the group $SO_4(j)$ have the following matrix form

$$X_1 = X_{01} = \begin{Vmatrix} 0 & -j_1^2 & 0 & 0 \\ 1 & 0 & 0 & 0 \\ 0 & 0 & 0 & 0 \\ 0 & 0 & 0 & 0 \end{Vmatrix}, \quad X_2 = X_{02} = \begin{Vmatrix} 0 & 0 & -j_1^2 j_2^2 & 0 \\ 0 & 0 & 0 & 0 \\ 1 & 0 & 0 & 0 \\ 0 & 0 & 0 & 0 \end{Vmatrix},$$

$$X_3 = X_{03} = \begin{Vmatrix} 0 & 0 & 0 & -j_1^2 j_2^2 j_3^2 \\ 0 & 0 & 0 & 0 \\ 0 & 0 & 0 & 0 \\ 1 & 0 & 0 & 0 \end{Vmatrix}, \quad X_4 = X_{12} = \begin{Vmatrix} 0 & 0 & 0 & 0 \\ 0 & 0 & -j_2^2 & 0 \\ 0 & 1 & 0 & 0 \\ 0 & 0 & 0 & 0 \end{Vmatrix},$$

$$X_5 = X_{13} = \begin{Vmatrix} 0 & 0 & 0 & 0 \\ 0 & 0 & 0 & -j_2^2 j_3^2 \\ 0 & 0 & 0 & 0 \\ 0 & 1 & 0 & 0 \end{Vmatrix}, \quad X_6 = X_{23} = \begin{Vmatrix} 0 & 0 & 0 & 0 \\ 0 & 0 & 0 & 0 \\ 0 & 0 & 0 & -j_3^2 \\ 0 & 0 & 1 & 0 \end{Vmatrix}, \tag{2.32}$$

and their nonzero commutators are given by the formula (2.6). The Casimir operators of the group $SO_4(j)$ may be easily found by the transitions method from the well known invariants of the rotation group SO(4) being expressed in

$$C_2(j) = j_2^2 j_3^2 X_{01}^2 + j_3^2 X_{02}^2 + X_{03}^2 + j_1^2 (j_3^2 X_{12}^2 + X_{13}^2 + j_2^2 X_{23}^2), \tag{2.33}$$

$$C'^{/_2}(j) = j_2^2 X_{01} X_{23} + X_{03} X_{12} - X_{02} X_{13}. \tag{2.34}$$

To write down the general element X(r,j) of the algebra $SO_4(j)$ in an appropriate form, let us introduce new notations for the group parameters r_4, r_5, r_6, namely, $r_4=s_3$, $r_5=s_2$, $r_6=-s_1$; then we have

$$X(\mathbf{r}, \mathbf{s}, \mathbf{j}) = \sum_{k=1}^{3} r_k X_{0k}(\mathbf{j}) - \sum_{k,m,p=1}^{3} \varepsilon_{kmp} s_k X_{mp}(\mathbf{j}) = \left\| \begin{matrix} 0 & -j_1^2 r_1 & -j_1^2 j_2^2 r_2 & -j_1^2 j_2^2 j_3^2 r_3 \\ r_1 & 0 & j_2^2 s_3 & -j_2^2 j_3^2 s_2 \\ r_2 & -s_3 & 0 & j_3^2 s_1 \\ r_3 & s_2 & -s_1 & 0 \end{matrix} \right\|, \tag{2.35}$$

where ε_{kmp} is completely antisymmetric unit tensor and $\varepsilon_{123}=1$. By means of the formula (2.13) we obtain the matrix of the finite rotation

(2.36)

where E is the four-dimensional identity matrix and the matrix X is given by the formula (2.35). The matrices X_1 and X^2 have the form

$$X_1 = \left\| \begin{matrix} 0 & -j_1^2 j_3^2 s_1 & -j_1^2 j_2^2 j_3^2 s_2 & -j_1^2 j_2^2 j_3^2 s_3 \\ j_3^2 s_1 & 0 & j_1^2 j_2^2 j_3^2 r_3 & -j_1^2 j_2^2 j_3^2 r_2 \\ j_3^2 s_2 & -j_1^2 j_3^2 r_3 & 0 & j_1^2 j_3^2 r_1 \\ s_3 & j_1^2 r_2 & -j_1^2 r_1 & 0 \end{matrix} \right\|, \tag{2.37}$$

$$X^2 = \left\| \begin{matrix} -r^2 & j_1^2 j_2^2 (\mathbf{r}\times\mathbf{s})_1 & j_1^2 j_2^2 (\mathbf{r}\times\mathbf{s})_2 & j_1^2 j_2^2 j_3^2 (\mathbf{r}\times\mathbf{s})_3 \\ j_2^2 (\mathbf{r}\times\mathbf{s})_1 & j_3^2 s_1^2 - j_1^2 r_1^2 - s^2 & j_2^2 j_3^2 s_1 s_2 - j_1^2 j_2^2 r_1 r_2 & j_2^2 j_3^2 s_1 s_3 - j_1^2 j_2^2 j_3^2 r_1 r_3 \\ (\mathbf{r}\times\mathbf{s})_2 & j_3^2 s_1 s_2 - j_1^2 r_1 r_2 & j_2^2 j_3^2 s_2^2 - j_1^2 j_2^2 r_2^2 - s^2 & j_2^2 j_3^2 s_2 s_3 - j_1^2 j_2^2 j_3^2 r_2 r_3 \\ (\mathbf{r}\times\mathbf{s})_3 & s_1 s_3 - j_1^2 r_1 r_3 & j_2^2 s_2 s_3 - j_1^2 j_2^2 r_2 r_3 & j_2^2 s_3^2 - j_1^2 j_2^2 j_3^2 r_3^2 - s^2 \end{matrix} \right\|. \tag{2.38}$$

The functions A, B, C, D in the formula (2.36) are given by the relations

$$A = z_1 \cos\sqrt{-z_2} - z_2 \cos\sqrt{-z_1}, \qquad B = z_1 \frac{\sin\sqrt{-z_1}}{\sqrt{-z_1}} - z_2 \frac{\sin\sqrt{-z_2}}{\sqrt{-z_2}},$$

$$C = \frac{\sin\sqrt{-z_1}}{\sqrt{-z_1}} - \frac{\sin\sqrt{-z_2}}{\sqrt{-z_2}}, \qquad D = \cos\sqrt{-z_1} - \cos\sqrt{-z_2}, \tag{2.39}$$

where

$$z_{1,2} = -\tfrac{1}{2}\{r^2 + s^2 \mp [(r^2+s^2)^2 - 4j_1^2 j_3^2 (\mathbf{r},\mathbf{s})^2]^{1/2}\}, \qquad z_1 z_2 = j_1^2 j_3^2 (\mathbf{r},\mathbf{s})^2,$$
$$z_1 - z_2 = [(r^2+s^2)^2 - 4j_1^2 j_3^2 (\mathbf{r},\mathbf{s})^2]^{1/2}, \qquad z_1 + z_2 = -(r^2+s^2). \tag{2.40}$$

We used the notations

$$r^2 = j_1^2 r_1^2 + j_1^2 j_2^2 r_2^2 + j_1^2 j_2^2 j_3^2 r_3^2, \qquad s^2 = j_3^2 s_1^2 + j_2^2 j_3^2 s_2^2 + j_2^2 s_3^2,$$
$$(\mathbf{r},\mathbf{s}) = r_1 s_1 + j_2^2 (r_2 s_2 + r_3 s_3), \qquad (\mathbf{r}\times\mathbf{s})_1 = r_2 s_3 - j_3^2 r_3 s_2,$$
$$(\mathbf{r}\times\mathbf{s})_2 = j_3^2 r_3 s_1 - r_1 s_3, \qquad (\mathbf{r}\times\mathbf{s})_3 = r_1 s_2 - r_2 s_1. \tag{2.41}$$

In the study [45] it was shown that for the groups $SO_4(j)$ with $j_1 \neq \iota_1$ and $j_3 \neq \iota_3$ one can construct the analogs of the parametrization introduced by F.I. Fedorov [46] for the

Lorentz group. If $j_1=\iota_1$ or $j_3=\iota_3$, there does not exist the parametrization of the groups $SO_4(j)$ which have the composition rule for the group multiplication as simple as for the Fedorov parametrization of the Lorentz group.

We have defined the orthogonal groups in Cayley-Klein spaces and shown that their generators, Casimir operators and other algebraic constructions characterizing these groups can be obtained by the transformation of the corresponding constructions of the classical orthogonal groups. This approach is natural, and it is appropriate because the classical groups and the algebraic constructions connected with these groups are well studied. But the question arises whether such an approach is the only one? Maybe it is possible as an initial group takes one of the groups in the Cayley-Klein space. The positive answer to these questions is given by the following theorem about the structure of the transitions between the groups. First we introduce some definitions.

The Cayley-Klein space will be called a nonfiber bundle if no one parameter from the set $j_k, k=1,2,\ldots,n$ is pure dual unit. The space $\Re_{n+1}(j)$ will be called $(k_1, k_2,\ldots, k_p)$ fiber bundle if the parameters $j_{k_1}, j_{k_2},\ldots, j_{k_p}$ are equal to pure dual units, i.e., $j_{k_1}=\iota_{k_1},\ldots, j_{k_p}=\iota_{k_p}$, $1\le K_1<K_2<\ldots<K_p\le n$. This space is characterized by the existence of the sequence of the projections $pr_1, pr_2, \ldots, pr_p$, the subspace $(X_0, X_1,\ldots, X_{k-1})$ being the base and the fiber being the subspace $(X_{k_1}, X_{k_1+1},\ldots,X_n)$ for pr_1; the space $(X_{k_1}, X_{k_1},\ldots,X_{k_2-1})$ being the base and the fiber being the subspace $(X_{k_2}, X_{k_2+1},\ldots,X_n)$ for pr_2, etc. From the mathematical point of view, the Cayley-Klein space as the fiber bundle is a trivial one. Its global properties are the same as the local ones. From the physical point of view, the fiber structure leads to appearance of the absolute physical observables. For example, the Galilean space which is realized on the sphere $S_4(\iota_1, \iota_2, 1,1)$ is characterized by the absolute time $t=x_1$ and by the absolute space $R_3=(X_2, X_3,X_4)$.

For the spaces of constant curvature, which are realized on the spheres $S_n(j)$ described by the equations (2.2), the parameter j_1 determines the curvature: at $j_1=1$ the curvature K=const>0, at $j_1=\iota 1$ the curvature K=0, at $j_1=\iota$ the curvature K=const<0. The dual values of the other parameters $j_1,\ldots,j_n$ give the system of fibers in the spaces of the constant curvatures.

For groups in the Cayley-Klein spaces, the pure dual value of the parameter $j_1=\iota$ leads to so called inhomogeneous groups which are the semidirect product of the translation subgroup and homogeneous group: $SO_{n+1}(\iota_1, j_2,\ldots,j_n)=R_n x SO_n(j_2,\ldots,j_n)$. Inhomogeneous classical groups have been considered in the study [48].

Let us formally define the transition from the space $\Re_{n+1}(j)$, the generators $X_{\mu\nu}(j)$ and the Casimir operators $C'_{2p}(j)$, $C'_{2p}(j)$, $C^{1/2}(j)$ of the group $SO_{n+1}(j)$ to the space $\Re_{n+1}(j')$, the operators $X_{\mu\nu}(j')$ and the Casimir operators $C'_{2p}(j')$, $C_{2p}(j')$, $C^{1/2}(j')$ of the group $SO_{n+1}(j')$ by the formulae [26, 27]

$$\psi': \Re_{n+1}(\mathbf{j}) \rightarrow \Re_{n+1}(\mathbf{j}'),$$

$$\psi' x_0 = x_0, \quad \psi' x_k = x_k \prod_{m=1}^{k} j'_m j_m^{-1}, \quad k = 1, 2, \ldots, n, \tag{2.42}$$

$$X_{\mu\nu}(\mathbf{j}') = \left(\prod_{m=\mu+1}^{\nu} j'_m j_m^{-1} \right) X_{\mu\nu}(\mathbf{j}; \rightarrow), \tag{2.43}$$

$$C'_{2p}(\mathbf{j}') = \prod_{k=1}^{p-1} (j'_k j'_{n-k+1} j_k^{-1} j_{n-k+1}^{-1})^{2k} \prod_{l=p}^{n-l+1} (j'_l j_l^{-1})^{2p} \cdot C'_{2p}(\mathbf{j}; \rightarrow); \tag{2.44}$$

$$C_{2p}(\mathbf{j}') = \prod_{k=1}^{n} (j_k' j_k^{-1})^{2p} \cdot C_{2p}(\mathbf{j}; \to), \tag{2.45}$$

$$C^{1/2}(\mathbf{j}') = (j_{(n+1)/2}' j_{(n+1)/2}^{-1})^{(n+1)/2} \prod_{k=1}^{(n-1)/2} (j_k' j_k^{-1} j_{n-k+1}' j_{n-k+1}^{-1})^{k} \cdot C^{1/2}(\mathbf{j}; \to). \tag{2.46}$$

The inverse transitions may be obtained from the formulae (2.42)-(2.46) by the replacements of the parameters j by j' and vice versa.

The constructed transitions do not have the sense for any groups and spaces (formality of the transformations (2.42)-(2.46) is connected namely with this fact) since the division by the pure dual unit is not defined. But the ratio of two equal dual units is defined. The analysis of the possible transitions [25,26,30] leads to the following theorem.

The theorem on the structure of the transitions. 1. Let $SO_{n+1}(j)$ be the group in nonfiber Cayley-Klein space $\Re_{n+1}(j)$, and $SO_{n+1}(j')$ be the group in arbitrary Cayley-Klein space $\Re_{n+1}(j')$. Then the transformations (2.42)-(2.46) give the transition of the group $SO_{n+1}(j)$ into the group $SO_{n+1}(j')$. If $\Re_{n+1}(j')$ is a nonfiber Cayley-Klein space, the inverse transformation gives the transition of the group $SO_{n+1}(j')$ into the group $SO_{n+1}(j)$.

2. Let $SO_{n+1}(j)$ be the group in $(k_1, k_2,\ldots,k_p)$–fiber Cayley-Klein space $\Re_{n+1}(j)$, and $SO_{n+1}(j')$ is the group in $(r_1, r_2,\ldots,r_q)$–fiber Cayley-Klein space $\Re_{n+1}(j')$. Then the transformations (2.42)-(2.46) give the transition of the group $SO_{n+1}(j)$ into the group $SO_{n+1}(j')$ if the set of numbers $(k_1, k_2,\ldots,k_p)$ is contained in the set of the numbers $(r_1,r_2,\ldots,r_q)$. The inverse transformations give the transition of the group $SO_{n+1}(j')$ into the group $SO_{n+1}(j)$ if $p=q$ and $k_1=r_2,\ldots,k_p=r_q$.

It follows from the theorem on the structure of the transitions that the group $SO_{n+1}(j)$ may be obtained for any set of the parameters j not only from the classical rotation group $SO_{n+1}(1)$, but from any group $SO_{n+1}(j')$ of the rotations in arbitrary nonfiber Cayley-Klein space. Of course, it is easier to take the group $SO_{n+1}(1)$ as the initial group. We have used this method.

3. JORDAN-SCHWINGER REPRESENTATION

In the previous section we have considered the matrix representation of the special orthogonal groups in the Cayley-Klein spaces. The generators of these groups were realized by the matrices of definite dimensionality. Another method used for description of the representations is the second quantization method. In the frame of this approach, the generators of a group representation are given either in terms of the boson creation and annihilation operators ("bosonization" procedure) or in terms of the fermion creation and annihilation operators ("fermionization" procedure). The representations of the rotation group by means of the creation and annihilation operators of the bose-oscillators have been constructed by Schwinger [1]. Earlier in the study [49], Jordan considered the representation of the kinematic symmetries in terms of field operators either of Bose or Fermi kind. Since the Jordan representation and the Schwinger representation are equivalent this method is called Jordan-Schwinger representation [50]. Some properties of the Jordan-Schwinger representation for the Lie algebra of n-dimensional matrices are considered in the study [50].

Let us remember the basic statements of the second quantization method in the appropriate form. Let three N-dimensional matrices A, B, C have the matrix elements A_{ik}, B_{ik}, C_{ik}, i, k=1,2,..., N and obey the commutation relation

$$[A,\ B] = C. \tag{3.1}$$

If one constructs three operators A, B, C acting in the space of states of N-dimensional quantum oscillator using the following formulae

$$\hat{A} = \sum_{i,k} A_{ik} a_i^+ a_k, \quad \hat{B} = \sum_{i,k} B_{ik} a_i^+ a_k, \quad \hat{C} = \sum_{i,k} C_{ik} a_i^+ a_k, \tag{3.2}$$

the commutation relation (3.1) may be reproduced, i.e. the following relation holds

$$[\hat{A},\ \hat{B}] = \hat{C}. \tag{3.3}$$

In this case, it does not matter whether the quantum oscillator is Bose-oscillator with the commutation relations

$$[a_i,\ a_k] = [a_i^-,\ a_k^-] = 0, \quad [a_i,\ a_k^+] = \delta_{ik}, \tag{3.4}$$

or Fermi-oscillator for which commutation relations (3.4) are replaced by the anticommutators formulae

$$[a_i,\ a_k]_- = [a_i^+,\ a_k^-]_+ = 0, \quad [a_i,\ a_k^+]_+ = \delta_{ik}. \tag{3.5}$$

The cross implies the hermitian conjugation. The statement may be proved by direct checking of the formula (3.3) using either the relations (3.4) or the relations (3.5). Let us note that one can describe the infinite dimensional representations taking in the formula (3.2)-(3.5), i,k=1,2,...,∞.

Before writing down the general formulae, we will consider the groups of the lowest dimensions [33]. For the one-parametric group $SO_2(j_1)$ the matrix generator of the rotation takes the form

$$X_{01}(j_1) = \begin{Vmatrix} 0 & -j_1^2 \\ 1 & 0 \end{Vmatrix}. \tag{3.6}$$

Let us construct the realization of this operator by means of the two pairs of the boson creation and annihilation operators $\alpha_o^+, \alpha_o, \alpha_1^+, \alpha_1$ which satisfy the commutation relation (3.4). The antihermitian operator

$$L_{01}(1) = \mathbf{a}^+ X_{01}(1)\,\mathbf{a} = (a_0^+, a_1^+) \begin{Vmatrix} 0 & -1 \\ 1 & 0 \end{Vmatrix} \begin{pmatrix} a_0 \\ a_1 \end{pmatrix} = a_1^+ a_0 - a_0^+ a_1, \tag{3.7}$$

corresponds to the compact (antisymmetric generator) $X_{01}(1)$ of the rotation group $SO_2(1)$. The operator

$$L_{01}(\iota_1) = \mathbf{a}^+ X_{01}(\iota_1)\,\mathbf{a} = (\mathbf{a}_0^+, a_1^+) \begin{Vmatrix} 0 & 0 \\ 1 & 0 \end{Vmatrix} \begin{pmatrix} a_0 \\ a_1 \end{pmatrix} = a_1^+ a_0, \tag{3.8}$$

corresponds to the generator $X_{01}(\iota_1)$ of the one-dimensional Galilean group $SO_2(\iota_1)$. The antihermitian operator

$$L_{01}(i) = \tilde{\varphi} X_{01}(i)\,\varphi = (a_0^+, -a_1) \begin{Vmatrix} 0 & 1 \\ 1 & 0 \end{Vmatrix} \begin{pmatrix} a_0 \\ a_1^+ \end{pmatrix} = a_0^+ a_1^+ - a_0 a_1. \tag{3.9}$$

corresponds to the symmetric noncompact generator $X_{01}(i)$ of the one-dimensional Lorentz group $SO_2(i)$. The last formula is constructed using the general rules of the bosonization of the group infinitesimal operators [42]. So, for both the antisymmetric matrix $X_{01}(1)$ and the symmetric matrix $X_{01}(i)$, we construct antihermitian operators (using the formulae (3.7) and (3.9)) $L_{01}(1)$ and $L_{01}(i)$ in terms of creation and annihilation operators of two bose-oscillators. When $j_1=\iota_1$ the matrix $X_{01}(\iota_1)$ is neither symmetric nor antisymmetric and the operator $L_{01}(\iota_1)$ is neither hermitian nor antihermitian, i.e. $L^{+}_{01}(\iota_1) \neq \pm L_{01}(\iota_1)$.

In order to write down the formulae (3.7)-(3.9) in the form which is appropriate for all three groups $SO_2(j_1)$ we define the transformations of the creation and annihilation operators α^{+}_{k}, α_k, k=0,1 by the map (2.15) inducing the transition from the rotation group SO_2 to the group $SO_2(j_1)$ as follows:

$$\psi \mathbf{a}^{+} = \tilde{\varphi} = (a_0^{+}, j_1^{-1} a_1^{+}), \qquad \psi \mathbf{a} = \varphi = (a_0, j_1 a_1) \tag{3.10}$$

We demand that at $j_1=1, \iota_1$ the transformation (3.10) be the identical one, i.e. $j_1^{-1}\alpha_1^{+}=\alpha_1^{+}$, $j_1\alpha_1=\alpha_1$. Then, at $j_1=i$ we have $j_1^{-1}\alpha_1^{+}=i\alpha_1^{+}=\alpha_1$ and $j_1\alpha_1=i\alpha_1=\alpha^{+}_1$. We have taken into account the usual relations for Bose-operators $i\alpha^{+}_k=\alpha_k$, $i\alpha_k=\alpha^{+}_k$, and, consequently, the transformation (3.10) determines the vectors $\tilde{\varphi}$ and φ which are contained in relation (3.9). It is worthy of note that the commutation relations of the operators $\psi\alpha^{+}_k$ and $\psi\alpha_k$ are the same as the commutation relations of the operators α^{+}_k and α_k. In fact, $[\psi\alpha_1, \psi\alpha^{+}_1]=[j_1\alpha_1, j^{-1}_1\alpha^{+}_1]=[\alpha_1,\alpha^{+}_1]=1$. So, the transformation (3.10) conserves the commutation relations of the boson creation and annihilation operators. Now the boson realization of the generators $X_{01}(j_1)$ may be written down in the unified form

$$L_{01}(j_1) = \psi \mathbf{a}^{+} X_{01}(j_1)\, \psi \mathbf{a} = (a_0^{+}, j_1^{-1} a_1^{+}) \begin{Vmatrix} 0 & -j_1^2 \\ 1 & 0 \end{Vmatrix} \begin{pmatrix} a_0 \\ j_1 a_1 \end{pmatrix} = a_0 (j_1^{-1} a_1^{+}) - j_1^2 a_0^{+} (j_1 a_1). \tag{3.11}$$

To write down the matrix generators $X_{01}(j_1)$ in terms of the creation and annihilation operators $b_0^{+}, b_0, b_1^{+}, b_1$ of two Fermi-oscillators, we define the transformation of these operators by the map (2.15) using the relation which is the analog of the formula (3.10) namely

$$\psi \mathbf{b}^{+} = (b_0^{+}, j_1^{-1} b_1^{+}), \qquad \psi \mathbf{b} = (b_0, j_1 b_1), \tag{3.12}$$

and we demand that for $j_1=1, \iota_1$ the formulae $j_1^{-1}b_1^{+}=b_1^{+}, j_1 b_1 = b_1$ hold. When $j_1=i$ the transformation (3.12) with the relations for Fermi-operators $ib_1=b^{+}_1$, $ib^{+}_1=-b_1$ which converse the commutation relations $[ib_1, i^{-1}b^{+}_1]_+=[b^{+}_1,b_1]_+=[b_1b^{+}_1]_+$, provides the formulae $\psi b^{+}=(b^{+}_o,b_1)$, $\psi b=(b_1,b^{+}_1)$. The realization of the generators $X_{01}(j_1)$ in terms of the Fermi-operators may be written down in the form

$$L'_{01}(j_1) = \psi \mathbf{b}^{+} X_{01}(j_1)\, \psi \mathbf{b} = b_0 (j_1^{-1} b_1^{+}) - j_1^2 b_0^{+} (j_1 b_1). \tag{3.13}$$

In the boson case as well as the antihermitian operator $L'_{01}(1)=b_ob^{+}_1-b^{+}_ob_1$ corresponds to the antisymmetric matrix $X_{01}(1)$ and the Hermitian operator $L'_{01}(i)=b_ob_1+b^{+}_ob^{+}_1$ corresponds to the symmetric matrix $X_{01}(i)$.

Let us write down the formulae (3.10)-(3.13) in the unified form. We denote $\hat{a}^{+}_k$, $\hat{a}_k$ the creation and annihilation operators of both Fermi and Bose type. The transformation of these operators corresponding to the multiplication by the imaginary unit is defined as follows:

$$i\hat{a}_k = \hat{a}_k^-, \qquad i\hat{a}_k^+ = \varepsilon\hat{a}_k, \tag{3.14}$$

where ε=+1 for bosons, and ε=-1 for fermions. We define the two-dimensional diagonal matrices $\psi_1(j_1)$, $\psi_2(j_1)$ by the relations: $(\psi_1(j_1))_{\infty}=1$, $(\psi_1(j_1))_{11}=1$ if $j_1=1, \iota_1$, and $(\psi_1(j_1))_{11}=0$, if $j_1=i$; $(\psi_2(j_1))_{\infty}=0$, $(\psi_2(j_1))_{\infty}=0$ if $j_1=1, \iota_1$, and $(\psi_2(j_1))_{11}=-1$, if $j_1=i$. From this definition the properties of the matrices follow: $\psi_1\psi_2=\psi_2\psi_1=0$, $\psi^2{}_1+\psi^2{}_2=E_2$, E_2 being the two-dimensional unity matrix. The transformations (3.10) and (3.12) of the creation and annihilation operators by the map (2.15) may be written down in the form

$$\begin{aligned} \psi\hat{\mathbf{a}} &= \psi_1(j_1)\,\hat{\mathbf{a}} - \psi_2(j_1)\,\hat{\mathbf{a}}^+, \\ \psi\hat{\mathbf{a}}^+ &= \psi_1(j_1)\,\hat{\mathbf{a}}^+ + \varepsilon\psi_2(j_1)\,\hat{\mathbf{a}}, \end{aligned} \tag{3.15}$$

that is equivalent to the relations

$$\begin{pmatrix} \psi\hat{\mathbf{a}} \\ \psi\hat{\mathbf{a}}^+ \end{pmatrix} = \begin{Vmatrix} \psi_1(j_1) & -\psi_2(j_1) \\ \varepsilon\psi_2(j_1) & \psi_1(j_1) \end{Vmatrix} \begin{pmatrix} \hat{\mathbf{a}} \\ \hat{\mathbf{a}}^+ \end{pmatrix} \equiv \Psi^{-1}(j_1) \begin{pmatrix} \hat{\mathbf{a}} \\ \hat{\mathbf{a}}^+ \end{pmatrix}. \tag{3.16}$$

For four-dimensional matrix $\Psi^{-1}(j_1)$ we have $\Psi(j_1)=(\Psi^{-1}(j_1))^T$. The formulae (3.11) and (3.13) for the generators of the group $SO_2(j_1)$ in the Jordan-Schwinger representation may be combined and described by one expression

$$\hat{L}_{01}(j_1) = \psi\hat{\mathbf{a}}^+ X_{01}(j_1)\,\psi\hat{\mathbf{a}}, \tag{3.17}$$

in which the left vector $\psi\hat{a}^+$ is the row-vector, the right vector yâ+ is the column vector, and the multiplication of these vectors and the matrix $X_{01}(j_1)$ is fulfilled according to the rules of the multiplication of the matrices. This rule will also be used in further derivations.

The consideration of the group $SO_2(j_1)$ gives us the possibility to connect the matrix generators with the operators $\hat{L}$ which are expressed in terms of the creation and annihilation operators both of the boson and fermion types. To clarify what commutation relations are satisfied by these operators $\hat{L}$, let us consider the groups $SO_3(j)$, $j=(j_1,j_2)$. Three pairs of the boson and fermion creation and annihilation operators are transformed by the map (2.8) as follows:

$$\psi\hat{\mathbf{a}} = (\hat{a}_0, j_1\hat{a}_1, j_1 j_2\hat{a}_2),\ \psi\hat{\mathbf{a}}^+ = (\hat{a}_0^+, j_1^{-1}\hat{a}_1^+, j_1^{-1}j_2^{-1}\hat{a}_2^+), \tag{3.18}$$

where the transformation (3.18) is the identity one when $j_k=1, \iota_k$, k=1,2. We introduce three-dimensional diagonal matrices $\psi_1(j)$, $\psi_2(j)$ with the elements: $(\psi_1(j))_{\infty}=1$, $(\psi_1(j))_{kk}=\pm1$ if $\prod_{m=1}^{k} j_m = \pm\gamma$ where γ is positive real or dual number, and $(\psi_1(j))_{kk}=0$ in the opposite case (i.e., if $\prod_{m=1}^{k} j_m = \pm i\gamma$). The diagonal elements of the matrix $\psi_2(j)$ are as follows: $(\psi_2(j))_{\infty}=0$, $(\psi_2(j))_{kk}=0$ if $(\psi_1(j))_{kk}=\pm1$ and $(\psi_2(j))_{kk}=\mp1$, if $\prod_{m=1}^{k} j_m = \pm i\gamma$ k=1,2. These matrices have the properties: $\psi_1(j)\psi_2(j)=\psi_2(j)\psi_1(j)=0$ and $\psi_1^2(j)+\psi_2^2(j)=E_3$, where E_3 is the three-dimensional unity matrix. The transformations (3.18) and the inverse transformations may be written down in the form

$$\begin{pmatrix} \psi\hat{\mathbf{a}} \\ \psi\hat{\mathbf{a}}^+ \end{pmatrix} = \Psi^{-1}(j) \begin{pmatrix} \hat{\mathbf{a}} \\ \hat{\mathbf{a}}^+ \end{pmatrix}, \qquad \begin{pmatrix} \hat{\mathbf{a}} \\ \hat{\mathbf{a}}^+ \end{pmatrix} = \Psi(j) \begin{pmatrix} \psi\hat{\mathbf{a}} \\ \psi\hat{\mathbf{a}}^+ \end{pmatrix}, \tag{3.19}$$

where the six-dimensional matrices $\Psi(j)$ and $\Psi^{-1}(j)$ have the form

$$\Psi(j) = \begin{Vmatrix} \psi_1(j) & \varepsilon\psi_2(j) \\ -\psi_2(j) & \psi_1(j) \end{Vmatrix}, \qquad \Psi^{-1}(j) = \Psi^T(j) = \begin{Vmatrix} \psi_1(j) & -\psi_2(j) \\ \varepsilon\psi_2(j) & \psi_1(j) \end{Vmatrix}. \tag{3.20}$$

The Jordan-Schwinger representation corresponding to the matrix generators (2.21) of the group $SO_3(j)$ is realized by the operators

$$\hat{L}_{\mu\nu}(j) = \psi\hat{a}^{+} X_{\mu\nu}(j)\,\psi\hat{a}, \tag{3.21}$$

which may be expressed explicitly

$$\hat{L}_{01}(j) = (j_1^{-1}\hat{a}_1^{+})\,\hat{a}_0 - j_1^2\hat{a}_0^{+}\,(j_1\hat{a}_1), \qquad \hat{L}_{02}(j) = (j_1^{-1}j_2^{-1}\hat{a}_2^{+})\,\hat{a}_0 - j_1^2 j_2^2\hat{a}_0^{+}\,(j_1 j_2\hat{a}_2),$$
$$\hat{L}_{12}(j) = (j_1^{-1}j_2^{-1}\hat{a}_2^{+})(j_1\hat{a}_1) - j_2^2\,(j_1^{-1}\hat{a}_1^{+})(j_1 j_2\hat{a}_2). \tag{3.22}$$

Using the commutation relations (3.4) of the Bose-operators and (3.5) of the Fermi-operators, which for the operators with heads may be written down in the form,

$$\hat{a}_i\hat{a}_k = \varepsilon\hat{a}_k\hat{a}_i, \qquad \hat{a}_i^{+}\hat{a}_k^{+} = \varepsilon\hat{a}_k^{+}\hat{a}_i^{+}, \qquad \hat{a}_i\hat{a}_k^{+} - \varepsilon\hat{a}_k^{+}\hat{a}_i = \delta_{ik}, \tag{3.23}$$

we find out the commutation relations of the operators $L_{\mu\nu}(j)$

$$[\hat{L}_{01}, \hat{L}_{02}] = j_1^2\hat{L}_{12}, \qquad [\hat{L}_{02}, \hat{L}_{12}] = j_2^2\hat{L}_{01}, \qquad [\hat{L}_{01}, \hat{L}_{12}] = -\hat{L}_{02} \tag{3.24}$$

which obviously coincide with the commutation relations (2.23) of the group $SO_3(j)$. Hence, the operators (3.22) realize the Jordan-Schwinger representation of the group $SO_3(j)$. The Casimir operators (2.25) look in this representation as follows

$$\hat{L}^2(j) = j_2^2\hat{L}_{01}^2(j) + \hat{L}_{02}^2(j) + j_1^2\hat{L}_{12}^2(j). \tag{3.25}$$

The generalization for the group $SO_{n+1}(j)$, $j=(j_1, j_2,\ldots, j_n)$ is not difficult [34]. The creation and annihilation operators of bosons or fermions transform as follows:

$$\psi\hat{a} = \left(\hat{a}_0, \hat{a}_k\prod_{m=1}^{k} j_m\right), \qquad \psi\hat{a}^{+} = \left(\hat{a}_0^{+}, \hat{a}_k^{+}\prod_{m=1}^{k} j_m^{-1}\right), \qquad k = 1, 2, \ldots, n. \tag{3.26}$$

When $j_k=1$, ι_k the transformation (3.26) is the identity one. The equivalent form of the transformation (3.26) may be described by the formulae (3.19) and (3.20) in which the matrices $\psi_1(j)$ and $\psi_2(j)$ have the dimensionality n+1; i.e., k=1,2,...,n. The generators $L_{\mu\nu}(j)$ of the group $SO_{n+1}(j)$ in the Jordan-Schwinger representation satisfying the commutation relations (2.6) correspond to the matrix generators (2.5) if one uses the formula (3.21).

Let us consider in detail, the sets of the operators $\psi\hat{a}$ and $\psi\hat{a}^{+}$ which are obtained by means of the transformation (3.26). If one considers the group contractions only when the parameters j_k take two values 1, i_k, then we have $\psi\hat{a}=\alpha$, $\psi\hat{a}^{+}=\alpha^{+}$, and each of the sets consist of the creation operators or the annihilation operators only. If not only the contractions are used, but the analytical continuations of the group parameters are used too (in this case one or several parameters, j_k are equal to imaginary units) the sets of operators $\psi\hat{a}$ and $\psi\hat{a}^{+}$ become the mixed ones, i.e. each of them contains both creation and annihilation operators. For example, in the case of Lorentz group $SO_2(i)$ we have $\psi a=(\alpha, i\alpha_1)=(\alpha_o, \alpha^{+}{}_1)$, $\psi\hat{a}^{+}=(\alpha^{+}, -i\alpha^{+}{}_1)=(\alpha^{+}{}_o, -\alpha_1)$.

Let us note that the mixed sets of the creation and annihilation operators (for bosons and fermions) and the representations of the matrix Lie algebras by these sets (analogs of the representation (3.21)) have been considered in study [50] without connection with the fundamental map of the type (2.3). In this study the mixed sets are ordered in such a manner that at first the annihilation operators are collected and then the creation operators and vice versa. From our transformation (3.26) it follows that, firstly, the mixed sets appear only in the case when at least one or several parameters

j_k take the imaginary values and, secondly, the creation and annihilation operators in relations (3.26) are not ordered, they are mixed in strictly determined manner. Namely, this fact gives us the possibility to study both the contractions and the analytical continuations of the groups in the Jordan-Schwinger representation.

4. GROUP QUANTUM SYSTEMS IN COHERENT STATE REPRESENTATION

Quantum system will be called the group quantum system if it has the Hamiltonian which in linear form is Lie group generators expressed in terms of creation and annihilation operators either of Bose-oscillators or Fermi-oscillators. It means that the Hamiltonian $\hat{H}$ of such a system is an element of the Lie algebra of the Lie group in Jordan-Schwinger representation. Choosing these groups to be the rotation groups in the Cayley-Klein spaces described in the previous sections in the frame of unified approach, we realize one of the possibilities to generalize the concept of the dynamical symmetry studying a set of several systems with different Hamiltonians as one system [12]. Such an approach also demonstrates the concrete realization of the program suggested in the study [24] to investigate simultaneously the sets of the physical theories characterized by the symmetry groups related by the limit transitions and analytical continuations of the group parameters.

Let us consider the quantum system corresponding to the group $SO_{n+1}(j)$ which is described by the Schrodinger type equation

$$i\hbar\partial\psi/\partial t = \hat{H}\psi \tag{4.1}$$

with the Hamiltonian

$$\hat{H}\,(\mathbf{j}) = \sum_{\lambda=1}^{n(n+1)/2} r_\lambda \hat{L}_\lambda\,(\mathbf{j}), \tag{4.2}$$

where the generators $\hat{L}_\lambda(j)$ are given by the formulae (3.21) in Jordan-Schwinger representation and the index λ is connected with the pair of indices μ, ν by the relation (2.14). The Hamiltonian (4.2) may be rewritten in the equivalent form

$$\hat{H}\,(\mathbf{j}) = \frac{1}{2}\,(\psi\hat{\mathbf{a}}, \psi\hat{\mathbf{a}}^+)\,\Delta\,(\mathbf{r}, \mathbf{j}) \begin{pmatrix} \psi\hat{\mathbf{a}} \\ \psi\hat{\mathbf{a}}^+ \end{pmatrix} = \frac{1}{2}\,(\hat{\mathbf{a}}, \hat{\mathbf{a}}^+)\,\mathcal{B}\,(\mathbf{r}, \mathbf{j}) \begin{pmatrix} \hat{\mathbf{a}} \\ \hat{\mathbf{a}}^+ \end{pmatrix}, \tag{4.3}$$

where the matrix $\Delta(r,j)$ is as follows

$$\Delta\,(\mathbf{r}, \mathbf{j}) = \begin{Vmatrix} 0 & \varepsilon X^T\,(\mathbf{r}, \mathbf{j}) \\ X\,(\mathbf{r}, \mathbf{j}) & 0 \end{Vmatrix}, \tag{4.4}$$

and the matrix $X(r,j)$ is a general element of the algebra $SO_{n+1}(j)$ described by the formula (2.12). Using the transformation (3.19) for the creation and annihilation operators due to the map ψ, we can find the relation of the matrices Δ and ß

$$\mathcal{B}\,(\mathbf{r}, \mathbf{j}) = \Psi\,(\mathbf{j})\,\Delta\,(\mathbf{r}, \mathbf{j})\,\Psi^{-1}\,(\mathbf{j}). \tag{4.5}$$

Let us note that if one considers only the contractions of the rotation group, i.e. $j_k=1, \iota_k$, the transformations are as follows:
$\Psi(j)\equiv 1$ and $ß(r,j)=\Delta(r,j)$.

The Hamiltonian $\hat{H}(j)$ is the quadratic form in boson or fermion creation and annihilation operators that gives the possibility to use for the description of the group quantum systems well-known theory of the general quadratic quantum systems [8]. The simplest method to find the propagators (the kernels of the evolution operators) of such systems is the effective method of the integrals of motion [12].

The linear integrals of the motion for the Hamiltonian (4.3) have the form

$$\hat{\mathbf{Q}}(t) = \Lambda(t)\,\hat{\mathbf{q}}, \quad \hat{\mathbf{q}} = \begin{pmatrix} \hat{\mathbf{a}} \\ \hat{\mathbf{a}}^{+} \end{pmatrix}, \quad \hat{\mathbf{Q}} = \begin{pmatrix} \hat{\mathbf{A}} \\ \hat{\mathbf{A}}_1 \end{pmatrix},$$

$$\Lambda = \begin{Vmatrix} \xi & \eta \\ \eta_1 & \xi_1 \end{Vmatrix}, \quad \hat{\mathbf{A}}(0) = \hat{\mathbf{a}}, \quad \hat{\mathbf{A}}_1(0) = \hat{\mathbf{a}}^{+}, \tag{4.6}$$

where the 2(n+1) dimensional matrix Λ is the solution to the following matrix equation

$$-i\hbar\dot{\Lambda} = \Lambda\Sigma\mathcal{B}(\mathbf{r}, \mathrm{j}), \quad \Lambda(0) = E_{2(n+1)}, \quad \Sigma = \begin{Vmatrix} 0 & E_{n+1} \\ -\varepsilon E_{n+1} & 0 \end{Vmatrix}. \tag{4.7}$$

Here E_k is k-dimensional unity matrix.

In a stationary case when the Hamiltonian (4.3) does not depend on the time, the solution to the equation (4.7) may be easily found

$$\Lambda(\mathbf{r}, \mathrm{j}, t) = \exp\left\{\frac{i}{\hbar} t\Sigma\mathcal{B}(\mathbf{r}, \mathrm{j})\right\}. \tag{4.8}$$

For the group $SO_{n+1}(j)$, which are obtained only by the contractions of the rotation group (when $j_k=1$, ι_k, k=1,2,...,n), the matrix $\mathcal{B}$(r,j) equals Δ(r,j) and it is given by the relation (4.4). The solution to the equation (4.7) has the form

$$\Lambda(\mathbf{r}, \mathrm{j}, t) = \exp\left\{\frac{i}{\hbar} t\Sigma\Delta(\mathbf{r}, \mathrm{j})\right\} = \begin{Vmatrix} \Xi(i\hbar^{-1}t\mathbf{r}, \mathrm{j}) & 0 \\ 0 & \Xi^{T}(-i\hbar^{-1}t\mathbf{r}, \mathrm{j}) \end{Vmatrix}, \tag{4.9}$$

where the matrix Ξ(r,j) described by the formula (2.3) is the matrix of the finite group transformation. Using the relations (4.6) and (4.9) one can conclude that $\eta=\eta_1=0$, $\varepsilon^{T}{}_1=\varepsilon^{-1}$ and the matrix $\varepsilon(t)$ is expressed as follows: $\varepsilon(t)=\Xi(i\hbar^{-1}t\mathbf{r},\mathrm{j})$, i.e. the transformation Λ does not connect the creation operators with the annihilation operators. Hence, the linear integrals of the motion $\hat{\Delta}(t)$ are expressed only in terms of the annihilation operators, and the integrals of the motion $\hat{\Delta}_1(t)$ are expressed only in terms of the creation operators.

$$\hat{\mathbf{A}}(t) = \Xi\left(\frac{i}{\hbar} t\mathbf{r}, \mathrm{j}\right)\hat{\mathbf{a}}, \quad \hat{\mathbf{A}}_1(t) = \Xi^{T}\left(-\frac{i}{\hbar} t\mathbf{r}, \mathrm{j}\right)\hat{\mathbf{a}}^{+}. \tag{4.10}$$

In the case of the boson systems, the coherent state representation [8] is the most appropriate one. We use the version of the representation in which the system state $|f\rangle$ is described by the analytical function of the complex argument $f(\alpha^*)=\langle\alpha^*|f\rangle\exp(|\alpha|^2/2)$, where $|\alpha\rangle$ is the normalized eigenstate of the annihilation operator $\hat{a}$ corresponding to the eigenvalue α, i.e. $\hat{\alpha}\,|\alpha\rangle=\alpha|\alpha\rangle$.

The most convenient representation in the case of fermion system is the representation based on using the Grassman anticommuting variables [51]. In both cases the following relations hold

$$\hat{a}f(\alpha^*) = \frac{\partial}{\partial\alpha^*} f(\alpha^*), \quad \hat{a}^{+}f(\alpha^*) = \alpha^* f(\alpha^*), \tag{4.11}$$

α^* being the Grassman variables in the case of Fermi-oscillator.

Knowing the quantum integrals of the motion (4.6), one can find [8] the evolution operation $\hat{U}(t)=\exp(-i\hbar^{-1}t\hat{H}(\mathrm{j}))$ of the system solving the operator equation $\hat{Q}(t)\hat{U}(t)=\hat{U}(t)\hat{q}$ which may be written down as the system of equation for the Green function (the kernel of the evolution operator) in the coherent representation

$$\left[\xi(t)\frac{\partial}{\partial\boldsymbol{\alpha}^*}+\eta(t)\boldsymbol{\alpha}^*\right]G(\boldsymbol{\alpha}^*,\boldsymbol{\beta},t)=\boldsymbol{\beta}G(\boldsymbol{\alpha}^*,\boldsymbol{\beta},t),$$
$$\left[\eta_1(t)\frac{\partial}{\partial\boldsymbol{\alpha}^*}+\xi_1(t)\boldsymbol{\alpha}^*\right]G(\boldsymbol{\alpha}^*,\boldsymbol{\beta},t)=\varepsilon\frac{\partial}{\partial\boldsymbol{\beta}}G(\boldsymbol{\alpha}^*,\boldsymbol{\beta},t). \tag{4.12}$$

The solution to this system of equations consistent with the equation (4.1) is given by the formula [8]

$$G(\boldsymbol{\alpha}^*,\boldsymbol{\beta},t)=[\det\xi(t)]^{-\varepsilon/2}\exp\{-{}^1/_2\boldsymbol{\alpha}^*\xi^{-1}(t)\eta(t)\boldsymbol{\alpha}^*+\boldsymbol{\alpha}^*\xi^{-1}(t)\boldsymbol{\beta}+$$
$$+{}^1/_2\varepsilon\boldsymbol{\beta}\eta_1(t)\xi^{-1}(t)\boldsymbol{\beta}\}=[\det\xi(t)]^{-\varepsilon/2}\exp\left\{-\frac{1}{2}(\boldsymbol{\alpha}^*,\boldsymbol{\beta})R(t)\binom{\boldsymbol{\alpha}^*}{\boldsymbol{\beta}}\right\}, \tag{4.13}$$

where the matrix R(t) has the form

$$R(t)=\begin{Vmatrix}\xi^{-1}(t)\eta(t) & -\xi^{-1}(t)\\ -\varepsilon\xi^{-1T}(t) & -\varepsilon\eta_1(t)\xi^{-1}(t)\end{Vmatrix}. \tag{4.14}$$

As it is known [8], the replacement $t=-i\hbar\tau$ in the formula (4.13) transforms the Green function (the propagator) of the stationary quantum system into the equilibrium density matrix of the system in the coherent state representation

$$\rho(\boldsymbol{\alpha}^*,\boldsymbol{\beta},\tau)=[\det\varphi(\tau)]^{-\varepsilon/2}\exp\{-{}^1/_2\boldsymbol{\alpha}^*\varphi^{-1}\varkappa\boldsymbol{\alpha}^*+\boldsymbol{\alpha}^*\varphi^{-1}\boldsymbol{\beta}+{}^1/_2\varepsilon\boldsymbol{\beta}\varkappa_1\varphi^{-1}\boldsymbol{\beta}\}, \tag{4.15}$$

where the following notations are introduced

$$\varphi(\tau)=\xi(-i\hbar\tau),\qquad \varkappa(\tau)=\eta(-i\hbar\tau),\qquad \varkappa_1(\tau)=\eta_1(-i\hbar\tau). \tag{4.16}$$

Such change of the time induces the transformation of the evolution operator $\hat{U}(t)=\exp(-i\hbar^{-1}t\hat{U})$ into the operator $\hat{U}_g(t)=\exp(-\tau\hat{H})$. The density matrix (4.15) is the kernel of this operator in the coherent state representation. At $\tau=1$ this operator

$$\hat{U}_g(\mathbf{j})=\exp(-\hat{H}(\mathbf{j}))=\exp\left(-\sum_{\lambda=1}^{n(n+1)/2}r_\lambda\hat{L}_\lambda(\mathbf{j})\right) \tag{4.17}$$

is the operator of the finite group transformation in the Jordan-Schwinger representation. Hence, the propagators, and consequently, all the properties of the stationary group quantum systems are determined completely by the kernel of the secondary quantized operator $\hat{U}_g$ of the finite group transformations. Because of this we will consider the kernel U(α*, β, r, j) of the operator (4.17) connected with the matrix elements $\langle\alpha|\hat{U}_g|\beta\rangle$ in the coherent state representation

$$U(\boldsymbol{\alpha}^*,\boldsymbol{\beta},\mathbf{r},\mathbf{j})=\langle\boldsymbol{\alpha}|\hat{U}_g|\boldsymbol{\beta}\rangle\exp({}^1/_2|\boldsymbol{\alpha}|^2+{}^1/_2|\boldsymbol{\beta}|^2). \tag{4.18}$$

This kernel is given by the formula

$$U(\boldsymbol{\alpha}^*,\boldsymbol{\beta},\mathbf{r},\mathbf{j})=[\det\xi]^{-\varepsilon/2}\exp\{-{}^1/_2\boldsymbol{\alpha}^*\xi^{-1}\eta\boldsymbol{\alpha}^*+\boldsymbol{\alpha}^*\xi^{-1}\boldsymbol{\beta}+{}^1/_2\varepsilon\boldsymbol{\beta}\eta_1\xi^{-1}\boldsymbol{\beta}\}=$$
$$=[\det\xi]^{-\varepsilon/2}\exp\left\{-\frac{1}{2}(\boldsymbol{\alpha}^*,\boldsymbol{\beta})R\binom{\boldsymbol{\alpha}^*}{\boldsymbol{\beta}}\right\}, \tag{4.19}$$

where the matrix

$$R=\begin{Vmatrix}\xi^{-1}\eta & -\xi^{-1}\\ -\varepsilon\xi^{-1T} & -\varepsilon\eta_1\xi^{-1}\end{Vmatrix}, \tag{4.20}$$

and the matrices ε, η, ε_1 are given by the relation

$$\Lambda(\mathbf{r}, \mathbf{j}) = \exp(\Sigma \mathfrak{R}(\mathbf{r}, \mathbf{j})) = \left\| \begin{matrix} \xi & \eta \\ \eta_1 & \xi_1 \end{matrix} \right\|. \tag{4.21}$$

The propagator (4.13) of the stationary group quantum system is connected with the kernel (4.18) of the operator of the finite group transformation: $G(\alpha^*, \beta, i\hbar^{-1}tr, j)=U(\alpha^*, \beta, i\hbar^{-1}tr, j)$.

Before discussing the concrete systems, let us remember another factor from the theory of the quantum systems with quadratic Hamiltonian [8] connected with Hermite polynomials. We will consider the case of bosons $\varepsilon=1$. Though the coherent states $|\alpha\rangle$ which are labeled by the continuous complex index α form the complete (even overcomplete) system of the states, they are not orthogonal

$$\langle \alpha \mid \beta \rangle = \exp\left(-{}^1/_2 \mid \alpha \mid^2 - {}^1/_2 \mid \beta \mid^2 + \alpha^* \beta\right) \neq 0. \tag{4.22}$$

The orthogonal and normalized basis in the space of the states of Bose-oscillators form the Fock states $|n\rangle$, i.e. the eigenstates of the particle number operator

$$a_k^+ a_k \mid \mathbf{n} \rangle = n_k \mid \mathbf{n} \rangle, \qquad k = 0, 1, 2, \ldots \tag{4.23}$$

Using the relation of these bases

$$|\alpha\rangle = e^{-1/_2 |\alpha|^2} \sum_{\mathbf{n}=0}^{\infty} \frac{\alpha^{\mathbf{n}}}{(\mathbf{n}!)^{1/_2}} \mid \mathbf{n} \rangle, \tag{4.24}$$

where $n=(n_o, n_1,\ldots,n_N)$, $\alpha=(\alpha_o,\alpha_1,\ldots,\alpha_N)$, $n_N!, \equiv n_o!n_1!\ldots n_N!$, $\alpha^n = \alpha_o^{n_o}\alpha_1^{n_1}\ldots\alpha_N^{n_N}$ we find that the matrix elements (the kernel) of the operator (4.17) in the basis of the coherent states are the generating functions for the matrix elements of this operator in the Fock basis

$$U(\alpha^*, \beta, \mathbf{r}, \mathbf{j}) = \sum_{\mathbf{m}, \mathbf{n}=0}^{\infty} \frac{\alpha^{*\mathbf{m}} \beta^{\mathbf{n}}}{(\mathbf{m}!\, \mathbf{n}!)^{1/_2}} \langle \mathbf{m} \mid U_g \mid \mathbf{n} \rangle. \tag{4.25}$$

The multidimensional Hermite polynomials $H_{m,n}^{(R)}(x)$ are determined by the generating function [52]

$$\exp\left(\frac{1}{2} \mathbf{a} R \mathbf{a} + \mathbf{a} R \mathbf{x}\right) = \sum_{\mathbf{k}=0}^{\infty} \frac{\mathbf{a}^{\mathbf{k}}}{\mathbf{k}!} H_{\mathbf{k}}^{\{R\}}(\mathbf{x}). \tag{4.26}$$

Hence, from the formula (4.19) one can derive the generating function for the Hermite polynomials at zero argument

$$[\det \xi]^{1/_2} U(\alpha^*, \beta, \mathbf{r}, \mathbf{j}) = \exp\left\{-\frac{1}{2}(\alpha^*, \beta) R \begin{pmatrix} \alpha^* \\ \beta \end{pmatrix}\right\} = \sum_{\mathbf{m}, \mathbf{n}=0}^{\infty} \frac{\alpha^{*\mathbf{m}} \beta^{\mathbf{n}}}{\mathbf{m}!\, \mathbf{n}!} H_{\mathbf{m}, \mathbf{n}}^{\{R\}}(0) \tag{4.27}$$

with the matrix R given by the relation (4.20). The relation of the matrix elements of the operator of the finite group transformation (4.17) in the Fock basis and the Hermite polynomials at zero argument is given by the formula

$$\langle \mathbf{m} \mid U_g \mid \mathbf{n} \rangle = (\det \xi \cdot \mathbf{m}!\, \mathbf{n}!)^{-1/_2} H_{\mathbf{m}, \mathbf{n}}^{\{R\}}(0). \tag{4.28}$$

The transition amplitudes from the Fock state into the coherent state and vice versa lead to the Hermite polynomials at nonzero argument [8]

$$\begin{aligned} \langle \mathbf{m} \mid U_g \mid \beta \rangle &= (\mathbf{m}!)^{-1/_2} U(0, \beta, \mathbf{r}, \mathbf{j}) H_{\mathbf{m}}^{\{\xi^{-1}\eta\}}(\eta^{-1}\xi\beta), \\ \langle \alpha^* \mid U_g \mid \mathbf{n} \rangle &= (\mathbf{n}!)^{-1/_2} U(\alpha^*, 0, \mathbf{r}, \mathbf{j}) H_{\mathbf{n}}^{\{-\eta_1\xi^{-1}\}}(-\xi\eta_1^{-1}\alpha^*). \end{aligned} \tag{4.29}$$

$$\Xi(\mathbf{r}'', \mathbf{j}) = \Xi(\mathbf{r}, \mathbf{j})\, \Xi(\mathbf{r}', \mathbf{j}), \qquad \mathbf{r}'' = \mathbf{r} \oplus \mathbf{r}', \tag{4.30}$$

The group rule of the composition which is valid also for the operators (4.17) produces the adding formulae for the Hermite polynomials. For the operators (4.17) in the Fock basis we have $\langle m | U_g(r'') | n \rangle = \langle m | U_g(r) U_g(r') | n \rangle$. Inserting the unity operator $1 = \sum_{k=0}^{\infty} | k \rangle\langle k |$ between the operators $U_g(r)$ and $U_g(r')$ we obtain

$$\langle \mathbf{m} | U_g(\mathbf{r}'') | \mathbf{n} \rangle = \sum_{k=0}^{\infty} \langle \mathbf{m} | U_g(\mathbf{r}) | \mathbf{k} \rangle \langle \mathbf{k} | U_g(\mathbf{r}') | \mathbf{n} \rangle. \tag{4.31}$$

Using the relation of the matrix elements of the operator in Fock basis and Hermite polynomials (4.28) one finds the adding formula for multidimensional Hermite polynomials at zero argument

$$H^{\{R(\mathbf{r}'')\}}_{\mathbf{m},\mathbf{n}}(0) = \left(\frac{\det \xi(\mathbf{r}'')}{\det \xi(\mathbf{r}) \cdot \det \xi(\mathbf{r}')} \right)^{1/2} \sum_{\mathbf{k}=0}^{\infty} \frac{1}{\mathbf{k}!} H^{\{R(\mathbf{r})\}}_{\mathbf{m},\mathbf{k}}(0)\, H^{\{R(\mathbf{r}')\}}_{\mathbf{k},\mathbf{n}}(0). \tag{4.32}$$

The matrix R is given by the formula (4.20).

In previous sections, it was clarified that the generators $\hat{L}_{\mu\nu}(j)$ in the boson Jordan-Schwinger representation due to the formula (3.21), are the antihermitian operators if no one from the parameters j is equal to pure dual number, i.e. the group $SO_{n+1}(j)$ is the notation group in nonfiber Cayley-Klein space $R_{n+1}(j)$. In this case, the operator (4.17) is a unitary one $\hat{U}_g \hat{U}_g^{+} = 1$. From the unitarity relation one obtains the sum rule for the Hermite polynomials at zero argument

$$\sum_{\mathbf{m}=0}^{\infty} \frac{1}{\mathbf{m}!} \left| H^{\{R(\mathbf{r},\,\mathbf{j})\}}_{\mathbf{n},\mathbf{m}}(0) \right|^2 = \mathbf{n}!\, |\det \xi|. \tag{4.33}$$

Let us consider now the quantum systems which correspond to the rotation groups in the Cayley-Klein spaces of the low dimensions.

4.1 One Parametric Rotation Group and Galilean Group

In this section the parameter j_1 takes two values $j_1=1, \iota_1$. According to (3.17) secondary quantized generators of the group $SO_2(j_1)$ is equal to $\hat{L}_{01}(j_1) = a_1^{+} a_o - j_1^2 a_o^{+} \hat{a}_1$ due to relations $\psi \hat{\alpha}^{+} = \hat{\alpha}^{+}\ \psi \hat{\alpha} = \hat{\alpha}$, where $\hat{a} = (\hat{\alpha}_0, \hat{\alpha}_1)$, $\hat{a}^{+} = (\hat{a}^{+}_0, \hat{a}^{+}_1)$. The Hamiltonian of the group quantum system is described by the formulae (4.3), (4.4) and it has the form

$$\hat{H}(j_1) = \frac{1}{2} (\hat{\mathbf{a}}, \hat{\mathbf{a}}^{+})\, \Delta(r_1, j_1) \begin{pmatrix} \hat{\mathbf{a}} \\ \hat{\mathbf{a}}^{+} \end{pmatrix} = \frac{1}{2} (\hat{\mathbf{a}}, \hat{\mathbf{a}}^{+}) \begin{Vmatrix} 0 & \varepsilon X^T(r_1, j_1) \\ X(r_1, j_1) & 0 \end{Vmatrix} \begin{pmatrix} \hat{\mathbf{a}} \\ \hat{\mathbf{a}}^{+} \end{pmatrix}, \tag{4.34}$$

where the matrix

$$X(r_1, j_1) = r_1 X_1(j_1) = \begin{Vmatrix} 0 & -j_1^2 r_1 \\ r_1 & 0 \end{Vmatrix},$$

due to the formulae (2.12). The equation (4.21) may be easily solved

$$= \exp \begin{Vmatrix} X(r_1, j_1) & 0 \\ 0 & X^T(-r_1, j_1) \end{Vmatrix} = \begin{Vmatrix} \Xi(r_1, j_1) & 0 \\ 0 & \Xi^T(-r_1, j_1) \end{Vmatrix} = \begin{Vmatrix} \xi & \eta \\ \eta_1 & \xi_1 \end{Vmatrix}, \tag{4.35}$$

i.e. $\eta=\eta_1=0$, $\varepsilon_1=\Xi^T(-r_1,j_1)$,$\varepsilon=\Xi^T(r^1,j^1)$ where the matrix $\Xi(r_1,j_1)$ is given by the formula (2.17) and $\det\xi=\det\Xi(r_1,j_1)=1$, $\xi^{-1}=\Xi^{-1}(r_1,j_1)=\Xi(-r_1,j_1)$. Using the formula (4.19) one obtains that kernel of the operator (4.17) of the finite group transformations in the coherent state basis

$$U(\alpha^*, \beta, r_1, j_1) = \exp\{\alpha^*\Xi(-r_1, j_1)\beta\} = \exp\{(\alpha_0^*\beta_0 + \alpha_1^*\beta_1)\cos j_1r_1 - - \alpha_1^*\beta_0 j_1^{-1}\sin j_1r_1 + \alpha_0^*\beta_1 j_1\sin j_1r_1\}. \quad (4.36)$$

This expression is correct in both boson and fermion cases. The only differences of these cases is that for bosons α_k^*, β_k, k=0,1 are the complex variables and for fermions these numbers are the Grassman anticommuting variables [51]. The kernel (4.36) may be written down in the form (4.19) with the matrix $R(r_1,j_1)$

$$R(r_1, j_1) = \begin{Vmatrix} 0 & -\Xi(-r_1, j_1) \\ -\varepsilon\Xi^T(-r_1, j_1) & 0 \end{Vmatrix}. \quad (4.37)$$

It follows from these expressions that in the boson case at $\varepsilon=1$ the kernel (4.36) of the operator of the finite group transformations of the group $SO_2(j_1)$ is the generating function for the Hermite polynomials of four zero-valued variables, i.e.

$$H_{\mathbf{m},\mathbf{n}}^{\{R(r_1, j_1)\}}(0) = \frac{\partial^{\mathbf{m}+\mathbf{n}}}{\partial\alpha^{*\mathbf{m}}\partial\beta^{\mathbf{n}}} U(\alpha^*, \beta, r_1, j_1)\,|_{\alpha^*=\beta=0}, \quad (4.38)$$

where $m=(m_o,m_1)$, $n=(n_o,n_1)$, $\partial\alpha^{*m}=\partial\alpha_o^{*m_o}\partial\alpha_1^{*m_1}$, $\partial\beta^n=\partial\beta_o^{n_o}\partial\beta_1^{n_1}$. For the completeness of the consideration, let us write down the expression for the matrix elements of the operator $U_g(r_1,j_1)$ in Fock basis in terms of Hermite polynomials

$$\langle\mathbf{m}|U_g(r_1, j_1)|\mathbf{n}\rangle = (\mathbf{m}!\,\mathbf{n}!)^{-1/2} H_{\mathbf{m},\mathbf{n}}^{\{R(r_1, j_1)\}}(0), \quad (4.39)$$

where the matrix $R(r_1,j_1)$ is given by the formula (4.37) at $\varepsilon=-1$. Using the formulae (4.36), (4.38) let us write down several first Hermite polynomials (at zero valued variables)

$$\begin{aligned}
&H_{0,0}(r_1, j_1) = 1, \quad H_{1,1}(r_1, j_1) = \cos^2 j_1r_1 - \sin^2 j_1r_1,\\
&H_{1,0;1,0}(r_1, j_1) = \cos j_1r_1, \quad H_{1,0;0,1}(r_1, j_1) = j_1\sin j_1r_1,\\
&H_{0,1;1,0}(r_1, j_1) = -j_1^{-1}\sin j_1r_1, \quad H_{0,1;0,1}(r_1, j_1) = \cos j_1r_1,\\
&H_{n,0;n,0}(r_1, j_1) = H_{0,n;0,n}(r_1, j_1) = \Gamma(n+1)(\cos j_1r_1)^n,\\
&H_{n,0;0,n}(r_1, j_1) = \Gamma(n+1)(j_1\sin j_1r_1)^n,\\
&H_{0,n;n,0}(r_1, j_1) = (-1)^n\Gamma(n+1)(j_1^{-1}\sin j_1r_1)^n,\\
&H_{2,1;2,1}(r_1, j_1) = H_{1,2;1,2}(r_1, j_1) = 2\cos j_1r_1(\cos^2 j_1r_1 - 2\sin^2 j_1r_1).\\
&H_{2,1;1,2}(r_1, j_1) = 2j_1\sin j_1r_1(2\cos^2 j_1r_1 - \sin^2 j_1r_1),\\
&H_{1,2;2,1}(r_1, j_1) = -2j_1^{-1}\sin j_1r_1(2\cos^2 j_1r_1 - \sin^2 j_1r_1).
\end{aligned} \quad (4.40)$$

In the case of the Galilean group $SO_2(\iota_1)$ the parameter $j_1=\iota_1$ and the part of the Hermite polynomials is equal to zero due to the contraction. In fact, $\cos\iota_1r_1=1$, $\sin\iota_1r_1=\iota_1r_1$, then $\Xi_{oo}=\Xi_{11}=1$, $\Xi_{01}=\iota_1\sin\iota_1r_1=0$, $\Xi_{10}=-\iota_1^{-1}\sin\iota_1r_1=-r_1$, hence

$$\begin{aligned}
&H_{n,0;n,0}(r_1, \iota_1) = H_{0,n;0,n}(r_1, \iota_1) = \Gamma(n+1), \quad H_{n,0;0,n}(r_1, \iota_1) = 0,\\
&H_{0,n;n,0}(r_1, \iota_1) = (-1)^n\Gamma(n+1)r_1^n, \quad H_{2,1;2,1}(r_1, \iota_1) = H_{1,2;1,2}(r_1, \iota_1) = 2,\\
&H_{1,2;2,1}(r_1, \iota_1) = -4r_1, \quad H_{2,1;1,2}(r_1, \iota_1) = 0.
\end{aligned} \quad (4.41)$$

Then for m=n the Hermite polynomials connected with the Galilean group do not depend on the group variable r_1 since the following equality takes place

$$H_{n,n}^{\{R(r_1,i_1)\}}(0) = \nu!, \tag{4.42}$$

where $\nu=\max(n_0, n_1)$, $\mu=\min(n_0,n_1)$ and all the dependence on the parameter r_1 is contained in the Hermite polynomial $H^{(R)}_{\mu,\nu,\nu,\mu}(0)$, $\mu<\nu$.

The group multiplication rule of the finite transformation operators creates the adding formulae for the matrix elements of these operators. One-parametric groups $SO_2(j_1)$ have very simple multiplication rule $\Xi(r_1,j_1)\Xi(r'_1,j_1)=\Xi(r''_1,j_1)$ where $r''_1=r_1+r'_1$. Because of this the adding formula (4.32) for the Hermite polynomials at zero-valued variables takes the form

$$H_{m,n}^{\{R(r_1+r'_1,j_1)\}}(0) = \sum_{k=0}^{\infty}\frac{1}{k!}H_{m\,k}^{\{R(r_1,j_1)\}}(0)\,H_{k,n}^{\{R(r'_1,j_1)\}}(0). \tag{4.43}$$

The sum rule (4.33) for Hermite polynomials is expressed by the relation

$$\sum_{m=0}^{\infty}\frac{1}{m!}\left|H_{n,m}^{\{R(r_1,j_1)\}}(0)\right|^2 = n!. \tag{4.44}$$

4.2 One-parametric Group of Hyperbolic Rotations

According to the relation (3.9) the symmetric matrix generator $X_{01}(i)=\begin{Vmatrix}0 & 1\\ 1 & 0\end{Vmatrix}$ of the hyperbolic rotation $SO_2(i)$ corresponds to the generator $\hat{L}_{01}(i)=\hat{\alpha}_o^+\hat{\alpha}_1^+-\varepsilon\hat{\alpha}_o\hat{\alpha}_1$ in the Jordan-Schwinger representation, and the Hamiltonian $\hat{H}(i)=r_1\hat{L}_{01}(i)$ may be rewritten in the form (4.3)

$$\hat{H}(i) = \frac{1}{2}(\hat{\mathbf{a}},\hat{\mathbf{a}}^+)\begin{Vmatrix}-r_1h^T & 0\\ 0 & r_1h\end{Vmatrix}\begin{pmatrix}\hat{\mathbf{a}}\\ \hat{\mathbf{a}}^+\end{pmatrix}, \quad h=\begin{Vmatrix}0 & 1\\ \varepsilon & 0\end{Vmatrix}. \tag{4.45}$$

The formula (4.21) provides the relation

$$\Lambda(r_1,i) = \exp\Sigma\mathcal{B}(r_0 i) = \exp\begin{Vmatrix}0 & r_1h\\ r_1h & 0\end{Vmatrix} = \begin{Vmatrix}E_2\operatorname{ch}r_1\sqrt{\varepsilon} & h\operatorname{sh}r_1\sqrt{\varepsilon}\\ h\operatorname{sh}r_1\sqrt{\varepsilon} & E_2\operatorname{ch}r_1\sqrt{\varepsilon}\end{Vmatrix}, \tag{4.46}$$

i.e.

$$\xi=\xi_1=E_2\operatorname{ch}r_1\sqrt{\varepsilon}, \qquad \eta=\eta_1=h\operatorname{sh}r_1\sqrt{\varepsilon}, \tag{4.47}$$

where E_2 is the two-dimensional unity matrix. From these relations we have the formulae

$$\xi^{-1}=E_2(\operatorname{ch}r_1\sqrt{\varepsilon})^{-1}, \qquad \xi^{-1}\eta=\eta_1\xi^{-1}=h\ \operatorname{th}r_1\sqrt{\varepsilon}, \qquad \det\xi=\operatorname{ch}^2r_1\sqrt{\varepsilon} \tag{4.48}$$

and, according to (4.19), the kernel of the hyperbolic rotation operator in the coherent state basis has the form

$$U(\alpha^*,\beta,r_1,i) = [\operatorname{ch}r_1\sqrt{\varepsilon}]^{-\varepsilon}\exp\{(\operatorname{ch}r_1\sqrt{\varepsilon})^{-1}[\alpha_0^*\beta_0+\alpha_1^*\beta_1+ \\ +(\beta_1\beta_0-\alpha_0^*\alpha_1^*)\operatorname{sh}r_1\sqrt{\varepsilon}]\}. \tag{4.49}$$

In the boson case ($\varepsilon=1$) this kernel is the generating function for the Hermite polynomials $H^{(R(r_1,i))}_{m,n}(0)$ determined by the matrix (4.20) which for the group $SO_2(i)$ has the form

$$R(r_1, i) = \left\| \begin{matrix} \xi^{-1}\eta & -\xi^{-1} \\ -\xi^{-1T} & -\eta_1\xi^{-1} \end{matrix} \right\| = (\mathrm{ch}\, r_1)^{-1} \left\| \begin{matrix} h\,\mathrm{sh}\, r_1 & -E_2' \\ -E_2 & -h\,\mathrm{sh}\, r_1 \end{matrix} \right\|. \tag{4.50}$$

Let us write down some of the Hermite polynomials

$$\begin{aligned} &H_{0,0} = 1, \quad H_{1,0;1,0} = H_{0,1;0,1} = (\mathrm{ch}\, r_1)^{-1}, \quad H_{1,0;0,1} = H_{0,1;1,0} = 0, \\ &H_{n,0;0,n} = H_{0,n;0,n} = (\mathrm{ch}\, r_1)^{-n}\,\Gamma(n+1), \quad H_{n,0;0,n} = H_{0,n;n,0} = 0, \\ &H_{1,1;1,1} = (\mathrm{ch}\, r_1)^{-2}(1 - \mathrm{sh}^2 r_1), \quad H_{2,1;2,1} = H_{1,2;1,2} = \\ &= 2(\mathrm{ch}\, r_1)^{-3}(1 - 2\,\mathrm{sh}^2 r_1), \quad H_{2,1;1,2} = H_{1,2;2,1} = 0. \end{aligned} \tag{4.51}$$

Since in this case the rule of the composition of the group parameters has the form $r_1'' = r_1 + r_1'$ and det $\xi(r_1) = \mathrm{ch}^2 r_1$, the relation (4.32) leads to the following adding formula for the Hermite polynomials corresponding to the group $SO_2(i)$

$$H_{\mathbf{m},\mathbf{n}}^{\{R(r_1+r_1', i)\}}(0) = \frac{\mathrm{ch}\,(r_1 + r_1')}{\mathrm{ch}\, r_1 \cdot \mathrm{ch}\, r_1'} \sum_{\mathbf{k}=0}^{\infty} \frac{1}{\mathbf{k}!} H_{\mathbf{m},\mathbf{k}}^{\{R(r_1, i)\}}(0)\, H_{\mathbf{k},\mathbf{n}}^{\{R(r_1', i)\}}(0), \tag{4.52}$$

and the sum rule (4.33) may be written in the form

$$\sum_{\mathbf{m}=0}^{\infty} \frac{1}{\mathbf{m}!} |H_{\mathbf{n},\mathbf{m}}^{\{R(r_1, i)\}}(0)|^2 = \mathbf{n}!\,\mathrm{ch}^2 r_1. \tag{4.53}$$

4.3 Rotation Group SO_3 and Its Contractions

Here we will consider the group $SO_3(\mathbf{j})$ when each of the parameters j_k takes two values $j_1 = 1, \iota_1, j_2 = 1, \iota_2$, i.e. we will study only the contractions of the rotation group SO_3. We will not study the analytical continuations of the group parameters and restrict ourselves by the boson case ($\varepsilon = 1$), then $\psi a^+ = a^+$, $\psi a = a$ and the formulae (3.21), (3.22) give

$$\begin{aligned} L_{01}(\mathbf{j}) &= \mathbf{a}^+ X_1(\mathbf{j})\,\mathbf{a} = a_0 a_1^+ - j_1^2 a_0^+ a_1 = L_1(\mathbf{j}), \\ L_{02}(\mathbf{j}) &= \mathbf{a}^+ X_2(\mathbf{j})\,\mathbf{a} = a_0 a_2^+ - j_1^2 j_2^2 a_0^+ a_2 = L_2(\mathbf{j}), \\ L_{12}(\mathbf{j}) &= \mathbf{a}^+ X_3(\mathbf{j})\,\mathbf{a} = a_1 a_2^+ - j_2^2 a_1^+ a_2 = L_3(\mathbf{j}). \end{aligned} \tag{4.54}$$

First we will obtain the matrix elements of the generators L_λ and the Casimir operator L^2 in the coherent state basis and in the Fock basis and then we will study the finite rotations. It may be easily done knowing the matrix elements for the creation and annihilation operators in both bases

$$\begin{aligned} &a_k |\boldsymbol{\beta}\rangle = \beta_k |\boldsymbol{\beta}\rangle, \quad \langle\boldsymbol{\alpha}| a_k^+ = \alpha_k^* \langle\boldsymbol{\alpha}|, \\ &a_k |\mathbf{n}\rangle = \sqrt{n_k}\,|n_k - 1\rangle, \quad a_k^+ |\mathbf{n}\rangle = \sqrt{n_k + 1}\,|n_k + 1\rangle. \end{aligned} \tag{4.55}$$

In the last formulae we take into account that the components of the vector $\mathbf{n}$-(n_0, n_1, n_2) which differ from the components n_k do not change. Then we have

$$\begin{aligned} &\mathcal{L}_1(\boldsymbol{\alpha}^*, \boldsymbol{\beta}) = (\alpha_1^* \beta_0 - j_1^2 \alpha_0^* \beta_1)\, e^{\boldsymbol{\alpha}^* \boldsymbol{\beta}}, \quad \mathcal{L}_2(\boldsymbol{\alpha}^*, \boldsymbol{\beta}) = (\alpha_2^* \beta_0 - j_1^2 j_2^2 \alpha_0^* \beta_2)\, e^{\boldsymbol{\alpha}^* \boldsymbol{\beta}}, \\ &\mathcal{L}_3(\boldsymbol{\alpha}^*, \boldsymbol{\beta}) = (\alpha_2^* \beta_1 - j_2^2 \alpha_1^* \beta_2)\, e^{\boldsymbol{\alpha}^* \boldsymbol{\beta}}, \end{aligned} \tag{4.56}$$

where $L_\lambda(\alpha^*, \beta)$ is the kernel of the operator L_λ in the coherent state representation. In the Fock basis we have

$$\begin{aligned}\langle \mathbf{m} | L_1 | \mathbf{n}\rangle &= \sqrt{n_1(n_2+1)}\,\delta_{m_1,n_1-1}\delta_{m_2,n_2+1}\delta_{m_3,n_3} - \\ &- j_1^2\sqrt{(n_1+1)n_2}\,\delta_{m_1,n_1+1}\delta_{m_2,n_2-1}\delta_{m_3,n_3},\\ \langle \mathbf{m} | L_2 | \mathbf{n}\rangle &= \sqrt{n_1(n_3+1)}\,\delta_{m_1,n_1-1}\delta_{m_2,n_2}\delta_{m_3,n_3+1} - \\ &- j_1^2 j_2^2\sqrt{(n_1+1)n_3}\,\delta_{m_1,n_1+1}\delta_{m_2,n_2}\delta_{m_3,n_3-1},\\ \langle \mathbf{m} | L_3 | \mathbf{n}\rangle &= \sqrt{n_2(n_3+1)}\,\delta_{m_1,n_1}\delta_{m_2,n_2-1}\delta_{m_3,n_3+1} - \\ &- j_2^2\sqrt{(n_2+1)n_3}\,\delta_{m_1,n_1}\delta_{m_2,n_2+1}\delta_{m_3,n_3-1}. \end{aligned} \tag{4.57}$$

The Casimir operation (3.25) is expressed in terms of the creation and annihilation operators in the form

$$\begin{aligned} L^2(\mathrm{j}) &= (a_2^+a_0)^2 + j_1^2(a_2^+a_1)^2 + j_2^2(a_1^+a_0)^2 + j_1^2j_2^2\{j_1^2(a_0^+a_1)^2 + j_1^2j_2^2(a_0^+a_2)^2 + \\ &+ j_2^2(a_1^+a_2)^2 - [(\sum_{k=0}^{2} a_k^+a_k)(2 + \sum_{k=0}^{2} a_k^+a_k) - \sum_{k=0}^{2}(a_k^+a_k)^2]\}. \end{aligned} \tag{4.58}$$

Its kernel $L^2(\alpha^*,\beta)$ in the coherent state basis may be obtained from the expression (4.58) if in the right side of the equality one replaces the operators α_k by the numbers β_k and the operators $\alpha^+{}_k$ by the numbers $\alpha^*{}_k$ adding the factor $\exp(\alpha^*,\beta)$. The matrix elements of the Casimir operator in the Fock basis are as follows

$$\begin{aligned} \langle \mathbf{m} | L^2(\mathrm{j}) | \mathbf{n}\rangle &= [n_0(n_0-1)(n_2+1)(n_2+2)]^{1/2}\langle \mathbf{m} | n_0-2, n_1, n_2+2\rangle + \\ &+ j_1^2[n_1(n_1-1)(n_2+1)(n_2+2)]^{1/2}\langle \mathbf{m} | n_0, n_1-2, n_2+2\rangle + \\ &+ j_2^2[n_0(n_0-1)(n_1+1)(n_1+2)]^{1/2}\langle \mathbf{m} | n_0-2, n_1+2, n_2\rangle + \\ &+ j_1^2j_2^2\{j_1^2[(n_0+1)(n_0+2)n_1(n_1-1)]^{1/2}\langle \mathbf{m} | n_0+2, n_1-2, n_2\rangle + \\ &+ j_1^2j_2^2[(n_0+1)(n_0+2)n_2(n_2-1)]^{1/2}\langle \mathbf{m} | n_0+2, n_1, n_2-2\rangle + \\ &+ j_2^2[(n_1+1)(n_1+2)n_2(n_2-1)]^{1/2}\langle \mathbf{m} | n_0, n_1+2, n_2-2\rangle\} - \\ &- j_1^2j_2^2[(n_0+n_1+n_2)(2+n_0+n_1+n_2) - n_0^2 - n_1^2 - n_2^2]^{1/2}\langle \mathbf{m} | \mathbf{n}\rangle. \end{aligned} \tag{4.59}$$

It follows from the expression that on the state $|\mathrm{n}\rangle = |1,1,1\rangle$ the Casimir operator is proportional to the unity operator

$$L^2(\mathrm{j})\,|1, 1, 1\rangle = -12 j_1^2 j_2^2\,|1, 1, 1\rangle. \tag{4.60}$$

The unitary operator $U_g(\mathbf{r},\mathrm{j}) = \exp(-\sum_{\lambda=1}^{3} r_\lambda L_\lambda(j))$ corresponds to the operators of the finite rotation (2.26) in the space $R_3(\mathrm{j})$ in the boson realization. The kernel of this operator in the coherent state basis is described by the formula (4.19) where the matrices $\eta = \eta_1 = 0$, and the matrix $\xi^{-1} = \Xi(-\mathbf{r},\mathrm{j})$. Explicitly it has the form

$$\begin{aligned} U(\alpha^*, \beta, \mathbf{r}, \mathrm{j}) &= \exp\{\alpha_0^*\beta_0(\cos r + j_2^2 r_3^2(1-\cos r)/r^2) + \alpha_0^*\beta_1 j_1^2(r_1(\sin r)/r - \\ &- j_2^2 r_2 r_3(1-\cos r)/r^2) + \alpha_0^*\beta_2 j_1^2 j_2^2(r_2(\sin r)/r + r_1 r_3(1-\cos r)/r^2) - \\ &- \alpha_1^*\beta_0(r_1(\sin r)/r + j_2^2 r_2 r_3(1-\cos r)/r^2) + \alpha_1^*\beta_1(\cos r + j_1^2 j_2^2 r_2^2 \times \\ &\times (1-\cos r)/r^2) + \alpha_1^*\beta_2 j_2^2(r_3(\sin r)/r - j_1^2 r_1 r_2(1-\cos r)/r^2) + \\ &+ \alpha_2^*\beta_0(-r_2(\sin r)/r + r_1 r_3(1-\cos r)/r^2) - \alpha_2^*\beta_1(r_3(\sin r)/r + \\ &+ j_1^2 r_1 r_2(1-\cos r)/r^2) + \alpha_2^*\beta_2(\cos r + j_1^2 r_1^2(1-\cos r)/r^2)\}. \end{aligned} \tag{4.61}$$

The matrix elements of the operator $U_g(\mathbf{r},\mathrm{j})$ in the Fock state basis may be found from the relation (4.61) using the formula (4.25). Writing down the relation (4.61) in the form (4.19) where the matrix R is as follows

$$R(\mathbf{r},\mathrm{j}) = \begin{Vmatrix} 0 & -\xi^{-1} \\ -\xi^{-1T} & 0 \end{Vmatrix} = \begin{Vmatrix} 0 & -\Xi(-\mathbf{r},\mathrm{j}) \\ -\Xi^T(-\mathbf{r},\mathrm{j}) & 0 \end{Vmatrix}, \tag{4.62}$$

and comparing it with the relation (4.26) one can conclude that the function $U(\alpha^*,\beta,r,j)$ is the generating function for the Hermite polynomials $H^{(R)}_{m,n}(0)$ of six variables $z=(z_1,\dots,z_6)$ calculated at $z=0$. These polynomials are obtained from the relation (4.61) using the formula (4.38) where $m=(m_0,m_1,m_2)$,

$$n=(n_0,n_1,n_2),\partial\alpha^{*m}=\partial\alpha_0^{*m_0}\partial\alpha_1^{*m_1}\partial\alpha_2^{*m_2},\partial\beta^n=\partial\beta_0^{n_0}\partial\beta_1^{n_1}\partial\beta_2^{n_2}$$

and the matrix R is given by the formula (4.62). Some first Hermite polynomials have the form

$$\begin{aligned} &H^{\{R\}}_{0,0}(0)=1,\ H^{\{R\}}_{\mathbf{m}_k,\mathbf{n}_s}(0)=[\Xi_{ks}(-\mathbf{r},\mathbf{j})]^n\,\Gamma(n+1),\\ &H^{\{R\}}_{1,1}(0)=\cos^2 r-\sin^2 r\cdot\cos r+((1-\cos r)^2/r^4)\,[6j_1^4j_2^4r_1^2r_2^2r_3^2(1-\cos r)/r^2+\\ &\quad+2j_1^2j_2^2(j_1^2r_1^2r_2^2+r_1^2r_3^2+j_2^2r_2^2r_3^2)\cos r+(1+\cos r)\times\\ &\quad\times[-j_1^4r_1^4-j_1^4j_2^4r_2^4-j_2^4r_3^4+2j_1^2j_2^2(j_1^2r_1^2r_2^2+r_1^2r_3^2+j_2^2r_2^2r_3^2)]], \end{aligned} \tag{4.63}$$

where $m_k(n_s)$ is the vector in which k-th (s-th) component is equal to the number n and all the other components are equal to zero. The formulae demonstrated describes at $j_1=j_2=1$ the boson realization of the rotation group SO_3. Taking the parameters j_1, j_2 to be pure dual units, we consider the different contractions of the rotation group and obtain the boson realization of these contracted groups.

At $j_1=\iota_1$ the rotation group transforms into the Euclidean group $SO_3(\iota_1, 1)$. For this group $r(\iota_1,j_2)=j_2r_3$ and the matrix $\xi^{-1}\Xi(-r,\iota_1,j_2)$ has the form

$$\Xi(-\mathbf{r},\iota_1,j_2)=\left\|\begin{matrix} 1 & 0 & 0\\ -r_1\dfrac{\sin j_2r_3}{j_2r_3}-\dfrac{r_2}{r_3}(1-\cos j_2r_3) & \cos j_2r_3 & j_2\sin j_2r_3\\ -r_2\dfrac{\sin j_2r_3}{j_2r_3}+r_1\dfrac{1-\cos j_2r_3}{j_2^2r_3} & -\dfrac{1}{j_2}\sin j_2r_3 & \cos j_2r_3 \end{matrix}\right\|. \tag{4.64}$$

We keep the parameter j_2 which is equal to unity for the Euclidean group in order to consider the transformation of the Euclidean group into the Galilean group $SO_3(\iota_1,\iota_2)$ due to the contraction. The kernel of the operator of finite group transformations is as follows

$$\begin{aligned} U(\alpha^*,\beta;\ \iota_1,j_2)&=\exp\{\alpha_0^*\beta_0-\alpha_1^*\beta_0[r_1(\sin j_2r_3)/j_2r_3+\\ &+r_2(1-\cos j_2r_3)/r_3]+\alpha_1^*\beta_1\cos j_2r_3+\alpha_1^*\beta_2j_2\sin j_2r_3+\\ &+\alpha_2^*\beta_0[-r_2(\sin j_2r_3)/j_2r_3+r_1(1-\cos j_2r_3)/r_3]-\alpha_2^*\beta_1j_2^{-1}\sin j_2r_3+\\ &+\alpha_2^*\beta_2\cos j_2r_3\}. \end{aligned} \tag{4.65}$$

Let us write down some first Hermite polynomials

$$\begin{aligned} &H^{\{R\}}_{0,0}(0)=1,\quad H^{\{R\}}_{\mathbf{m}_k,\mathbf{n}_s}(0)=[\Xi_{ks}(-\mathbf{r};\iota_1,j_2)]^n\,\Gamma(n+1),\\ &H^{\{R\}}_{\mathbf{m}_k,\mathbf{n}_s}(0)=0,\quad k=0,s=1,2,\quad H^{\{R\}}_{\mathbf{m}_0,\mathbf{n}_0}(0)=\Gamma(n+1),\\ &H^{\{R\}}_{1,1}(0)=\cos^2 j_2r_3-\sin^2 j_2r_3. \end{aligned} \tag{4.66}$$

Due to the contraction at $j_2=\iota_2$ the rotation group transforms into the semispherical group $SO_3(1,\iota_2)$. In this case $r(j_1,\iota_2)=j_1r_1$, the matrix $\xi^{-1}=\Xi(-r,j_1,\iota_2)$ and

$$\Xi(-\mathbf{r},j_1,\iota_2)=\left\|\begin{matrix} \cos j_1r_1 & j_1\sin j_1r_1 & 0\\ -j_1^{-1}\sin j_1r_1 & \cos j_1r_1 & 0\\ -r_2\dfrac{\sin j_1r_1}{j_1r_1}+r_3\dfrac{1-\cos j_1r_1}{j_1^2r_1} & -r_3\dfrac{\sin j_1r_1}{j_1r_1}-r_2\dfrac{1-\cos j_1r_1}{r_1} & 1 \end{matrix}\right\|. \tag{4.67}$$

We keep the parameter j_1 which is equal to unity for the semispherical group to consider the transformation of this group into the Galilean group $SO_3(\iota_1,\iota_2)$ due to the contraction. We will write down the kernel of the operator (4.17) of the finite group transformations in the coherent state basis

$$U(\alpha^*, \beta, j_1, \iota_2) = \exp\{\alpha_0^*\beta_0 \cos j_1 r_1 + \alpha_0^*\beta_1 j_1 \sin j_1 r_1 - \alpha_1^*\beta_0 j_1^{-1} \sin j_1 r_1 + \\ + \alpha_1^*\beta_1 \cos j_1 r_1 + \alpha_2^*\beta_2 + \alpha_2^*\beta_0\left(-r_2 \frac{\sin j_1 r_1}{j_1 r_1} + r_3 \frac{1-\cos j_1 r_1}{j_1^2 r_1}\right) - \\ - \alpha_2^*\beta_1\left(r_3 \frac{\sin j_1 r_1}{j_1 r_1} + r_2 \frac{1-\cos j_1 r_1}{r_1}\right)\} = U_2(\alpha^*, \beta, j_1) \exp \times \\ \times \left\{\alpha_2^*\left[\beta_2 + \beta_0\left(-r_2 \frac{\sin j_1 r_1}{j_1 r_1} + r_3 \frac{1-\cos j_1 r_1}{j_1^2 r_1^2}\right) - \right.\right. \\ \left.\left. - \beta_1\left(r_3 \frac{\sin j_1 r_1}{j_1 r_1} + r_2 \frac{1-\cos j_1 r_1}{r_1}\right)\right]\right\}, \tag{4.68}$$

where the function $U_2(\alpha^*,\beta,j_1)$ is the kernel of the operator of the finite rotations of the group $SO_2(j_1)$ (see also (4.36)). The Hermite polynomials, connected with the semispherical group, i.e., with the matrix (4.62) where $\Xi(-r,j)$ is given by the formula (4.67), have the form

$$H_{0,0}^{\{R\}}(0) = 1, \quad H_{m_k, n_s}^{\{R\}}(0) = 0, \quad k = 0, 1, \quad s = 2, \\ H_{m_2, n_2}^{\{R\}}(0) = \Gamma(n+1), \quad H_{1,1}^{\{R\}}(0) = \cos^2 j_1 r_1 - \sin^2 j_1 r_1. \tag{4.69}$$

The simplest formulae are obtained for $j_1=\iota_1$, $j_2=\iota_2$, i.e. for the case when the rotation group is transformed by the contraction into the Galilean group $SO_3(\iota)$. Then $r(\iota)=0$, $(\sin r)/r=1$, $(1-\cos r)/r^2=1/2$ and the formula (2.26) provides for the matrix $\Xi(-r,\iota)$ the expression

$$\Xi(-\mathbf{r}, \iota) = \begin{Vmatrix} 1 & 0 & 0 \\ -r_1 & 1 & 0 \\ -r_2 + {}^1/_2 r_1 r_3 & -r_3 & 1 \end{Vmatrix}. \tag{4.70}$$

For the Galilean group the kernel of the operator of the finite transformations in the Jordan-Schwinger representation has the form

$$U(\alpha^*, \beta, \mathbf{r}, \iota) = \exp\{\alpha_0^*\beta_0 + \alpha_1^*\beta_1 + \alpha_2^*\beta_2 - \alpha_1^*\beta_0 r_1 - \alpha_2^*\beta_1 r_3 + \\ + \alpha_2^*\beta_0(-r_2 + {}^1/_2 r_1 r_3)\}, \tag{4.71}$$

and the Hermite polynomials are as follows:

$$H_{0,0}^{\{R\}}(0) = 1, \quad H_{m_k, n_s}^{\{R\}}(0) = 0, \quad k = 0,\ s = 1, 2; \quad k = 1,\ s = 2, \\ H_{m_k, n_k}^{\{R\}}(0) = \Gamma(n+1), \quad k = 0, 1, 2, \quad H_{1,1}^{\{R\}}(0) = 1, \\ H_{m_1, n_0}^{\{R\}}(0) = (-1)^n r_1^n \Gamma(n+1), \quad H_{m_2, n_1}^{\{R\}}(0) = (-1)^n r_3^n \Gamma(n+1), \\ H_{m_2, n_0}^{\{R\}}(0) = (-r_2 + {}^1/_2 r_1 r_3)^n \Gamma(n+1). \tag{4.72}$$

If one obtains the Galilean group $SO_3(\iota)$ either from the Euclidean group $SO_3(\iota_1,j_2)$ (taking $j_2=\iota_2$ in the formulae (4.64)-(4.66) or from the semispherical group $SO_3(j_1,\iota_2)$ (taking $j_1=\iota_1$ in the formulae (4.67)-(4.69)) the same expressions (4.70)-(4.72) will be found. In other words, subsequent one-dimensional contractions (in one parameter) provide the same result that may be obtained due to the multidimensional contraction (in our case two-dimensional one).

Contraction transforms some nondiagonal elements of the matrix Ξ into zeros and it provides the equality of some of the Hermitian polynomials to zero. Some of the diagonal elements of the matrix Ξ become equal to unity and, consequently, some of the Hermite polynomials do not, depending in this case, on the group parameters r. For the Galilean group the Hermite polynomials $H^{(R)}_{m,m}(0)$ do not depend on r, the explicit form of the polynomials being as follows:

$$H^{\{R\}}_{m,m}(0) = \Gamma(\nu + 1), \qquad \nu = \max(m_0, m_1, m_2). \tag{4.73}$$

Due to the group composition rule Ξ(r",j)=Ξ(r,j)Ξ(r',j'), r"=r⊕r', and taking into account that detξ=detΞ=1 one obtains the adding formula for the Hermite polynomials

$$\sum_{k=0}^{\infty} \frac{1}{k!} H^{\{R(r)\}}_{m,k}(0)\, H^{\{R(r')\}}_{k,n}(0) = H^{\{R(r'')\}}_{m,n}(0), \tag{4.74}$$

where $k!=k_0!k_1!k_2!$. The simplest form of the group multiplication rule takes place in the case of the Galilean group

$$r'' = r \oplus r' = (r_1 + r_1', r_2 + r_2' - {}^1\!/_2(r_3 r_1' - r_1 r_3'), r_3 + r_3'). \tag{4.75}$$

It is worthy of note that the sum rule of the type (4.33) holds for the Hermite polynomials related to the rotation group $SO_3(1,1)$. If one considers the contractions of the rotation group (i.e., one takes $j_1=\iota_1$, or $j_2=\iota_2$, or $j_2=\iota_1$, $j_2=\iota_2$) the Hermite polynomials related to the contracted groups do not obey the sum rules since the operator (4.17) of the finite group transformations becomes a nonunitary one.

4.4 Lobachevski Group and Semihyperbolic Group in Boson Realization

We have considered the rotation group SO_3 and the groups to which this group transforms due to the contractions. The Lobachevski group $SO_3(i,1)$ may be obtained from the rotation group by the analytical continuation which means in terms of the parameters j_k that $j_1=i$. In order to conserve the rule of the boson realization when the matrix generator X, which may be symmetrical or an antisymmetrical one is corresponded to the antihermitian operator L, the bosonization procedure of the generator X should be done according to the formula (3.21). Due to this change of the bosonization rule, one can not simultaneously consider all nine groups $SO_3(j)$. The groups with some of the parameters j_k equal to imaginary unity must be considered separately. Other parameters are equal either unity or pure dual units. The last case leads to the group contractions.

Let parameter $j_1=i$ and the second parameter $j_2=1$, ι_2 that corresponds to the Lobachevski group $SO_3(\iota,1)$ and the semihyperbolic group $SO_3(i,\iota_2)$. The matrix generators of the group $SO_3(i,j_2)$ are given by the formula (2.21) at $j_1=i$ and they have the form

$$X_1 = \begin{Vmatrix} 0 & 1 & 0 \\ 1 & 0 & 0 \\ 0 & 0 & 0 \end{Vmatrix}, \quad X_2 = \begin{Vmatrix} 0 & 0 & j_2^2 \\ 0 & 0 & 0 \\ 1 & 0 & 0 \end{Vmatrix}, \quad X_3 = \begin{Vmatrix} 0 & 0 & 0 \\ 0 & 0 & -j_2^2 \\ 0 & 1 & 0 \end{Vmatrix}. \tag{4.76}$$

The generators satisfy the commutation relations

$$[X_1, X_2] = -X_3, \qquad [X_2, X_3] = j_2^2 X_1, \qquad [X_1, X_3] = -X_2. \tag{4.77}$$

At $j_1=i$ the formula (3.18) give $\psi a^+=(\alpha^+_0,-\alpha_1,-\alpha_2)$, $\psi a=(\alpha_0,\alpha^+_1,\alpha^+_2)$. Using the formula (3.21) we construct the boson representation of the generator (4.76) and obtain

$$L_1(i, j_2) = \overset{+}{a_0}\overset{+}{a_1} - a_0 a_1, \quad L_2(i, j_2) = j_2^2 \overset{+}{a_0}\overset{+}{a_2} - a_0 a_2, \quad L_3(i, j_2) = j_2^2 a_1 \overset{+}{a_2} - \overset{+}{a_1} a_2. \tag{4.78}$$

The Hamiltonian $H(i,j_2) = \sum_{\lambda=1}^{3} r_\lambda L_\lambda(i,j_2)$ of the quantum system corresponding to the group $SO_3(i,j_2)$ may be written down in the equivalent form (4.3) with the matrix

$$\mathcal{B}(\mathbf{r}, i, j_2) = \begin{Vmatrix} -c & -b \\ -b^T & d \end{Vmatrix}, \tag{4.79}$$

where the matrices b, c, d are as follows:

$$b = \begin{Vmatrix} 0 & 0 & 0 \\ 0 & 0 & -j_2^2 r_3 \\ 0 & r_3 & 0 \end{Vmatrix}, \quad c = \begin{Vmatrix} 0 & r_1 & r_2 \\ r_1 & 0 & 0 \\ r_2 & 0 & 0 \end{Vmatrix}, \quad d = \begin{Vmatrix} 0 & r_1 & j_2^2 r_2 \\ r_1 & 0 & 0 \\ j_2^2 r_2 & 0 & 0 \end{Vmatrix}. \tag{4.80}$$

Then, the equation (4.7) in which the variable t is replaced by the variable $\tau = i\hbar^{-1}\tau$, leads to the system of differential equations for the matrices, ξ, ξ_1, η, η_1

$$\dot{\xi} = -\xi b^T + \eta c, \quad \xi(0) = E; \quad \dot{\xi}_1 = \xi_1 b + \eta_1 d, \quad \xi_1(0) = E;$$

$$\dot{\eta} = \xi d + \eta b, \quad \eta(0) = 0; \quad \dot{\eta}_1 = \xi_1 c - \eta_1 b^T, \quad \eta_1(0) = 0. \tag{4.81}$$

Before solving this system, let us consider the following system of the first order differential equations with the constant coefficients

$$\dot{x} = 0 + \alpha y + \beta z, \quad \dot{y} = \gamma x + 0 + \delta z, \quad \dot{z} = \mu x + \nu y + 0, \tag{4.82}$$

where the real coefficients of this system satisfy the conditions: $\alpha\mu\delta + \beta\gamma\nu = 0$, $\Delta \equiv \mu^2\delta = \gamma^2\nu \neq 0$. Then the characteristic polynomial of the system (4.82) has the form $s(s^2 - v^2) = 0$ where the notation $v^2 = \alpha\gamma + \beta\mu + \delta\nu$ is introduced and the roots of the polynomial are $s_1 = 0$, $s_2 = -v$, $s_3 = v$. Thus, every solution to the system (4.82) is expressed in terms of three linearly independent solutions: C_1=const, $C_2\varepsilon^{vt}$, $C_3 e^{-vt}$. It is more convenient to choose as the independent solutions C_1, $C_2 \operatorname{sh} vt$, $C_3 \operatorname{ch} vt$. The general solution to the system (4.82) may be written down in the form

$$x(t) = A \operatorname{sh} vt + B \operatorname{ch} vt + C,$$

$$y(t) = \frac{\delta\mu A + v\gamma B}{v^2 - \delta\nu} \operatorname{sh} vt + \frac{v\gamma A + \delta\mu B}{v^2 - \delta\nu} \operatorname{ch} vt - \frac{\mu}{\nu} C,$$

$$z(t) = \frac{\gamma\nu A + v\mu B}{v^2 - \delta\nu} \operatorname{sh} vt + \frac{v\mu A + \gamma\nu B}{v^2 - \delta\nu} \operatorname{ch} vt - \frac{\gamma}{\delta} C, \tag{4.83}$$

where A, B, C are the constant determined by the initial conditions.

The initial conditions $x(0)=1$, $y(0)=0$, $z(0)=0$ give the solution

$$x(t) = \operatorname{ch} vt - \delta\nu v^{-2} (\operatorname{ch} vt - 1), \quad y(t) = \gamma v^{-1} \operatorname{sh} vt + \mu\delta v^{-2} (\operatorname{ch} vt - 1), \quad z(t) = \mu v^{-1} \operatorname{sh} vt + \gamma\nu v^{-2} (\operatorname{ch} vt - 1). \tag{4.84}$$

If $x(0)=0$, $y(0)=1$, $z(0)=0$ the solution is

$$x(t) = \Delta^{-1}\nu (v^2 - \delta\nu) \left(-\gamma \frac{\operatorname{sh} vt}{v} + \delta\mu \frac{\operatorname{ch} vt - 1}{v^2} \right),$$

$$y(t) = \operatorname{ch} vt - \Delta^{-1}\delta\mu^2 (v^2 - \delta\nu) \frac{\operatorname{ch} vt - 1}{v^2},$$

$$z(t) = \nu \frac{\operatorname{sh} vt}{v} - \Delta^{-1}\gamma\mu\nu (v^2 - \delta\nu) \frac{\operatorname{ch} vt - 1}{v^2}. \tag{4.85}$$

For $x(0)=0$, $y(0)=0$, $z(0)=1$ the solution is

$$x(t) = \Delta^{-1}\delta(v^2 - \delta\nu)\left(\mu\frac{\mathrm{sh}\, vt}{v} - \gamma\nu\frac{\mathrm{ch}\, vt - 1}{v^2}\right),$$

$$y(t) = \delta\frac{\mathrm{sh}\, vt}{v} + \Delta^{-1}\delta\gamma\mu(v^2 - \delta\nu)\frac{\mathrm{ch}\, vt - 1}{v^2},$$

$$z(t) = \mathrm{ch}\, vt + \Delta^{-1}\gamma^2\nu(v^2 - \delta\nu)\frac{\mathrm{ch}\, vt - 1}{v^2}. \tag{4.86}$$

For zero initial conditions one has zero solution.

It may be checked that the system (4.81) with the coefficients (4.80) is split giving independent subsystems of equations of the type (4.82). For example, for $\xi_{00}, \eta_{01}, \eta_{02}$, for $\xi_{01}, \xi_{02}, \eta_{00}$, etc. Using now the formulae (4.84)-(4.86) we obtain the matrices

$$\xi(\tau) = \begin{Vmatrix} \mathrm{ch}\, r\tau + j_2^2 r_3^2\frac{\mathrm{ch}\, r\tau - 1}{r^2} & 0 & 0 \\ 0 & \mathrm{ch}\, r\tau - j_2^2 r_2^2\frac{\mathrm{ch}\, r\tau - 1}{r^2} & -r_3\frac{\mathrm{sh}\, r\tau}{r} + r_1 r_2\frac{\mathrm{ch}\, r\tau - 1}{r^2} \\ 0 & j_2^2\left(r_3\frac{\mathrm{sh}\, r\tau}{r} + r_1 r_2\frac{\mathrm{ch}\, r\tau - 1}{r^2}\right) & \mathrm{ch}\, r\tau - r_1^2\frac{\mathrm{ch}\, r\tau - 1}{r^2} \end{Vmatrix}, \tag{4.87}$$

$$\xi_1(\tau) = \begin{Vmatrix} \mathrm{ch}\, r\tau + j_2^2 r_3^2\frac{\mathrm{ch}\, r\tau - 1}{r^2} & 0 & 0 \\ 0 & \mathrm{ch}\, r\tau - j_2^2 r_2^2\frac{\mathrm{ch}\, r\tau - 1}{r^2} & -r_3\frac{\mathrm{sh}\, r\tau}{r} + r_1 r_2\frac{\mathrm{ch}\, r\tau - 1}{r^2} \\ 0 & r_3\frac{\mathrm{sh}\, r\tau}{r} + r_1 r_2\frac{\mathrm{ch}\, r\tau - 1}{r^2} & \mathrm{ch}\, r\tau - r_1^2\frac{\mathrm{ch}\, r\tau - 1}{r^2} \end{Vmatrix}, \tag{4.88}$$

$$\eta(\tau) = \begin{Vmatrix} 0 & r_1\frac{\mathrm{sh}\, r\tau}{r} + j_2^2 r_2 r_3\frac{\mathrm{ch}\, r\tau - 1}{r^2} & j_2^2\left(r_2\frac{\mathrm{sh}\, r\tau}{r} - r_1 r_3\frac{\mathrm{ch}\, r\tau - 1}{r^2}\right) \\ r_1\frac{\mathrm{sh}\, r\tau}{r} - j_2^2 r_2 r_3\frac{\mathrm{ch}\, r\tau - 1}{r^2} & 0 & 0 \\ j_2^2\left(r_2\frac{\mathrm{sh}\, r\tau}{r} + r_1 r_3\frac{\mathrm{ch}\, r\tau - 1}{r^2}\right) & 0 & 0 \end{Vmatrix}, \tag{4.89}$$

$$\eta_1(\tau) = \begin{Vmatrix} 0 & r_1\frac{\mathrm{sh}\, r\tau}{r} + j_2^2 r_2 r_3\frac{\mathrm{ch}\, r\tau - 1}{r^2} & r_2\frac{\mathrm{sh}\, r\tau}{r} - r_1 r_3\frac{\mathrm{ch}\, r\tau - 1}{r^2} \\ r_1\frac{\mathrm{sh}\, r\tau}{r} - j_2^2 r_2 r_3\frac{\mathrm{ch}\, r\tau - 1}{r^2} & 0 & 0 \\ r_2\frac{\mathrm{sh}\, r\tau}{r} + r_1 r_3\frac{\mathrm{ch}\, r\tau - 1}{r^2} & 0 & 0 \end{Vmatrix}. \tag{4.90}$$

The relations $\xi_1=\xi$, $\eta_1=\eta$ which are valid for the unitary operator are fulfilled also at $j_2=1$. After contraction $j_2=\iota_2$ the operator of the finite group transformations transforms into nonunitary operator and these relations are not correct.

Let us find out the matrices ξ^{-1}, $\zeta^{-1}\eta$, $\eta_1\xi^{-1}$ for $\tau=1$. We have

$$\det \xi(\mathbf{r}) = (\xi_{00})^2 = \left[\mathrm{ch}\, r + j_2^2 r_3^2\frac{\mathrm{ch}\, r - 1}{r^2}\right]^2, \qquad r^2 = r_1^2 + j_2^2 r_2^2 - j_2^2 r_3^2, \tag{4.91}$$

$$\xi^{-1} = (\det\xi)^{-1/2}\begin{Vmatrix} 1 & 0 & 0 \\ 0 & \mathrm{ch}\, r - r_1^2\frac{\mathrm{ch}\, r - 1}{r^2} & r_3\frac{\mathrm{sh}\, r}{r} - r_1 r_2\frac{\mathrm{ch}\, r - 1}{r^2} \\ 0 & -j_2^2\left(r_3\frac{\mathrm{sh}\, r}{r} + r_1 r_2\frac{\mathrm{ch}\, r - 1}{r^2}\right) & \mathrm{ch}\, r - j_2^2 r_2^2\frac{\mathrm{ch}\, r - 1}{r^2} \end{Vmatrix}, \tag{4.92}$$

$$\xi^{-1}\eta = (\det \xi)^{-1/2} \times$$
$$\times \begin{Vmatrix} 0 & r_1 \frac{\operatorname{sh} r}{r} - j_2^2 r_2 r_3 \frac{\operatorname{ch} r - 1}{r^2} & r_2 \frac{\operatorname{sh} r}{r} + r_1 r_3 \frac{\operatorname{ch} r - 1}{r^2} \\ r_1 \frac{\operatorname{sh} r}{r} + j_2^2 r_2 r_3 \frac{\operatorname{ch} r - 1}{r^2} & 0 & 0 \\ j_2^2 \left(r_2 \frac{\operatorname{sh} r}{r} - r_1 r_3 \frac{\operatorname{ch} r - 1}{r^2} \right) & 0 & 0 \end{Vmatrix}, \quad (4.93)$$

$$\eta_1 \xi^{-1} = (\det \xi)^{-1/2} \times$$
$$\times \begin{Vmatrix} 0 & r_1 \frac{\operatorname{sh} r}{r} - j_2^2 r_2 r_3 \frac{\operatorname{ch} r - 1}{r^2} & r_2 \frac{\operatorname{sh} r}{r} + r_1 r_3 \frac{\operatorname{ch} r - 1}{r^2} \\ r_1 \frac{\operatorname{sh} r}{r} - j_2^2 r_2 r_3 \frac{\operatorname{ch} r - 1}{r^2} & 0 & 0 \\ r_2 \frac{\operatorname{sh} r}{r} + r_1 r_3 \frac{\operatorname{ch} r - 1}{r^2} & 0 & 0 \end{Vmatrix}. \quad (4.94)$$

The matrices $\xi^{-1}\eta$ and $\eta_1\xi^{-1}$ are symmetrical ones including the case $j_2=\iota_2$, i.e. for the semihyperbolic group $SO_3(i,\iota_2)$.

Substituting the relations (4.92)-(4.94) into the formula (4.19) we obtain the kernel of the operator of the finite transformations in the coherent state basis for the groups $SO_3(i,j_2)$, i.e. for Lobachevski group and semispherical one

$$U(\alpha^*, \beta, \mathbf{r}, j_2) = (\det \xi)^{-1/2} \exp\left\{ \left(\operatorname{ch} r + j_2^2 r_3 \frac{\operatorname{ch} r - 1}{r^2} \right)^{-1} \left[-\alpha_0^* \alpha_1^* \left(r_1 \frac{\operatorname{sh} r}{r} + \right.\right.\right.$$
$$\left. + j_2^2 r_2 r_3 \frac{\operatorname{ch} r - 1}{r^2} \right) - j_2^2 \alpha_0^* \alpha_2^* \left(r_2 \frac{\operatorname{sh} r}{r} - j_2^2 r_1 r_3 \frac{\operatorname{ch} r - 1}{r^2} \right) + \alpha_0^* \beta_0 +$$
$$+ \alpha_1^* \beta_1 \left(\operatorname{ch} r - r_1^2 \frac{\operatorname{ch} r - 1}{r^2} \right) + \alpha_1^* \beta_2 \left(r_3 \frac{\operatorname{sh} r}{r} - r_1 r_2 \frac{\operatorname{ch} r - 1}{r^2} \right) -$$
$$- j_2^2 \alpha_2^* \beta_1 \left(r_3 \frac{\operatorname{sh} r}{r} + r_1 r_2 \frac{\operatorname{ch} r - 1}{r^2} \right) + \alpha_2^* \beta_2 \left(\operatorname{ch} r - j_2^2 r_2^2 \frac{\operatorname{ch} r - 1}{r^2} \right) +$$
$$\left.\left. + \beta_0 \beta_1 \left(r_1 \frac{\operatorname{sh} r}{r} - j_2^2 r_2 r_3 \frac{\operatorname{ch} r - 1}{r^2} \right) + \beta_0 \beta_2 \left(r_2 \frac{\operatorname{sh} r}{r} + r_1 r_3 \frac{\operatorname{ch} r - 1}{r^2} \right) \right] \right\}. \quad (4.95)$$

This kernel is the generating function for the Hermite polynomials $H_{m,n}^{(R)}(0)$ with the matrix R of the type (4.20) where the matrices ξ^{-1}, $\xi^{-1}\eta$, $\eta_1\zeta^{-1}$ are given by the formulae (4.91)-(4.94). For these Hermite polynomials the adding formula of the type (4.32) is valid, the group parameter r" in this formula being expressed in terms of r and r' by the relation $\Xi(r'')=\Xi(r)\Xi(r')$. The sum rule of the type (4.33) is correct only for the Hermite polynomials connected with the Lobachevski group.

4.5 Anti-de-Sitter and Poincare Groups

Anti-de-Sitter group $SO_3(1,i)$ and Poincare group $SO_3(\iota_1,i)$ may be obtained from the group $SO_3(j)$ at $j_2=i$, $j_1=1,\iota_1$. The matrix generators of these groups

$$X_1 = \begin{Vmatrix} 0 & -j_1^2 & 0 \\ 1 & 0 & 0 \\ 0 & 0 & 0 \end{Vmatrix}, \quad X_2 = \begin{Vmatrix} 0 & 0 & j_1^2 \\ 0 & 0 & 0 \\ 1 & 0 & 0 \end{Vmatrix}, \quad X_3 = \begin{Vmatrix} 0 & 0 & 0 \\ 0 & 0 & 1 \\ 0 & 1 & 0 \end{Vmatrix} \quad (4.96)$$

satisfy the commutation relations

$$[X_1, X_2] = j_1^2 X_3, \quad [X_2, X_3] = -X_1, \quad [X_1, X_3] = -X_2. \quad (4.97)$$

From (3.18) at $j_2=i$ we obtain $\psi a^+=(\alpha_0, \alpha^+_1, -\alpha_2)$, $\psi a=(\alpha_0,\alpha_1,\alpha^+_2)$ and the generators expressed in terms of boson operators are

$$L_1(j_1, i) = a_0 a_1^+ - j_1^2 a_1 a_0^+, \qquad L_2(j_1, i) = j_1^2 a_0^+ a_2^+ - a_0 a_2,$$
$$L_3(j_1, i) = a_1^+ a_2^+ - a_1 a_2. \tag{4.98}$$

The operator of finite group rotations $U_g(r,j_1,i)=\exp(-\sum_{\lambda=1}^{\xi} r_\lambda L_\lambda(j_1,i))$, $r_\lambda \in R$ is unitary one (at $j_1=1$), and its kernel in the coherent state basis is given by the formula (4.19) where the matrices ξ, ξ_1, η, η_1 satisfy the system of equations (4.18) in which the matrix coefficients may be found from the Hamiltonian $H(j_1,i)=\sum_{\lambda=1}^{3} r_\lambda L_\lambda(j_1,i))$ written in the form (4.3). These coefficients are

$$-b^T = \begin{Vmatrix} 0 & -j_1^2 r_1 & 0 \\ r_1 & 0 & 0 \\ 0 & 0 & 0 \end{Vmatrix}, \quad c = \begin{Vmatrix} 0 & 0 & r_2 \\ 0 & 0 & r_3 \\ r_2 & r_3 & 0 \end{Vmatrix}, \quad d = \begin{Vmatrix} 0 & 0 & j_1^2 r_2 \\ 0 & 0 & r_3 \\ j_1^2 r_2 & r_3 & 0 \end{Vmatrix}. \tag{4.99}$$

The matrix B(r,j,i) in the Hamiltonian is given by the formula (4.79). Solving the obtained system of equations by means of the relations (4.82)-(4.86) we have (at $\tau=1$)

$$\xi = \begin{Vmatrix} \operatorname{ch} r - r_3^2 \frac{\operatorname{ch} r - 1}{r^2} & j_1^2\left(-r_1 \frac{\operatorname{sh} r}{r} + r_2 r_3 \frac{\operatorname{ch} r - 1}{r^2}\right) & 0 \\ r_1 \frac{\operatorname{sh} r}{r} + r_2 r_3 \frac{\operatorname{ch} r - 1}{r^2} & \operatorname{ch} r - j_1^2 r_2^2 \frac{\operatorname{ch} r - 1}{r^2} & 0 \\ 0 & 0 & \operatorname{ch} r + j_1^2 r_1^2 \frac{\operatorname{ch} r - 1}{r^2} \end{Vmatrix}, \tag{4.100}$$

$$\xi_1 = \begin{Vmatrix} \operatorname{ch} r - r_3^2 \frac{\operatorname{ch} r - 1}{r^2} & -r_1 \frac{\operatorname{sh} r}{r} + r_2 r_3 \frac{\operatorname{ch} r - 1}{r^2} & 0 \\ j_1^2\left(r_1 \frac{\operatorname{sh} r}{r} + r_2 r_3 \frac{\operatorname{ch} r - 1}{r^2}\right) & \operatorname{ch} r - j_1^2 r_2^2 \frac{\operatorname{ch} r - 1}{r^2} & 0 \\ 0 & 0 & \operatorname{ch} r + j_1^2 r_1^2 \frac{\operatorname{ch} r - 1}{r^2} \end{Vmatrix}, \tag{4.101}$$

$$\eta = \begin{Vmatrix} 0 & 0 & j_1^2\left(r_2 \frac{\operatorname{sh} r}{r} - r_1 r_3 \frac{\operatorname{ch} r - 1}{r^2}\right) \\ 0 & 0 & r_3 \frac{\operatorname{sh} r}{r} + j_1^2 r_1 r_2 \frac{\operatorname{ch} r - 1}{r^2} \\ j_1^2\left(r_2 \frac{\operatorname{sh} r}{r} + r_1 r_3 \frac{\operatorname{ch} r - 1}{r^2}\right) & r_3 \frac{\operatorname{sh} r}{r} - j_1^2 r_1 r_2 \frac{\operatorname{ch} r - 1}{r^2} & 0 \end{Vmatrix}, \tag{4.102}$$

$$\eta_1 = \begin{Vmatrix} 0 & 0 & r_2 \frac{\operatorname{sh} r}{r} - r_1 r_3 \frac{\operatorname{ch} r - 1}{r^2} \\ 0 & 0 & r_3 \frac{\operatorname{sh} r}{r} + j_1^2 r_1 r_2 \frac{\operatorname{ch} r - 1}{r^2} \\ r_2 \frac{\operatorname{sh} r}{r} + r_1 r_3 \frac{\operatorname{ch} r - 1}{r^2} & r_3 \frac{\operatorname{sh} r}{r} - j_1^2 r_1 r_2 \frac{\operatorname{ch} r - 1}{r^2} & 0 \end{Vmatrix}, \tag{4.103}$$

where

$$r^2 = -j_1^2 r_1^2 - j_1^2 r_2^2 + r_3^2. \tag{4.104}$$

The matrices ξ^{-1}, $\xi^{-1}\eta$, $\eta_1\xi^{-1}$ have the form

$$\xi^{-1} = (\det \xi)^{-1/2} \begin{Vmatrix} \operatorname{ch} r - j_1^2 r_2^2 \dfrac{\operatorname{ch} r - 1}{r^2} & j_1^2 \left(r_1 \dfrac{\operatorname{sh} r}{r} - r_2 r_3 \dfrac{\operatorname{ch} r - 1}{r^2} \right) & 0 \\ - r_1 \dfrac{\operatorname{sh} r}{r} - r_2 r_3 \dfrac{\operatorname{ch} r - 1}{r^2} & \operatorname{ch} r - r_3^2 \dfrac{\operatorname{ch} r - 1}{r^2} & 0 \\ 0 & 0 & 1 \end{Vmatrix}, \tag{4.105}$$

$$\xi^{-1}\eta = (\det \xi)^{-1/2} \times$$

$$\times \begin{Vmatrix} 0 & 0 & j_1^2 \left(r_2 \dfrac{\operatorname{sh} r}{r} + r_1 r_3 \dfrac{\operatorname{ch} r - 1}{r^2} \right) \\ 0 & 0 & r_3 \dfrac{\operatorname{sh} r}{r} - j_1^2 r_1 r_2 \dfrac{\operatorname{ch} r - 1}{r^2} \\ j_1^2 \left(r_2 \dfrac{\operatorname{sh} r}{r} + r_1 r_3 \dfrac{\operatorname{ch} r - 1}{r^2} \right) & r_3 \dfrac{\operatorname{sh} r}{r} - j_1^2 r_1 r_2 \dfrac{\operatorname{ch} r - 1}{r^2} & 0 \end{Vmatrix}, \tag{4.106}$$

$$\eta_1 \xi^{-1} = (\det \xi)^{-1/2} \times$$

$$\times \begin{Vmatrix} 0 & 0 & r_2 \dfrac{\operatorname{sh} r}{r} - r_1 r_3 \dfrac{\operatorname{ch} r - 1}{r^2} \\ 0 & 0 & r_3 \dfrac{\operatorname{sh} r}{r} + j_1^2 r_1 r_2 \dfrac{\operatorname{ch} r - 1}{r^2} \\ r_2 \dfrac{\operatorname{sh} r}{r} - r_1 r_3 \dfrac{\operatorname{ch} r - 1}{r^2} & r_3 \dfrac{\operatorname{sh} r}{r} + j_1^2 r_1 r_2 \dfrac{\operatorname{ch} r - 1}{r^2} & 0 \end{Vmatrix}. \tag{4.107}$$

Here

$$\det \xi = (\xi_{22})^2 = [\operatorname{ch} r + j_1^2 r_1^2 (\operatorname{ch} r - 1)/r^2]^2. \tag{4.108}$$

At $j_1=1$ and $j_1=\iota_1$ the matrices $\xi^{-1}\eta$ and $\eta_1\xi^{-1}$ are symmetrical ones.

The kernel of the operator of the finite transformations in the coherent state basis for anti-de-Sitter group and Poincare group may be found from the formula (4.19) and it is as follows:

$$\begin{aligned} U(\alpha^*, \beta, \mathbf{r}, j_1) &= (\det \xi)^{-1/2} \exp \Big\{ \Big(\operatorname{ch} r + j_1^2 r_1^2 \frac{\operatorname{ch} r - 1}{r^2} \Big)^{-1} \Big[- \alpha_0^* \alpha_2^* j_1^2 \Big(r_2 \frac{\operatorname{sh} r}{r} + \\ &+ r_1 r_3 \frac{\operatorname{ch} r - 1}{r^2} \Big) - \alpha_1^* \alpha_2^* \Big(r_3 \frac{\operatorname{sh} r}{r} - j_1^2 r_1 r_2 \frac{\operatorname{ch} r - 1}{r^2} \Big) + \\ &+ \alpha_0^* \beta_0 \Big(\operatorname{ch} r - j_1^2 r_2^2 \frac{\operatorname{ch} r - 1}{r^2} \Big) + \alpha_2^* \beta_2 + \alpha_0^* \beta_1 j_1^2 \Big(r_1 \frac{\operatorname{sh} r}{r} - r_2 r_3 \frac{\operatorname{ch} r - 1}{r^2} \Big) - \\ &- \alpha_1^* \beta_0 \Big(r_1 \frac{\operatorname{sh} r}{r} + r_2 r_3 \frac{\operatorname{ch} r - 1}{r^2} \Big) + \alpha_1^* \beta_1 \Big(\operatorname{ch} r - r_3^2 \frac{\operatorname{ch} r - 1}{r^2} \Big) + \\ &+ \beta_0 \beta_2 \Big(r_2 \frac{\operatorname{sh} r}{r} - r_1 r_3 \frac{\operatorname{ch} r - 1}{r^2} \Big) + \beta_1 \beta_2 \Big(r_3 \frac{\operatorname{sh} r}{r} + j_1^2 r_1 r_2 \frac{\operatorname{ch} r - 1}{r^2} \Big) \Big] \Big\}. \end{aligned} \tag{4.109}$$

For the Poincare group $SO_3(\iota_1, i)$ the kernel has the form

$$\begin{aligned} U(\alpha^*, \beta, \mathbf{r}, \iota_1) &= (\operatorname{ch} r_3)^{-1} \exp \Big\{ (\operatorname{ch} r_3)^{-1} \Big[\alpha_0^* \beta_0 \operatorname{ch} r_3 + \alpha_1^* \beta_1 + \alpha_2^* \beta_2 - \alpha_1^* \alpha_2^* \operatorname{sh} r_3 + \\ &+ \beta_1 \beta_2 \operatorname{sh} r_3 - \alpha_1^* \beta_0 \Big(r_1 \frac{\operatorname{sh} r_3}{r_3} + r_2 \frac{\operatorname{ch} r_3 - 1}{r_3} \Big) + \beta_0 \beta_2 \Big(r_2 \frac{\operatorname{sh} r_3}{r_3} - r_1 \frac{\operatorname{ch} r_3 - 1}{r_3} \Big) \Big] \Big\}. \end{aligned} \tag{4.110}$$

For the Hermite polynomials created by anti-de Sitter and Poincare groups the adding formula (4.32) is valid. The sum rule of the type (4.33) are correct only for the Hermite polynomials corresponding to the anti-de-Sitter group.

4.6 De Sitter Group in Boson Representation

The de Sitter group is contained in the set of the motion groups of two-dimensional spaces (planes) of constant curvature at $j_1=\iota_1$, $j_2=i$. The matrix generators of the group $SO_3(i)$ are easily obtained from the formula (2.21)

$$X_1 = \begin{Vmatrix} 0 & 1 & 0 \\ 1 & 0 & 0 \\ 0 & 0 & 0 \end{Vmatrix}, \quad X_2 = \begin{Vmatrix} 0 & 0 & -1 \\ 0 & 0 & 0 \\ 1 & 0 & 0 \end{Vmatrix}, \quad X_3 = \begin{Vmatrix} 0 & 0 & 0 \\ 0 & 0 & 1 \\ 0 & 1 & 0 \end{Vmatrix}. \tag{4.111}$$

They satisfy the commutation relations

$$[X_1, X_2] = -X_3, \quad [X_2, X_3] = -X_1, \quad [X_1, X_3] = -X_2. \tag{4.112}$$

According to the formula (3.21) we construct the boson realization of these generators for which $\psi a^+=(\alpha^+{}_0, -\alpha_1, \alpha^+{}_2)$, $\psi a=(\alpha_0,\alpha^+{}_1,\alpha_2)$. It has the form

$$L_1(\mathrm{i}) = a_0^+ a_1^+ - a_0 a_1, \quad L_2(\mathrm{i}) = a_0 a_2^+ - a_0^+ a_2, \quad L_3(\mathrm{i}) = a_1^+ a_2^+ - a_1 a_2. \tag{4.113}$$

The commutation relations of operators L_λ are given by the expression (4.112). It means that these operators realize the boson representation of de Sitter group.

The unitary operator of the finite group transformations $U_g(r,i)=\exp(\sum_{\lambda=1}^{3} r_\lambda L_\lambda(i))$, $r_\lambda \in R$ in the coherent state basis has the kernel $U(\alpha^*,\beta,r,i)$ which may be found from the formula (4.12) with the matrices ξ, η, ξ_1, η_1, these matrices being the solution to the system (4.81). The matrix coefficients of this system of equations are determined by the Hamiltonian $H(r,i)=\sum_{\lambda=1}^{3} r_\lambda L_\lambda(i))$ and for de Sitter group they have the form

$$b = \begin{Vmatrix} 0 & 0 & -r_2 \\ 0 & 0 & 0 \\ -r_2 & 0 & 0 \end{Vmatrix}, \quad c = d = \begin{Vmatrix} 0 & r_1 & 0 \\ r_1 & 0 & r_3 \\ 0 & r_3 & 0 \end{Vmatrix}. \tag{4.114}$$

Solving the system of the differential equations under discussion we have (at $\tau=1$)

$$\xi = \xi_1 = \begin{Vmatrix} \mathrm{ch}\, r - r_3^2 \dfrac{\mathrm{ch}\, r - 1}{r^2} & 0 & -r_2 \dfrac{\mathrm{sh}\, r}{r} + r_1 r_3 \dfrac{\mathrm{ch}\, r - 1}{r^2} \\ 0 & \mathrm{ch}\, r + r_2^2 \dfrac{\mathrm{ch}\, r - 1}{r^2} & 0 \\ r_2 \dfrac{\mathrm{sh}\, r}{r} + r_1 r_3 \dfrac{\mathrm{ch}\, r - 1}{r^2} & 0 & \mathrm{ch}\, r - r_1^2 \dfrac{\mathrm{ch}\, r - 1}{r^2} \end{Vmatrix}, \tag{4.115}$$

$$\eta = \eta_1 = \begin{Vmatrix} 0 & r_1 \dfrac{\mathrm{sh}\, r}{r} - r_2 r_3 \dfrac{\mathrm{ch}\, r - 1}{r^2} & 0 \\ r_1 \dfrac{\mathrm{sh}\, r}{r} + r_2 r_3 \dfrac{\mathrm{ch}\, r - 1}{r^2} & 0 & r_3 \dfrac{\mathrm{sh}\, r}{r} - r_1 r_2 \dfrac{\mathrm{ch}\, r - 1}{r^2} \\ 0 & r_3 \dfrac{\mathrm{sh}\, r}{r} + r_1 r_2 \dfrac{\mathrm{ch}\, r - 1}{r^2} & 0 \end{Vmatrix}, \tag{4.116}$$

where

$$r^2 = r_1^2 - r_2^2 + r_3^2. \tag{4.117}$$

The determinant of the matrix ξ differs from zero

$$\det \xi = (\xi_{11})^2 = [\mathrm{ch}\, r + r_2^2 (\mathrm{ch}\, r - 1)/r^2]^2, \tag{4.118}$$

and due to this inverse matrix exists

$$\xi^{-1} = (\det \xi)^{-1/2} \begin{Vmatrix} \mathrm{ch}\, r - r_1^2 \dfrac{\mathrm{ch}\, r - 1}{r^2} & 0 & r_2 \dfrac{\mathrm{sh}\, r}{r} - r_1 r_3 \dfrac{\mathrm{ch}\, r - 1}{r^2} \\ 0 & 1 & 0 \\ -r_2 \dfrac{\mathrm{sh}\, r}{r} - r_1 r_3 \dfrac{\mathrm{ch}\, r - 1}{r^2} & 0 & \mathrm{ch}\, r - r_3^2 \dfrac{\mathrm{ch}\, r - 1}{r^2} \end{Vmatrix}. \tag{4.119}$$

Let us find out the matrices $\xi^{-1}\eta$, $\eta_1\xi^{-1}$

$$\xi^{-1}\eta = (\det \xi)^{-1/2} \times$$
$$\times \begin{Vmatrix} 0 & r_1 \frac{\operatorname{sh} r}{r} + r_2 r_3 \frac{\operatorname{ch} r - 1}{r^2} & 0 \\ r_1 \frac{\operatorname{sh} r}{r} + r_2 r_3 \frac{\operatorname{ch} r - 1}{r^2} & 0 & r_3 \frac{\operatorname{sh} r}{r} - r_1 r_2 \frac{\operatorname{ch} r - 1}{r^2} \\ 0 & r_3 \frac{\operatorname{sh} r}{r} - r_1 r_2 \frac{\operatorname{ch} r - 1}{r^2} & 0 \end{Vmatrix}, \tag{4.120}$$

$$\eta_1\xi^{-1} = (\det \xi)^{-1/2} \times$$
$$\times \begin{Vmatrix} 0 & r_1 \frac{\operatorname{sh} r}{r} - r_2 r_3 \frac{\operatorname{ch} r - 1}{r^2} & 0 \\ r_1 \frac{\operatorname{sh} r}{r} - r_2 r_3 \frac{\operatorname{ch} r - 1}{r^2} & 0 & r_3 \frac{\operatorname{sh} r}{r} + r_1 r_2 \frac{\operatorname{ch} r - 1}{r^2} \\ 0 & r_3 \frac{\operatorname{sh} r}{r} + r_1 r_2 \frac{\operatorname{ch} r - 1}{r^2} & 0 \end{Vmatrix}. \tag{4.121}$$

These matrices are symmetrical ones.

The kernel of the finite transformations for de Sitter group in the coherent state basis is as follows:

$$U(\alpha^*, \beta, \mathbf{r}, i) = (\det \xi)^{-1/2} \exp\left\{\left(\operatorname{ch} r + r_2^2 \frac{\operatorname{ch} r - 1}{r^2}\right)^{-1} \left[-\alpha_0^*\alpha_1^*\left(r_1 \frac{\operatorname{sh} r}{r} + \right.\right.\right.$$
$$\left. + r_2 r_3 \frac{\operatorname{ch} r - 1}{r^2}\right) - \alpha_1^*\alpha_2^*\left(r_3 \frac{\operatorname{sh} r}{r} - r_1 r_2 \frac{\operatorname{ch} r - 1}{r^2}\right) +$$
$$+ \alpha_0^*\beta_0\left(\operatorname{ch} r - r_1^2 \frac{\operatorname{ch} r - 1}{r^2}\right) + \alpha_0^*\beta_2\left(r_2 \frac{\operatorname{sh} r}{r} - r_1 r_3 \frac{\operatorname{ch} r - 1}{r^2}\right) +$$
$$+ \alpha_1^*\beta_1 - \alpha_2^*\beta_0\left(r_2 \frac{\operatorname{sh} r}{r} + r_1 r_3 \frac{\operatorname{ch} r - 1}{r^2}\right) + \alpha_2^*\beta_2\left(\operatorname{ch} r - r_3^2 \frac{\operatorname{ch} r - 1}{r^2}\right) +$$
$$\left.\left. + \beta_0\beta_1\left(r_1 \frac{\operatorname{sh} r}{r} - r_2 r_3 \frac{\operatorname{ch} r - 1}{r^2}\right) + \beta_1\beta_2\left(r_3 \frac{\operatorname{sh} r}{r} + r_1 r_2 \frac{\operatorname{ch} r - 1}{r^2}\right)\right]\right\}. \tag{4.122}$$

This kernel is the generating function for the Hermite polynomials $H_{m,n}^{(r)}(0)$ with the symmetrical matrix R of the form (4.20) where the matrices $\xi^{-1}\eta$, ξ^{-1}, $\eta_1\xi^{-1}$ are given by the relations (4.119)-(4.121). For these Hermite polynomials the adding formula (4.32) and the sum rules of the type (4.33) are valid.

4.7 Matrix Elements of Finite Group Transformations and Analytical Continuation of Groups

In Section 4.4-4.6 we have found the matrix elements of the finite group transformations in the coherent state basis and in the Fock state basis for the groups which are obtained from the rotation group by the analytical continuation of the real group parameters into the region of the pure imaginary values determining the matrices ξ, ξ_1, η, η_1 by the solution of the system of equations (4.7) in the form (4.81). We will show that it is not necessary to solve this system. The matrices ξ, ξ_1, η, η_1 may be expressed through the matrix of the finite rotations $\Xi(r,j)=\exp(\sum_\lambda r_\lambda X_\lambda(j))$ determining the matrix elements of the operator of the finite transformations of the groups which may be obtained from the rotation group only by the transactions.

Let us consider the groups $SO_{n+1}(j)$, $j_k=1, \iota_k$, $k=1,2,\dots,n$, i.e. the rotation group SO_{n+1} and the groups which are obtained from it by the contractions. The Jordan-Schwinger representation of the generators for these groups is described by the formula (3.21) in which $\psi\hat{a}^+ = \hat{a}^+, \psi\hat{a} = \hat{a}$. Because of this the operator (4.17) of the finite group transformations is written down in the form

$$\hat{U}_g(\mathbf{r}, \mathrm{j}) = \exp\Big(-\sum_{\lambda=1}^{n(n+1)/2} r_\lambda \hat{L}_\lambda(\mathrm{j})\Big) = \exp\left\{-\frac{1}{2}(\hat{\mathbf{a}}, \hat{\mathbf{a}}^+)\,\Delta(\mathrm{j})\begin{pmatrix}\hat{\mathbf{a}}\\ \hat{\mathbf{a}}^+\end{pmatrix}\right\}, \tag{4.123}$$

where the matrix $\Delta(\mathrm{j})$ is given by the formula (4.4). The matrix $\Lambda(j) = \begin{Vmatrix}\xi & \eta\\ \eta_1 & \xi_1\end{Vmatrix}$ satisfies the equation

$$\dot{\Lambda} = \Lambda\Sigma\Delta(\mathrm{j}), \qquad \Lambda(0) = E_{2(n+1)}, \qquad \Sigma = \begin{Vmatrix} 0 & E_{n+1}\\ -\varepsilon E_{n+1} & 0\end{Vmatrix}, \tag{4.124}$$

which may be obtained from the equation (4.7) if one uses the replacement of the variable t by the variable $\tau = it/\hbar$. Its solution may be easily found and at $\tau=1$ this solution gives

$$\Lambda(\mathrm{j}) = \exp\{\Sigma\Delta(\mathrm{j})\} = \begin{Vmatrix}\Xi(\mathbf{r}, \mathrm{j}) & 0\\ 0 & \Xi^T(-\mathbf{r}, \mathrm{j})\end{Vmatrix}, \tag{4.125}$$

i.e. $\eta=\eta_1=0$, $\xi=\Xi(\mathrm{r,j})$, $\xi_1=\Xi^t(-\mathrm{r,j})$.

Let us consider the groups $SO_{n+1}(\mathrm{j})$ where $j_k=1, \iota_k$, i, k=1,2,...,n, i.e. the groups which may be obtained from the rotation group SO_{n+1} both by the contractions and by the analytical continuations of the group parameters. We introduce the parameters $\tilde{j}_k$, each of these parameters takes two values $\tilde{j}_k = 1, \iota_k$. These parameters are connected with the parameters j_k as follows. If some of the parameters j_m are equal to the imaginary unity, then $j_m = i\tilde{j}_m$. If the parameters j_k are equal, either real unity or dual units, then $j_k = \tilde{j}_k$. The aim of this new definition of the parameters is to take into account explicitly the analytical continuations and to keep the possibility of considering the contractions (for dual values of the parameters $\tilde{j}_k$) of these analytically continued groups. The operator of the finite transformations of the group $SO_{n+1}(\mathrm{j})$ takes the form

$$\hat{U}_g(\mathrm{j}) = \exp\left\{-\frac{1}{2}(\hat{\mathbf{a}}, \hat{\mathbf{a}}^+)\,\mathcal{B}(\mathrm{j})\begin{pmatrix}\hat{\mathbf{a}}\\ \hat{\mathbf{a}}^+\end{pmatrix}\right\}, \tag{4.126}$$

where according to formula (4.5) the matrix $B(\mathrm{j})$ is connected with the matrix $\Delta(\mathrm{j})$ by the relation

$$\mathcal{B}(\mathrm{j}) = \Psi(\mathrm{j})\,\Delta(\mathrm{j})\,\Psi^{-1}(\mathrm{j}). \tag{4.127}$$

The matrix elements of the operator (4.126) may be found by means of the matrix

$\tilde{\Lambda} = \begin{Vmatrix}\zeta & \eta\\ \eta_1 & \zeta_1\end{Vmatrix}$ which is the solution to the equation

$$\dot{\tilde{\Lambda}} = \tilde{\Lambda}\Sigma\mathcal{B}(\mathrm{j}), \qquad \tilde{\Lambda}(0) = E, \tag{4.128}$$

taken at $\tau=1$. Let us substitute into this equation the expression (4.127) for the matrix B through Δ and multiply from the right side the equation obtained by $\Psi(\mathrm{j})$ and from the left side by $\Psi^{-1}(\mathrm{j})$. Using the relation $\Psi\Psi^{-1}=E$ we have

$$\Psi^{-1}\dot{\tilde{\Lambda}}\Psi = \Psi^{-1}\tilde{\Lambda}\Sigma\Psi\Delta(j). \tag{4.129}$$

Inserting $E=\Psi\Psi^{-1}$ between $\tilde{\Lambda}$ and Σ we obtain the equation for the matrix $\Psi^{-1}\Lambda\Psi$

$$\Psi^{-1}\dot{\tilde{\Lambda}}\Psi = (\Psi^{-1}\tilde{\Lambda}\Psi)(\Psi^{-1}\Sigma\Psi)\,\Delta(j), \qquad \Psi^{-1}\tilde{\Lambda}(0)\,\Psi = E, \tag{4.130}$$

which differs from the equation (4.124) for the matrix $\Lambda(j)$ only by the replacement of the matrix Σ in formula (4.130) by the matrix $\Psi^{-1}\Sigma\Psi$. Let us prove that

$$\Psi^{-1}\Sigma\Psi = \Sigma. \tag{4.131}$$

In fact, from the definition (3.20) of the matrices Ψ, Ψ^{-1} and from the definition (3.19) of the vectors $\Psi\hat{a}$, $\Psi\hat{a}^+$ it follows that the action of the matrix Ψ on the column-vector $\begin{pmatrix}\psi\hat{a}\\ \psi\hat{a}^+\end{pmatrix}$ is reduced to the permutations of the components with equal labels of the vector $\psi\hat{a}$ and $\psi\hat{a}^+$, the component of the vector $\psi\hat{a}^+$ being simultaneously multiplied by -1. The permutations take place for all the components $\psi\hat{\alpha}_k$ and $\psi\hat{\alpha}^+{}_k$ which have the property that the product $\prod_{m=1}^{k} j_m$ is pure imaginary number. Analogously the matrix Ψ^{-1} acts on the row-vector $(\psi\hat{a}, \psi\hat{a}^+)$. Then the transformation (4.131) is reduced to the permutations of the rows and columns of the matrix Σ with corresponding multiplication by -1. It is easy to verify that this transformation does not change the form of the matrix Σ.

After we have proved the relation (4.131) and substituted it into equation (4.130), we can see that the equations and the initial conditions for the matrices Λ and $\Psi^{-1}\Lambda\Psi$ coincide, i.e. we obtain the formula

$$\tilde{\Lambda}(\mathbf{r}, j) = \Psi(j)\,\Lambda(\mathbf{r}, j)\,\Psi^{-1}(j), \tag{4.132}$$

analogous to the formula (4.127). Taking the expression (4.125) for the matrix Λ and using (3.20) we have

$$\begin{aligned}
\xi &= \psi_1\Xi(\mathbf{r}, j)\,\psi_1 + \psi_2\Xi^T(-\mathbf{r}, j)\,\psi_2,\\
\xi_1 &= \psi_2\Xi(\mathbf{r}, j)\,\psi_2 + \psi_1\Xi^T(-\mathbf{r}, j)\,\psi_1,\\
\eta &= -\psi_1\,\Xi(\mathbf{r}, j)\psi_2 + \varepsilon\psi_2\Xi^T(-\mathbf{r}, j)\psi_1,\\
\eta_1 &= -\psi_2\Xi(\mathbf{r}, j)\psi_1 + \varepsilon\psi_1\Xi^T(-\mathbf{r}, j)\psi_2.
\end{aligned} \tag{4.133}$$

Hence, knowing the finite group transformations of the group $SO_{n+1}(j)$, we obtain the integrals of the motion for all the quantum systems corresponding to the groups $SO_{n+1}(j)$ either using the formulae (4.125) if one considers only the contractions of the rotation group, or using the formulae (4.132), (4.133) if one also considers the analytical continuations of this group.

The next step consists of finding the matrix elements of the operator of the finite group transformations in the coherent state basis. As it follows from the formulae (4.19) to find the elements, it is necessary to calculate the matrices ξ^{-1}, $\xi^{-1}\eta$, $\eta_1\xi^{-1}$. The nontrivial operation here is to obtain the inverse matrix ξ^{-1}. Let us analyze the expression (4.133) for the matrix ξ. Since the matrix ψ_1 is a diagonal one, the part of diagonal elements which is equal to zero the first term $\psi_1\Xi(r)\psi_1$ is the matrix with zero columns, and rows the labels of which coincide with the labels of the zero diagonal elements of the matrix ψ_1. Let us remove these zero rows and columns and denote the matrix obtained as $(\psi_1\Xi(r)\psi_1)$. This matrix is a nondegenerate one. We find the inverse matrix $(\psi_1\Xi(r)\psi_1)^{-1}$ and reconstruct the removed zero rows and columns. Let us denote the matrix obtained as $[(\psi_1\Xi(r)\psi_1)^{-1}]$. Repeating this procedure with the second term in the expression (4.133) for the matrix ξ, we find the inverse matrix

$$\xi^{-1} = [(\psi_1\Xi\,(\mathbf{r})\psi_1)^{-1}] + [(\psi_2\Xi^T\,(-\mathbf{r})\psi_2)^{-1}]. \tag{4.134}$$

So, we reduce the problem of finding the inverse matrix ξ to the problem of finding two inverse matrices of smaller dimensionalities. Then taking into account the relations (4.133) we easily find the other matrices

$$\xi^{-1}\eta = -[(\psi_1\Xi\,(\mathbf{r})\,\psi_1)^{-1}]\,\psi_1\Xi\,(\mathbf{r})\,\psi_2 + \varepsilon\,[(\psi_2\Xi^T\,(-\mathbf{r})\,\psi_2)^{-1}]\,\psi_2\Xi^T\,(-\mathbf{r})\,\psi_1,$$

$$\eta_1\xi^{-1} = -\,\psi_2\Xi\,(\mathbf{r})\,\psi_1\,[(\psi_1\Xi\,(\mathbf{r})\,\psi_1)^{-1}] + \varepsilon\psi_1\Xi^T\,(-\mathbf{r})\,\psi_2\,[(\psi_2\Xi^T\,(-\mathbf{r})\,\psi_2)^{-1}]. \tag{4.135}$$

From the general properties of the quantum systems with the Hamiltonians which are quadratic in creation and annihilation operators, it follows [8] that the matrices (4.135) are symmetrical ones in the boson case (ξ=1) and antisymmetrical ones in the fermion case (ξ=-1). Because of this, it is necessary to calculate only the first terms in the formula (4.135) and the other matrices may be obtained by the transposition. This trick gives the possibility to reduce the calculations which are simple but very cumbersome. We will demonstrate the developed approach on the example of the quantum systems corresponding to the groups $SO_4(j)$.

4.8 The Rotation Group SO4 and Its Contractions

We consider the groups $SO_4(j)$ when each of the parameters $j=(j_1, j_2, j_3)$ may take two values $j_k=1, \iota_k$, $k=1,2,3$. Using the formula (3.21) the generators $\hat{L}_{\mu\nu}(j)$ in the Jordan-Schwinger representation correspond to the matrix generators (2.32). In the formula (3.21) the map is given by the relations $\psi\hat{a}=\hat{a}$, $\psi\hat{a}^+=\hat{a}^+$. We have the generators

$$\hat{L}_{01}\,(\mathrm{j}) = \hat{a}_1^+\hat{a}_0 - j_1^2\hat{a}_0^+\hat{a}_1, \qquad \hat{L}_{02}\,(\mathrm{j}) = \hat{a}_2^+\hat{a}_0 - j_1^2j_2^2\hat{a}_0^+\hat{a}_2,$$
$$\hat{L}_{03}\,(\mathrm{j}) = \hat{a}_3^+\hat{a}_0 - j_1^2j_2^2j_3^2\hat{a}_0^+\hat{a}_3,$$
$$\hat{L}_{12}\,(\mathrm{j}) = \hat{a}_2^+\hat{a}_1 - j_2^2\hat{a}_1^+\hat{a}_2, \qquad \hat{L}_{13}\,(\mathrm{j}) = \hat{a}_3^+\hat{a}_1 - j_2^2j_3^2\hat{a}_1^+\hat{a}_3, \qquad \hat{L}_{23}\,(\mathrm{j}) = \hat{a}_3^+\hat{a}_2 - j_3^2\hat{a}_2^+\hat{a}_3, \tag{4.136}$$

satisfying the same commutation relations (2.6) of $SO_4(j)$ group generators. The Casimir operators of the group $SO_4(j)$ which are determined in accordance with study [41] as the sum of the principle minors of the antisymmetrical matrix constructed from the generators $X_{\mu\nu}(j)$, may be easily found by the transition method using the formula (2.9)-(2.11) from the well-known invariant operators of the rotation group SO_4. In Jordan-Schwinger representation they are

$$\hat{L}^2\,(\mathrm{j}) = j_2^2j_3^2\hat{L}_{01}^2 + j_3^2\hat{L}_{02}^2 + \hat{L}_{03}^2 + j_1^2j_3^2\hat{L}_{12}^2 + j_1^2\hat{L}_{13}^2 + j_1^2j_2^2\hat{L}_{23}^2,$$
$$\hat{L}^{1/2}\,(\mathrm{j}) = j_2^2\hat{L}_{01}\hat{L}_{23} + \hat{L}_{03}\hat{L}_{12} - \hat{L}_{02}\hat{L}_{13}. \tag{4.137}$$

The finite rotation $\Xi(r, s, j)$ in the Cayley-Klein space $\Re_4(j)$ described by the formulae (2.35)-(2.41) is represented in Jordan-Schwinger scheme by the operator

$$\hat{U}_g\,(\mathbf{r},\,\mathbf{s},\,\mathrm{j}) = \exp\Big\{-\sum_{k=1}^{3} r_k\hat{L}_{0k}\,(\mathrm{j}) + \sum_{k,\,m,\,p=1}^{3} \varepsilon_{kmp}s_k\hat{L}_{mp}\,(\mathrm{j})\Big\} =$$
$$= \exp\left\{-\frac{1}{2}\,(\hat{\mathbf{a}},\,\hat{\mathbf{a}}^+)\,\Delta\,(\mathbf{r},\,\mathbf{s},\,\mathrm{j})\begin{pmatrix}\hat{\mathbf{a}}\\ \hat{\mathbf{a}}^+\end{pmatrix}\right\}, \tag{4.138}$$

where ε_{kmp} is a completely antisymmetric unit tensor and the matrix Δ is

$$\Delta(\mathbf{r}, \mathbf{s}, \mathbf{j}) = \begin{Vmatrix} 0 & \varepsilon X^T(\mathbf{r}, \mathbf{s}, \mathbf{j}) \\ X(\mathbf{r}, \mathbf{s}, \mathbf{j}) & 0 \end{Vmatrix}. \tag{4.139}$$

Here the matrix X(r, s, j) is given by the formula (2.35). The linear integrals of the motion may be found with the help of the matrix

$$\Lambda(\mathbf{r}, \mathbf{s}, \mathbf{j}) = \exp\{\Sigma\Delta(\mathbf{r}, \mathbf{s}, \mathbf{j})\} = \begin{Vmatrix} \Xi(\mathbf{r}, \mathbf{s}, \mathbf{j}) & 0 \\ 0 & \Xi^T(-\mathbf{r}, -\mathbf{s}, \mathbf{j}) \end{Vmatrix}, \tag{4.140}$$

i.e. $\eta=\eta_1=0$, $\xi=\Xi(r, s, j)$, $\xi_1=\Xi^T(-r, -s, j)$, where Ξ is the matrix of the finite group transformation (2.36) and $\det\Xi=1$. According to the Eq. (4.19) the kernel of the operator (4.138) in the coherent state basis has the form

$$\begin{aligned} &-\alpha_0^*\beta_k \prod_{m=1}^{k} j_m^2\Big) - s_1(\alpha_3^*\beta_2 - j_3^2\alpha_2^*\beta_3) + s_2(\alpha_3^*\beta_1 - j_2^2 j_3^2\alpha_1^*\beta_3) - \\ &- s_3(\alpha_2^*\beta_1 - j_2^2\alpha_1^*\beta_2)\Big] - C(\mathbf{r}, \mathbf{s})\,[j_3^2 s_1(\alpha_1^*\beta_0 - j_1^2\alpha_0^*\beta_1) + j_3^2 s_2(\alpha_2^*\beta_0 - j_1^2 j_2^2\alpha_0^*\beta_2) + \\ &+ s_3(\alpha_3^*\beta_0 - j_1^2 j_2^2 j_3^2\alpha_0^*\beta_3) - j_1^2 r_1(\alpha_3^*\beta_2 - j_3^2\alpha_2^*\beta_3) + j_1^2 r_2(\alpha_3^*\beta_1 - j_2^2 j_3^2\alpha_1^*\beta_3) - \\ &- j_1^2 j_3^2 r_3(\alpha_2^*\beta_1 - j_2^2\alpha_1^*\beta_2)] + D\Big[-r^2\alpha_0^*\beta_0 - \sum_{k=1}^{3}\alpha_k^*\beta_k\Big(s^2 + r_k^2\prod_{m=1}^{k} j_m^2\Big) + \\ &+ j_3^2 s_1^2\alpha_1^*\beta_1 + j_2^2 j_3^2 s_2^2\alpha_2^*\beta_2 + j_2^2 s_3^2\alpha_3^*\beta_3 + j_2^2(\mathbf{r}\times\mathbf{s})_1(\alpha_1^*\beta_0 + j_1^2\alpha_0^*\beta_1) + \\ &+ (\mathbf{r}\times\mathbf{s})_2(\alpha_2^*\beta_0 + j_1^2 j_2^2\alpha_0^*\beta_2) + (\mathbf{r}\times\mathbf{s})_3(\alpha_3^*\beta_0 + j_1^2 j_2^2 j_3^2\alpha_0^*\beta_3) + \\ &+ (j_3^2 s_1 s_2 - j_1^2 r_1 r_2)(\alpha_2^*\beta_1 + j_2^2\alpha_1^*\beta_2) + (s_1 s_3 - j_1^2 r_1 r_3)(\alpha_3^*\beta_1 + j_2^2 j_3^2\alpha_1^*\beta_3) + \\ &+ j_2^2(s_2 s_3 - j_1^2 r_2 r_3)(\alpha_3^*\beta_2 + j_3^2\alpha_2^*\beta_3)\Big]\Big]\Big\}, \end{aligned} \tag{4.141}$$

where the functions A, B, C, D are determined by the formulae (2.39). These expressions are valid in a boson case when α_k^*, β_k are the complex variables as well as in a fermion case when α_k^*, β_k are Grassman anticommuting variables.

Taking different values of the parameters j_k in the formula (4.140) we obtain the expressions for the kernels of the operators of finite group rotations corresponding to the group $SO_4(j)$. Let us consider, for example, the group $SO_4(\iota_1, \iota_2, j_3)$ which is isomorphic at $j_3=1$ to the motion group of the Galilean space with one time and two space axes, i.e. two-dimensional usual space. The matrix

$$\Xi(-\mathbf{r}, -\mathbf{s}) = E - \frac{\sin j_3 s_1}{j_3 s_1} X - \frac{1}{j_3^2}\frac{r_1}{s_1}\Big(1 - \frac{\sin j_3 s_1}{j_3 s_1}\Big) X_1 + \frac{1 - \cos j_3 s_1}{j_3^2 s_1^2} X^2, \tag{4.142}$$

where

$$X = \begin{Vmatrix} 0 & 0 & 0 & 0 \\ r_1 & 0 & 0 & 0 \\ r_2 & -s_3 & 0 & j_3^2 s_1 \\ r_3 & s_2 & -s_1 & 0 \end{Vmatrix}, \quad X_1 = \begin{Vmatrix} 0 & 0 & 0 & 0 \\ j_3^2 s_1 & 0 & 0 & 0 \\ j_3^2 s_2 & 0 & 0 & 0 \\ s_3 & 0 & 0 & 0 \end{Vmatrix},$$

$$X^2 = \begin{Vmatrix} 0 & 0 & 0 & 0 \\ 0 & 0 & 0 & 0 \\ (\mathbf{r}\times\mathbf{s})_2 & j_3^2 s_1 s_2 & -j_3^2 s_1^2 & 0 \\ (\mathbf{r}\times\mathbf{s})_3 & s_1 s_3 & 0 & -j_3^2 s_1^2 \end{Vmatrix}, \tag{4.143}$$

and the kernel of the operator has the form

$$U(\alpha^*, \beta, \mathbf{r}, \mathbf{s}) = \exp\Big\{\alpha_0^*\beta_0 + \alpha_1^*\beta_1 + (\alpha_2^*\beta_2 + \alpha_3^*\beta_3)\cos j_3 s_1 - r_1\alpha_1^*\beta_0 - $$
$$- \beta_0(\alpha_2^* r_2 + \alpha_3^* r_3)\frac{\sin j_3 s_1}{j_3 s_1} - \beta_0(j_3^2\alpha_2^* s_2 + \alpha_3^* s_3)\frac{r_1}{j_3^2 s_1}\left(1 - \frac{\sin j_3 s_1}{j_3 s_1}\right) +$$
$$+ \beta_0[\alpha_2^*(\mathbf{r}\times\mathbf{s})_2 + \alpha_3^*(\mathbf{r}\times\mathbf{s})_3]\frac{1-\cos j_3 s_1}{j_3^2 s_1^2} + \beta_1(\alpha_2^* s_3 - \alpha_3^* s_2)\frac{\sin j_3 s_1}{j_3 s_1} +$$
$$+ \beta_1(j_3^2\alpha_2^* s_2 + \alpha_3^* s_3)\, s_1\frac{1-\cos j_3 s_1}{j_3^2 s_1^2} + \alpha_3^*\beta_2\frac{1}{j_3}\sin j_3 s_1\Big\}. \qquad (4.144)$$

According to Eq. (4.27), the function (4.141) is the generating function for the Hermite polynomials with zero-valued argument $H_{mn}^{(R)}(0)$, $m=(m_0, m_1, m_2, m_3)$, $n=(n_0, n_1, n_2, n_3)$ where the matrix R is

$$R(\mathbf{r}, \mathbf{s}) = \begin{Vmatrix} 0 & -\Xi(-\mathbf{r}, -\mathbf{s}) \\ -\Xi^T(-\mathbf{r}, -\mathbf{s}) & 0 \end{Vmatrix}. \qquad (4.145)$$

From the group composition rule $\Xi(r, s)\Xi(r', s')=\Xi(r'', s'')$ one obtains the adding formula for Hermite polynomials of the type (4.32). For the group $SO_4(\iota)\equiv SO_4(\iota_1, \iota_2, \iota_3)$ the composition rule is the simplest one

$$s_1'' = s_1 + s_1', \quad s_2'' = s_2 + s_2' + (\mathbf{s}'\times\mathbf{s})_2, \quad s_3'' = s_3 + s_3',$$
$$r_1'' = r_1 + r_1', \quad r_2'' = r_2 + r_2' + {}^1/_2(\mathbf{r}'\times\mathbf{s} - \mathbf{r}\times\mathbf{s}')_2,$$
$$r_3'' = r_3 + r_3' + {}^1/_2(\mathbf{r}'\times\mathbf{s})_3 + {}^1/_2(\mathbf{s}'\times\mathbf{r})_3 + {}^1/_{12}(s_3 + s_3')[(\mathbf{r}', \mathbf{s}) + (\mathbf{r}, \mathbf{s}')] -$$
$$- {}^1/_6 s_3(\mathbf{r}', \mathbf{s}') - {}^1/_6 s_3'(\mathbf{r}, \mathbf{s}). \qquad (4.146)$$

Here the scalar and vector products should be understood in the sense of Eq. (2.41) at $j=\iota$.

So the group quantum systems of the type (4.1)-(4.3) with the Hamiltonians which are the linear forms in the generators of the group $SO_4(j)$ expressed in terms of the creation and annihilation operators of Bose or Fermi – oscillators are completely described by the finite group transformations $\Xi(r, s, j)$. According to Eqs. (4.15)-(4.17) this transformation provides at real values of the group parameters the density matrix of the system under consideration, the analytical continuation of the kernel (4.141) into the region of the Green function (4.13) of the system. Let us point out that the replacement of all the real group parameters by the pure imaginary variables does not change the group $SO_4(j)$. This change of variables is not connected with the discussed analytical continuation of some from the real group parameters into the region of the pure imaginary variables which induces the transition of one group into a different group. The result of this analytical continuation has a special explicit form determined by the property that some of the parameters j describing the group structure become equal to pure imaginary units.

4.9. Groups Obtained From SO_4 By Analytical Continuations And Contractions

In this section we consider the group quantum systems with Hamiltonians

$$\hat{H}(j) = \sum_{k=1}^{3} r_k \hat{L}_{0k}(j) - \sum_{k,m,p=1}^{3} \varepsilon_{kmp} s_k \hat{L}_{mp}(j), \qquad (4.147)$$

which are linear forms in generators of the group $SO_4(j)$ as well as in Eq. (4.138) realized in terms of the Bose or Fermi-creation and annihilation operators. But now the parameters j_k may also take pure imaginary values. Due to this, the transformation

$\Psi^{-1}(j)$ given by Eq. (3.19) is not an identical one, the secondary quantized generators $\hat{L}_\lambda(j)$ are described by the expression (3.21), the matrix B in the Hamiltonian (4.147) rewritten in the equivalent form

$$\hat{H}(\mathbf{j}) = \frac{1}{2}(\hat{\mathbf{a}}, \hat{\mathbf{a}}^+)\, \mathcal{B}(\mathbf{r}, \mathbf{s}, \mathbf{j}) \begin{pmatrix} \hat{\mathbf{a}} \\ \hat{\mathbf{a}}^+ \end{pmatrix}, \tag{4.148}$$

is expressed through the matrix $\Delta(\mathrm{r, s, j})$ by the relation

$$\mathcal{B}(\mathbf{r}, \mathrm{s}, \mathrm{j}) = \Psi(\mathrm{j})\, \Delta(\mathbf{r}, \mathrm{s}, \mathrm{j})\, \Psi^{-1}(\mathrm{j}). \tag{4.149}$$

The linear integrals of the motion for the system (4.148) may be found with the help of the matrix $\Lambda(\mathrm{r,s,j})$ which is obtained from the matrix $\Lambda(\mathrm{r,s,j})$ given by the formula (4.140) by the transformation

$$\tilde{\Lambda}(\mathbf{r}, \mathbf{s}, \mathrm{j}) = \Psi(\mathrm{j})\, \Lambda(\mathbf{r}, \mathbf{s}, \mathrm{j})\, \Psi^{-1}(\mathrm{j}). \tag{4.150}$$

As an example, let us consider the quantum system with the Hamiltonian (4.147) which is a linear form in generators of the group $SO_4(j=i\tilde{j})$ where $\tilde{j}_k=1, \iota_k$, k=1,2,3. From Eq. (3.19) we obtain the projectors ψ_1, ψ_2:

$$\psi_1 = \begin{Vmatrix} 1 & 0 & 0 & 0 \\ 0 & 0 & 0 & 0 \\ 0 & 0 & -1 & 0 \\ 0 & 0 & 0 & 0 \end{Vmatrix}, \quad \psi_2 = \begin{Vmatrix} 0 & 0 & 0 & 0 \\ 0 & -1 & 0 & 0 \\ 0 & 0 & 0 & 0 \\ 0 & 0 & 0 & 1 \end{Vmatrix}. \tag{4.151}$$

and according to Eq. (3.20) construct the transformations Ψ and Ψ^{-1}. The matrix X(r, s, ij) may be obtained from the expression (2.35) in which the parameters $\tilde{j}_k$ are replaced by the parameters $i\tilde{j}_k$ and it has the form

$$X(\mathbf{r}, \mathrm{s}, i\tilde{\mathbf{j}}) = \begin{Vmatrix} 0 & \tilde{j}_1^2 r_1 & -\tilde{j}_1^2\tilde{j}_2^2 r_2 & \tilde{j}_1^2\tilde{j}_2^2\tilde{j}_3^2 r_3 \\ r_1 & 0 & -\tilde{j}_2^2 s_3 & -\tilde{j}_2^2\tilde{j}_3^2 s_2 \\ r_2 & -s_3 & 0 & -\tilde{j}_3^2 s_1 \\ r_3 & s_2 & -s_1 & 0 \end{Vmatrix} \equiv \tilde{X}(\mathbf{r}, \mathrm{s}). \tag{4.152}$$

Substituting $\Delta(i\tilde{j}) = \begin{Vmatrix} 0 & \varepsilon\tilde{X}^T \\ \tilde{X} & 0 \end{Vmatrix}$ into the formula (4.149) one finds the Hamiltonian of the system

$$\mathcal{B}(\mathbf{r}, \mathrm{s}, i\tilde{\mathbf{j}}) = \begin{Vmatrix} \psi_1\tilde{X}^T\psi_2 + \varepsilon\psi_2\tilde{X}\psi_1 & \varepsilon(\psi_1\tilde{X}^T\psi_1 - \psi_2\tilde{X}\psi_2) \\ \psi_1\tilde{X}\psi_1 - \psi_2\tilde{X}^T\psi_2 & -\psi_1\tilde{X}\psi_2 - \varepsilon\psi_2\tilde{X}^T\psi_1 \end{Vmatrix} = \begin{Vmatrix} D_1 & \varepsilon D_0^T \\ D_0 & D_2 \end{Vmatrix}, \tag{4.153}$$

where

$$D_0 = \begin{Vmatrix} 0 & 0 & \tilde{j}_1^2\tilde{j}_2^2 r_2 & 0 \\ 0 & 0 & 0 & s_2 \\ -r_2 & 0 & 0 & 0 \\ 0 & -\tilde{j}_2^2\tilde{j}_3^2 s_2 & 0 & 0 \end{Vmatrix}, \quad D_1 = \begin{Vmatrix} 0 & -r_1 & 0 & r_3 \\ -\varepsilon r_1 & 0 & -\varepsilon\tilde{j}_2^2 s_3 & 0 \\ 0 & -\tilde{j}_2^2 s_3 & 0 & s_1 \\ \varepsilon r_3 & 0 & \varepsilon s_1 & 0 \end{Vmatrix},$$

$$D_2 = \begin{Vmatrix} 0 & \tilde{j}_1^2 r_1 & 0 & -\tilde{j}_1^2\tilde{j}_2^2\tilde{j}_3^2 r_3 \\ \varepsilon\tilde{j}_1^2 r_1 & 0 & \varepsilon s_3 & 0 \\ 0 & s_3 & 0 & -\tilde{j}_3^2 s_1 \end{Vmatrix}. \tag{4.154}$$

The matrices D_1 and D_2 are symmetrical ones in the boson case and antisymmetrical ones in the fermion case as it follows from the general form of the Hamiltonian which is quadratic form in creation and annihilation operators [8].

Replacing the parameters jk by $i\tilde{j}_k$, k=1,2,3 in the formulae (2.36)-(2.41) describing the matrix Ξ, we find the matrix $\Xi(r, s, i\tilde{j})$. Introducing the notations $Q=\Xi(r, s, i\tilde{j})$, $R=\Xi(-r, -s, i\tilde{j})$ we have $\Lambda(r,s,i\tilde{j})=\begin{Vmatrix} Q & 0 \\ 0 & R \end{Vmatrix}$ and the formula (4.150) gives the matrix $\tilde{\Lambda}=\begin{Vmatrix} \xi & \eta \\ \eta_1 & \xi_1 \end{Vmatrix}$ where according to Eq. (4.133) ξ, ξ_1, η, η_1 have the form

$$\xi = \psi_1 Q \psi_1 + \psi_2 R^T \psi_2 = \begin{Vmatrix} Q_{00} & 0 & -Q_{02} & 0 \\ 0 & R_{11} & 0 & -R_{31} \\ -Q_{20} & 0 & Q_{22} & 0 \\ 0 & -R_{13} & 0 & R_{33} \end{Vmatrix}, \tag{4.155}$$

$$\xi_1 = \psi_2 Q \psi_2 + \psi_1 R^T \psi_1 = \begin{Vmatrix} R_{00} & 0 & -R_{20} & 0 \\ 0 & Q_{11} & 0 & -Q_{13} \\ -R_{02} & 0 & R_{22} & 0 \\ 0 & -Q_{31} & 0 & Q_{33} \end{Vmatrix}, \tag{4.156}$$

$$\eta = -\psi_1 Q \psi_2 + \varepsilon \psi_2 R^T \psi_1 = \begin{Vmatrix} 0 & Q_{01} & 0 & -Q_{03} \\ -\varepsilon R_{01} & 0 & \varepsilon R_{21} & 0 \\ 0 & -Q_{21} & 0 & Q_{23} \\ \varepsilon R_{03} & 0 & -\varepsilon R_{23} & 0 \end{Vmatrix}, \tag{4.157}$$

$$\eta_1 = -\psi_2 Q \psi_1 + \varepsilon \psi_1 R^T \psi_2 = \begin{Vmatrix} 0 & -\varepsilon R_{10} & 0 & \varepsilon R_{30} \\ Q_{10} & 0 & -Q_{12} & 0 \\ 0 & \varepsilon R_{12} & 0 & -\varepsilon R_{32} \\ -Q_{30} & 0 & Q_{32} & 0 \end{Vmatrix}. \tag{4.158}$$

Let us calculate the inverse matrix ξ. From Eq. (4.155) one can see that the term $\psi_1 Q \psi_1$ has the zero rows and columns labeled by the numbers 1 and 3. Removing them we have nondegenerate matrix

$$(\psi_1 Q \psi_1) = \begin{Vmatrix} Q_{00} & -Q_{02} \\ -Q_{20} & Q_{22} \end{Vmatrix}, \tag{4.159}$$

the determinant of which $h_1=\det(\psi_1 Q \psi_1)=Q_{00}Q_{22}-Q_{02}Q_{20}$ differs from zero, then the inverse matrix is

$$(\psi_1 Q \psi_1)^{-1} = h_1^{-1} \begin{Vmatrix} Q_{22} & Q_{02} \\ Q_{20} & Q_{00} \end{Vmatrix}. \tag{4.160}$$

Reconstructing removed zero rows and columns labeled by the numbers 1 and 3 we obtain

$$[(\psi_1 Q \psi_1)^{-1}] = h_1^{-1} \begin{Vmatrix} Q_{22} & 0 & Q_{02} & 0 \\ 0 & 0 & 0 & 0 \\ Q_{20} & 0 & Q_{00} & 0 \\ 0 & 0 & 0 & 0 \end{Vmatrix}. \tag{4.161}$$

Analogously we calculate the inverse second matrix in (4.155)

$$[(\psi_2 R^T \psi_2)^{-1}] = h_2^{-1} \begin{Vmatrix} 0 & 0 & 0 & 0 \\ 0 & R_{33} & 0 & R_{31} \\ 0 & 0 & 0 & 0 \\ 0 & R_{13} & 0 & R_{11} \end{Vmatrix}, \tag{4.162}$$

where $h_2=\det(\psi_2 R^T \psi_2)=R_{11}R_{33}-R_{13}R_{31}$ and using the formula (4.134) we find the inverse matrix ξ^{-1}

$$\xi^{-1} = (h_1 h_2)^{-1} \begin{Vmatrix} h_2Q_{22} & 0 & h_2Q_{02} & 0 \\ 0 & h_1R_{33} & 0 & h_1R_{31} \\ h_2Q_{20} & 0 & h_2Q_{00} & 0 \\ 0 & h_1R_{13} & 0 & h_1R_{11} \end{Vmatrix}. \tag{4.163}$$

Let us note that if one permutates the first and the second rows and columns in Eq. (4.155) we will have the block-diagonal matrix

$$\xi' = \begin{Vmatrix} Q_{00} & -Q_{02} & 0 & 0 \\ -Q_{20} & Q_{22} & 0 & 0 \\ 0 & 0 & R_{11} & -R_{31} \\ 0 & 0 & -R_{13} & R_{33} \end{Vmatrix}. \tag{4.164}$$

If one finds the matrix $(\xi')^{-1}$ and permutates the rows and the columns in the inverse order, then the matrix (4.163) will be obtained exactly. The way we have chosen for finding inverse matrix ξ is equivalent to the method last discussed but it gives the possibility to avoid the extra operation of the permutations of the rows and columns.

Multiplying Eq. (4.157) from the left side by the expression (4.163) and Eq. (4.158) from the right side by the expression (4.163), we find the matrices $\xi^{-1}\eta$ and $\eta_1\xi^{-1}$. It is reasonable for reducing the calculations to find out only the first terms in Eq. (4.135) and then to use the symmetry properties of these matrices in boson case and antisymmetry properties in fermion case. We have

$$\psi_1 Q \psi_2 = \begin{Vmatrix} 0 & -Q_{01} & 0 & Q_{03} \\ 0 & 0 & 0 & 0 \\ 0 & Q_{21} & 0 & -Q_{23} \\ 0 & 0 & 0 & 0 \end{Vmatrix}, \tag{4.165}$$

$$-[(\psi_1 Q \psi_1)^{-1}]\,\psi_1 Q \psi_2 = h_1^{-1} \begin{Vmatrix} 0 & Q_{01}Q_{22} - Q_{02}Q_{21} & 0 & Q_{02}Q_{23} - Q_{22}Q_{03} \\ 0 & 0 & 0 & 0 \\ 0 & Q_{01}Q_{02} - Q_{00}Q_{21} & 0 & Q_{00}Q_{23} - Q_{20}Q_{03} \\ 0 & 0 & 0 & 0 \end{Vmatrix}, \tag{4.166}$$

$$\xi^{-1}\eta = h_1^{-1} \begin{Vmatrix} 0 & Q_{01}Q_{22} - Q_{02}Q_{21} & 0 & Q_{02}Q_{23} - Q_{22}Q_{03} \\ \varepsilon\,(Q_{01}Q_{22} - Q_{02}Q_{21}) & 0 & \varepsilon\,(Q_{01}Q_{02} - Q_{00}Q_{21}) & 0 \\ 0 & Q_{01}Q_{02} - Q_{00}Q_{21} & 0 & Q_{00}Q_{23} - Q_{20}Q_{03} \\ \varepsilon\,(Q_{02}Q_{23} - Q_{22}Q_{03}) & 0 & \varepsilon\,(Q_{00}Q_{23} - Q_{20}Q_{03}) & 0 \end{Vmatrix}. \tag{4.167}$$

Analogously we find

$$\psi_2 Q \psi_1 = \begin{Vmatrix} 0 & 0 & 0 & 0 \\ -Q_{10} & 0 & Q_{12} & 0 \\ 0 & 0 & 0 & 0 \\ Q_{30} & 0 & -Q_{32} & 0 \end{Vmatrix}, \tag{4.168}$$

$$-\psi_2 Q\psi_1 [(\psi_1 Q\psi_1)^{-1}] = h_1^{-1} \left\| \begin{matrix} 0 & 0 & 0 & 0 \\ Q_{10}Q_{22} - Q_{12}Q_{20} & 0 & Q_{10}Q_{02} - Q_{00}Q_{12} & 0 \\ 0 & 0 & 0 & 0 \\ Q_{32}Q_{20} - Q_{22}Q_{30} & 0 & Q_{00}Q_{32} - Q_{30}Q_{02} & 0 \end{matrix} \right\|, \tag{4.169}$$

$$\eta_1\xi^{-1} = h_1^{-1} \left\| \begin{matrix} 0 & \varepsilon(Q_{10}Q_{22} - Q_{12}Q_{20}) & 0 & \varepsilon(Q_{32}Q_{20} - Q_{22}Q_{30}) \\ Q_{10}Q_{22} - Q_{12}Q_{20} & 0 & Q_{10}Q_{02} - Q_{00}Q_{12} & 0 \\ 0 & \varepsilon(Q_{10}Q_{02} - Q_{00}Q_{12}) & 0 & \varepsilon(Q_{00}Q_{32} - Q_{30}Q_{02}) \\ Q_{32}Q_{20} - Q_{22}Q_{30} & 0 & Q_{00}Q_{32} - Q_{30}Q_{02} & 0 \end{matrix} \right\| \tag{4.170}$$

and using the formula (4.19) we obtain the kernel of the operator of the finite group transformations in the coherent state basis

$$\begin{aligned} U(\alpha^*, \beta, \mathbf{r}, \mathbf{s}, i\tilde{j}) &= (h_1 h_2)^{-\varepsilon/2} \exp \{(h_1 h_2)^{-1}[h_2 Q_{22}\alpha_0^*\beta_0 + h_1 R_{33}\alpha_1^*\beta_1 + \\ &+ h_2 Q_{00}\alpha_2^*\beta_2 + h_1 R_{11}\alpha_3^*\beta_3 + h_2 Q_{02}\alpha_0^*\beta_2 + h_2 Q_{20}\alpha_2^*\beta_0 + h_1 R_{31}\alpha_1^*\beta_3 + \\ &+ h_1 R_{13}\alpha_3^*\beta_1 + h_2 [(-Q_{01}Q_{22} + Q_{02}Q_{21})\alpha_0^*\alpha_1^* + (Q_{22}Q_{03} - Q_{02}Q_{23})\alpha_0^*\alpha_3^* + \\ &+ (Q_{00}Q_{21} - Q_{01}Q_{02})\varepsilon\alpha_1^*\alpha_2^* + (Q_{20}Q_{03} - Q_{00}Q_{23})\alpha_2^*\alpha_3^* + (Q_{10}Q_{22} - \\ &- Q_{12}Q_{20})\beta_0\beta_1 + (Q_{32}Q_{20} - Q_{22}Q_{30})\beta_0\beta_3 + (Q_{10}Q_{02} - Q_{00}Q_{12})\varepsilon\beta_1\beta_2 + \\ &+ (Q_{00}Q_{32} - Q_{30}Q_{02})\beta_2\beta_3]]\}. \end{aligned} \tag{4.171}$$

In the boson case (ε=1) from the composition rule for the finite transformations of the group $SO_4(i\tilde{j})$ one obtains the adding formula for the Hermite polynomials at zero-valued variables of the type (4.32). Not considering the contractions of the group $SO_4(i\tilde{j})$ from the unitarity condition for the operator $\hat{U}_g=\exp(-\hat{H})$ in the Fock basis, one obtains the sum rule for the Hermite polynomials at zero-valued variables of the type (4.33).

We changed the parameter j_k by $i\tilde{j}_k$ in order to have the possibility to consider the contractions (at $\tilde{j}_k=\iota_k$) of the group $SO_4(i\tilde{j})$ and, consequently, to study the quantum systems into which the system (4.147) transforms at the limit values of the parameters characterizing this system. We consider in detail the most degenerate case when all the parameters $\tilde{j}_k=\iota_k$, k=1,2,3, i.e. the system corresponding to the group $SO_4(i\iota)$. From the formula (2.35) it follows that X(r, s, $i\iota$)=X(r, s, ι), i.e. $SO_4(i\iota)$ and $SO_4(\iota)$ are same groups. The finite group transformation of this group may be obtained from Eq. 4 (4.142) at $j_3=\iota_3$ and it has the form

$$\Xi(\mathbf{r}, \mathbf{s}, \iota) = E + X + {}^1/_6(\mathbf{r}, \mathbf{s}) X_1 + {}^1/_2 X^2 =$$
$$= \left\| \begin{matrix} 1 & 0 & 0 & 0 \\ r_1 & 1 & 0 & 0 \\ r_2 + {}^1/_2(\mathbf{r}\times\mathbf{s})_2 & -s_3 & 1 & 0 \\ r_3 + {}^1/_2(\mathbf{r}\times\mathbf{s})_3 + {}^1/_6 s_3(\mathbf{r}, \mathbf{s}) & s_2 + {}^1/_2 s_1 s_3 & -s_1 & 1 \end{matrix} \right\|, \tag{4.172}$$

where $(r\times s)_2=-r_1s_3$, $(r\times s)_3=r_1s_2-r_2s_1$, $(r\times s)=r_1s_1$. The group composition rule is given by the formulae (4.146). But the Hamiltonians (4.138) and (4.153) of the systems corresponding to the groups $SO_4(\iota)$ and $SO_4(i\iota)$ are not equal. Because of this, we will keep the two notations for one group and for the same group in order to distinguish these two systems.

Taking j=ι in the relations Q=Ξ(r, s, $i\tilde{j}$), R=Ξ(-r, -s, $i\tilde{j}$) and using Eq. (4.172) we find the matrix elements of the matrices Q and R

$$Q_{km}=0,\ k<m,\quad Q_{kk}=1,\ k=0,1,2,3,\quad Q_{10}=r_1,\quad Q_{20}=r_2+{}^1/_2(\mathbf{r}\times\mathbf{s})_2,$$
$$Q_{31}=s_2+{}^1/_2\,s_1s_3,\quad Q_{30}=r_3+{}^1/_2\,(\mathbf{r}\times\mathbf{s})_3+{}^1/_6\,s_3\,(\mathbf{r},\mathbf{s}),\quad Q_{21}=-s_3,\quad Q_{32}=-s_1;$$
$$R_{km}=0,\quad k<m,\quad R_{kk}=1,\quad k=0,1,2,3,\quad R_{10}=-r_1,\quad R_{21}=s_3,$$
$$R_{20}=-r_2+{}^1/_2\,(\mathbf{r}\times\mathbf{s})_2,\ R_{30}=-r_3+{}^1/_2\,(\mathbf{r}\times\mathbf{s})_3-{}^1/_6 s_3\,(\mathbf{r},\mathbf{s}),\quad R_{31}=-s_2+$$
$$+{}^1/_2\,s_1s_3,\quad R_{32}=s_1. \tag{a}$$

Substituting these matrix elements into the formulae (4.163), (4.167), (4.170) we obtain the matrices $\xi^{-1}, \xi^{-1}\eta, \eta_1\xi^{-1}$

$$\xi^{-1}=\begin{Vmatrix} 1 & 0 & 0 & 0\\ 0 & 1 & 0 & -s_2+{}^1/_2\,s_1s_3\\ r_2+{}^1/_2\,(\mathbf{r}\times\mathbf{s})_2 & 0 & 1 & 0\\ 0 & 0 & 0 & 1\end{Vmatrix},$$
$$\xi^{-1}\eta=\begin{Vmatrix} 0 & 0 & 0 & 0\\ 0 & 0 & \varepsilon s_3 & 0\\ 0 & s_3 & 0 & 0\\ 0 & 0 & 0 & 0\end{Vmatrix},\quad \eta_1\xi^{-1}=\begin{Vmatrix} 0 & \varepsilon r_1 & 0 & -\varepsilon\mathcal{D}\\ r_1 & 0 & 0 & 0\\ 0 & 0 & 0 & -\varepsilon s_1\\ -\mathcal{D} & 0 & -s_1 & 0\end{Vmatrix}, \tag{4.173}$$

where $D=r_3+(r_1s_2+r_2s_1)/2-s_3(r,s)/3$. The kernel $U(\alpha^*, \beta, i\iota)$ of the operator of finite group transformation $\hat{U}_g\exp(-\hat{H})$ in the coherent state basis may be found with the help of the formula (4.19). It has the form

$$U(\tilde{\alpha}^*,\tilde{\beta},i\iota)=\exp\{\sum_{k=0}^{3}\tilde{\alpha}_k^*\tilde{\beta}_k+(-\tilde{s}_2+{}^1/_2\tilde{s}_1\tilde{s}_3)\,\tilde{\alpha}_1^*\tilde{\beta}_3+$$
$$+(\tilde{r}_2+{}^1/_2\,(\tilde{\mathbf{r}}\times\tilde{\mathbf{s}})_2)\,\tilde{\alpha}_2^*\tilde{\beta}_0-\varepsilon\tilde{s}_3\tilde{\alpha}_1^*\tilde{\alpha}_2^*+\tilde{r}_1\tilde{\beta}_0\tilde{\beta}_1-\tilde{s}_1\tilde{\beta}_2\tilde{\beta}_3-$$
$$-[\tilde{r}_3+{}^1/_2\,(\tilde{r}_1\tilde{s}_2+\tilde{r}_2\tilde{s}_1)-{}^1/_3\tilde{s}_3\,(\tilde{\mathbf{r}},\tilde{\mathbf{s}})]\,\tilde{\beta}_0\tilde{\beta}_3\}. \tag{4.174}$$

In this formula the variables related to the group $SO_4(i\iota)$ have the label tilda.

Let us write down the matrix elements of the operator $\hat{U}_g=\exp(-\hat{H})$ corresponding to the system with the Hamiltonian (4.138), i.e. the group $SO_4(\iota)$. Using Eqs. (4.172) and (4.140) we have

$$U(\alpha^*,\beta,\iota)=\exp\{\sum_{k=0}^{3}\alpha_k^*\beta_k+(-s_2+{}^1/_2s_1s_3)\,\alpha_3^*\beta_1+$$
$$+(-r_2+{}^1/_2\,(\mathbf{r}\times\mathbf{s})_2)\,\alpha_2^*\beta_0+s_3\alpha_2^*\beta_1-r_1\alpha_1^*\beta_0+s_1\alpha_3^*\beta_2+$$
$$+[-r_3+{}^1/_2\,(r_1s_2-r_2s_1)-{}^1/_6s_3\,(\mathbf{r},\mathbf{s})]\,\alpha_3^*\beta_0\}. \tag{4.175}$$

The matrix elements (4.174) and (4.175) are different. Nevertheless we will show that they are connected by some relation. Let us consider the transformation of the variables labeling the coherent state of the form

$$\begin{pmatrix}\alpha^*\\ \beta\end{pmatrix}=X(i\iota)\begin{pmatrix}\tilde{\alpha}^*\\ \tilde{\beta}\end{pmatrix}=\begin{Vmatrix}\chi_1 & \varepsilon\chi_2\\ \chi_2 & \chi_1\end{Vmatrix}\begin{pmatrix}\tilde{\alpha}^*\\ \tilde{\beta}\end{pmatrix}, \tag{4.176}$$

where the diagonal matrices x_1 and x_2 are

$$\chi_1=\begin{Vmatrix}1&0&0&0\\0&0&0&0\\0&0&1&0\\0&0&0&0\end{Vmatrix},\quad \chi_2=\begin{Vmatrix}0&0&0&0\\0&1&0&0\\0&0&0&0\\0&0&0&1\end{Vmatrix}, \tag{4.177}$$

and they have the matrix elements equal to the modulus of the matrix elements of the matrices ψ_1 and ψ_2 respectively. Let us also consider the following transformation of the group parameters

$$\mathbf{r} = T(\tilde{s})\tilde{r}, \qquad \mathbf{s} = \mathcal{R}\tilde{\mathbf{s}}, \tag{4.178}$$

where the transformation matrices have the form

$$T(\tilde{\mathbf{s}}) = \begin{Vmatrix} -1 & 0 & 0 \\ 0 & -1 & 0 \\ -{}^1/_6\tilde{s}_1\tilde{s}_3 & 0 & 1 \end{Vmatrix}, \qquad \mathcal{R} = \begin{Vmatrix} -1 & 0 & 0 \\ 0 & 1 & 0 \\ 0 & 0 & -1 \end{Vmatrix}. \tag{4.179}$$

Substituting into the formula (4.175) the variables α^*, β and group parameters r and s expressed in terms of the variables $\tilde{\alpha}^*,\tilde{\beta}$ and parameters $\tilde{r}$ and $\tilde{s}$ according to Eqs. (4.176)-(4.179) one obtains the formula (4.174) for the kernel of the finite transformations of the group $SO_4(i\iota)$ (taking into account the Grassman properties of the variables α^*, β in the fermion case). So, the following relation holds

$$U((\boldsymbol{\alpha}^*, \boldsymbol{\beta}) = \mathrm{X}(\tilde{\boldsymbol{\alpha}}^*, \tilde{\boldsymbol{\beta}}); \mathbf{r} = T(\tilde{\mathbf{s}})\tilde{\mathbf{r}}; \mathbf{s} = \mathcal{R}\tilde{\mathbf{s}}, \iota) = U(\tilde{\boldsymbol{\alpha}}^*, \tilde{\boldsymbol{\beta}}, \mathbf{r}, \mathbf{s}, i\iota), \tag{4.180}$$

and we also have the inverse relation

$$U((\tilde{\boldsymbol{\alpha}}^*, \tilde{\boldsymbol{\beta}}) = \mathrm{X}^{-1}(\boldsymbol{\alpha}^*, \boldsymbol{\beta}); \tilde{\mathbf{r}} = T(\mathbf{s})\mathbf{r}; \tilde{\mathbf{s}} = \mathcal{R}\mathbf{s}; i\iota) = U(\boldsymbol{\alpha}^*, \boldsymbol{\beta}, \mathbf{r}, \mathbf{s}, \iota), \tag{4.181}$$

where the inverse transformation (4.176) has the form

$$\mathrm{X}^{-1}(i\iota) = \begin{Vmatrix} \chi_1 & \chi_2 \\ \varepsilon\chi_2 & \chi_1 \end{Vmatrix}. \tag{4.182}$$

This close relation of two expressions for the kernels of the operators of the finite group transformations in the coherent state basis (and the close relation of two quantum systems) is connected with two different representations in terms of the creation and annihilation operators of the generators of the same $SO_4(\iota)$ group obtained from the rotation group with help of the multidimensional (in our case three-dimensional) contraction. In fact, the matrix generators (2.32) of the group $SO_4(\iota)$ are expressed in terms of the creation and annihilation operators by the formulae

$$\hat{L}_{01} = \hat{a}_1^+\hat{a}_0, \quad \hat{L}_{02} = \hat{a}_2^+\hat{a}_0, \quad \hat{L}_{03} = \hat{a}_3^+\hat{a}_0, \quad \hat{L}_{12} = \hat{a}_2^+\hat{a}_1, \quad \hat{L}_{13} = \hat{a}_3^+\hat{a}_1,$$
$$\hat{L}_{23} = \hat{a}_3^+\hat{a}_2, \tag{4.183}$$

for the case j=iι these matrix generators have the following form in Jordan-Schwinger representation

$$\hat{L}'_{01} = -\varepsilon\hat{a}_1\hat{a}_0, \quad \hat{L}'_{02} = \hat{a}_2^+\hat{a}_0, \quad \hat{L}'_{03} = -\varepsilon\hat{a}_3\hat{a}_0,$$
$$\hat{L}'_{12} = \hat{a}_2^+\hat{a}_1^+, \quad \hat{L}'_{13} = -\varepsilon\hat{a}_3\hat{a}_1^+, \quad \hat{L}'_{23} = -\varepsilon\hat{a}_3\hat{a}_2. \tag{4.184}$$

Here a, a^+ are either boson or fermion operators. Both sets of the operators $\hat{L}_{km}$ and $\hat{L}'_{km}$ satisfy the same commutation relations of the group $SO_4(\iota)$.

In the general case, the semi-simple group (algebras) $SO_{n+1}(j)$ for which no one parameter j is equal to pure dual unit, have the only Jordan-Schwinger representation expressed by the formula (3.21).

Each group (algebra) $SO_{n+1}(j)$ for which one of the parameters, for example the first one is equal to pure dual unit $j_1=\iota_1$ permits two Jordan-Schwinger representations corresponding to the parameters $j=(\iota_1, j_1,\ldots,j_n)$ and $j'=(i\iota_1, j_2, j_3,\ldots, j_n)$.

Each group for which two parameters, for example, the first and the second ones, are equal to pure dual units $j_1=\iota_1$, $j_2=\iota_2$ have four Jordan-Schwinger representations corresponding to the parameters $j=(\iota_1, \iota_2, j_3,\ldots,j_n)$, $j'=(i\iota_1, \iota_2, j_3,\ldots,j_n)$, $j''=(\iota_1, i\iota_2, j_3,\ldots,j_n)$, $j'''=(i\iota_1, i\iota_2, j_3,\ldots, j_n)$. In general, if the group $SO_{n+1}(j)$ has k pure dual parameters according to the formula (3.21) it has 2^k different Jordan-Schwinger representations realized by 2^k different mixed sets of the creation and annihilation operators. Since we consider the quantum systems with Hamiltonians which are linear functions in operators $\hat{L}_{\mu\nu}(j)$, there are 2^k different quantum systems corresponding to each of the groups $SO_{n+1}(j)$ with k dual parameters. As the considered example of the groups $SO_4(\iota)$ and $SO_4(i\iota)$ demonstrates, the kernels of the operators of the finite group transformations (consequently, the Green functions) of such systems may be transformed one into the other. This problem needs further study.

5. GROUP QUANTUM SYSTEMS IN CORRELATED STATE REPRESENTATION

The aim of this section is to construct the operators transforming well known expressions for the linear integrals of the motion and the propagators of the multidimensional quadratic quantum systems (on examples of the Hamiltonians connected with Lie groups) in the coherent state representation into the expression for the linear integrals of the motion and the propagators of the same systems in the correlated state representation. For completeness, in Section 5.1 we describe the correlated states. In Sections 5.2, 5.3 for the Hamiltonians, propagators and integrals of the motion of the quadratic quantum systems we study the relation of the Glauber coherent states and correlated states. In Section 5.4 we consider the special subset of the quadratic systems in the correlated state representation. The Hamiltonians of these systems are linear functions in generators of the rotation groups in the Cayley-Klein spaces.

5.1 Correlated Coherent States

The coherent states introduced in study [2] turned out to be very useful in many branches of the quantum theory. The most important physical property of the Glauber coherent states that created the effective application of these states for solving concrete physical problems is the fact that they minimize the uncertainty relation for the operators of canonically conjugate variables, i.e. in a definite sense, the quantum systems in these states are as close as possible to their classical analogs. The coherent states $|\alpha,\mu\rangle$ are the eigenvalues of the annihilation operator

$$a(\mu) = (2\mu)^{-1/2}(\hat{q} + i\mu\hat{p}), \quad \mu > 0,$$
$$a(\mu)|\alpha, \mu\rangle = \alpha\,|\alpha, \mu\rangle, \tag{5.1}$$

where $\hat{q}$ and $\hat{p}$ are the canonically conjugate variables ($i=[\hat{q},\hat{p}]$). In the coordinate representation, the normalized eigenstate of the annihilation operator corresponding to the complex eigenvalue α may be written down in the form

$$\psi_\alpha(x) \equiv \langle x \mid \alpha, \mu\rangle = (\pi\mu)^{-1/4}\exp\{-(x(2\mu)^{-1/2} - \alpha)^2 + {}^1/_2\alpha^2 - \\ - {}^1/_2\,|\alpha|^2\}. \tag{5.2}$$

For the states (5.2) the quantum-mechanical uncertainties of the operators $\hat{q}$ and $\hat{p}$ are minimal $\sigma_q\sigma_p=1/4$ (the Heisenberg relation) where $\sigma_q=\langle(\Delta\hat{q})^2\rangle\sigma_p=\langle(\Delta\hat{p})2\rangle$ are the operator dispersions, $\Delta\hat{q}=\hat{q}-<\hat{q}>$, $\Delta\hat{p}=\hat{p}-<\hat{p}>$ and the parenthesis $<...>$ imply the quantum-mechanical averages of the operators. In this case the correlation moment $\sigma_{pq}=<\Delta\hat{q}\Delta\hat{p}+\Delta\hat{p}\Delta\hat{q}>/2$ is equal to zero and corresponds to the statistical independence of the Gaussian wave packets (eigenfunctions) of the variables $\hat{q}$ and $\hat{p}$.

If there is additional information about the quantum system, for example, it is known that the correlation moment σ_{qp} of the canonically conjugate variables differs from zero. The uncertainty relation is written down in the form [9-12]

$$\sigma_q\sigma_p\,(1-R^2)\geqslant 1/4, \quad |R|<1, \tag{5.3}$$

where $R=\sigma_{qp}(\sigma_q\sigma_p)^{-1/2}$ is the correlated coefficient of the physical observables $\hat{q}$ and $\hat{p}$. The state of the quantum system which minimizes the uncertainty relation (5.3) was called the correlated coherent state [9-12] and as it was shown in these studies, it is the eigenstate of the boson annihilation operator

$$\begin{aligned} b\,(\nu,\chi,\varphi) &= e^{i\varphi}\,(2\nu\cos\chi)^{-1/2}\,(\hat{q}+i\nu e^{i\chi}\hat{p}), \quad \nu>0, \ -\pi/2<\chi<\pi/2, \\ b\,(\nu,\chi,\varphi)|\,\beta',\nu,\chi,\varphi\rangle &= \beta'\,|\,\beta',\nu,\chi,\varphi\rangle. \end{aligned} \tag{5.4}$$

In the coordinate representation the correlated coherent states are described by the formula

$$\langle x\,|\,\beta',\nu,\chi,\varphi\rangle = = \left(\frac{\cos\chi}{\pi\nu}\right)^{1/4}\exp\left\{-e^{-i\chi}\left(\frac{x}{(2\nu)^{1/2}}-(\cos\chi)^{1/2}\beta' e^{-i\varphi}\right)^2+\frac{\beta'^2}{2}e^{-2i\varphi}-\frac{|\beta'|^2}{2}\right\}. \tag{5.5}$$

In this parametrization the correlation coefficient of the observables $\hat{q}$ and $\hat{p}$ is equal to $R=\sin\chi$.

The boson creation and annihilation operators b^+, b create the coherent states $|\beta>$ may be considered in differing parametrizations [11]. Let us consider one of these parametrizations which we will use in further sections. Let Glauber coherent states be the eigenstates of the boson annihilation operators $\alpha(\mu)$, $\alpha^+(\mu)$. Let us construct the operators b and b^+ with help of the Bogolubov transform

$$\begin{pmatrix} b \\ b^+ \end{pmatrix} = \begin{Vmatrix} u & v \\ v^* & u^* \end{Vmatrix} \begin{pmatrix} a\,(\mu) \\ a^+\,(\mu) \end{pmatrix} \equiv V \begin{pmatrix} a\,(\mu) \\ a^+\,(\mu) \end{pmatrix}. \tag{5.6}$$

Here u and v are the complex numbers satisfying the condition $\det V=|u|^2-|v|^2=1$, the matrix V belongs to the group SU(1,1), and the operators $\alpha(\mu)$ and $\alpha^+(\mu)$ satisfy the commutation relation $[\alpha,\alpha^+]=1$. Then the commutation relations for the operators b and b^+ is $[b,b^+]=1$. The eigenvectors $|\beta, u, v>$ of the annihilation operator $b(u, v)$ are the correlated coherent states in the coordinate representation and they may be written in the form [11]

$$\langle x\,|\,\beta,u,v\rangle = (\pi\mu\,|\,u-v\,|^2)^{-1/4}\times \times\exp\left\{-\frac{1}{2\mu}\frac{u+v}{u-v}x^2+\frac{2\,(2\mu)^{-1/2}}{u-v}\beta x-\frac{1}{2}\beta^2\frac{u^*-v^*}{u-v}-\frac{1}{2}|\beta|^2\right\} \tag{5.7}$$

with the correlation coefficient of the coordinate and momentum

$$R=\mathrm{Im}\,(uv^*)[(1/2+|\,v\,|^2)^2-(\mathrm{Re}\,uv^*)^2]^{-1/2}. \tag{5.8}$$

It is worthy of note that the correlated coherent state possess all the analytica properties of the Glauber coherent states. They form the complete (overcomplete) set o states and the scalar product is given by the formula

$$\langle \gamma \mid \beta \rangle = \exp \{\gamma^* \beta - {}^1/_2 (\mid \beta \mid^2 + \mid \gamma \mid^2)\}. \tag{5.9}$$

The expansion coefficients of the correlated states in powers of the complex parameters β are the eigenstates of the number operator of the correlated particles $\hat{N} = b^+b$

$$\mid \beta \rangle = \exp \left\{ -\frac{1}{2} \mid \beta \mid^2 \right\} \sum_{n=0}^{\infty} \frac{\beta^n}{\sqrt{n!}} \mid n \rangle,$$

$$\mid n \rangle = \frac{(b^+)^n}{\sqrt{n!}} \mid 0 \rangle, \quad b \mid n \rangle = \sqrt{n} \mid n-1 \rangle, \quad b^+ \mid n \rangle = \sqrt{n+1} \mid n+1 \rangle, \tag{5.10}$$

where the ground correlated state is determined by the equation $b \mid 0 \rangle = 0$.

The transformation (5.6) of the Glauber creation and annihilation operators α^+ and α into the correlated creation and annihilation operators b^+ and b as well as the inverse transformation

$$\begin{pmatrix} a \\ a^+ \end{pmatrix} = \begin{Vmatrix} u^* & -v \\ -v^* & u \end{Vmatrix} \begin{pmatrix} b \\ b^+ \end{pmatrix} \equiv V^{-1} \begin{pmatrix} b \\ b^+ \end{pmatrix} \tag{5.11}$$

also provide the connection of the matrix elements of these operators on the basis of correlated coherent states

$$\begin{pmatrix} \langle \delta \mid b \mid \gamma \rangle \\ \langle \delta \mid b^+ \mid \gamma \rangle \end{pmatrix} = V \begin{pmatrix} \langle \delta \mid a \mid \gamma \rangle \\ \langle \delta \mid a^+ \mid \gamma \rangle \end{pmatrix}, \quad \begin{pmatrix} \langle \delta \mid a \mid \gamma \rangle \\ \langle \delta \mid a^+ \mid \gamma \rangle \end{pmatrix} = V^{-1} \begin{pmatrix} \langle \delta \mid b \mid \gamma \rangle \\ \langle \delta \mid b^+ \mid \gamma \rangle \end{pmatrix}, \tag{5.12}$$

the matrix V^{-1} belonging to the group SU(1,1) too.

5.2 Correlated and Coherent Hamiltonians of Quantum Systems

Let us consider the Hamiltonian systems which are multidimensional homogeneous quadratic forms in the boson creation and annihilation operators. The general form of Hamiltonians for such systems is as follows [8]

$$H_{\mathbf{a}} = \frac{1}{2} (\mathbf{a}, \mathbf{a}^+) \Delta_{\mathbf{a}} \begin{pmatrix} \mathbf{a} \\ \mathbf{a}^+ \end{pmatrix} + (\mathbf{f}, \mathbf{f}^*) \begin{pmatrix} \mathbf{a} \\ \mathbf{a}^+ \end{pmatrix}, \quad \mathbf{a} = (a_1, \ldots, a_N), \tag{5.13}$$

where α^+_k, α_k are Glauber creation and annihilation operators with the commutation relations $[\alpha_k, \alpha^+_k] = 1$, k=1,2,...,N and Δ_a is the 2N-dimensional symmetrical matrix. The theory of such systems is well known. In study [8] the expressions for the propagators (Green functions) of the Schrodinger equation and the equilibrium density matrices related to the propagators have been obtained in different representations: in coordinate, momentum, coherent state and Wigner-Weyl representations. Here we consider the representation of the correlated coherent states. In this representation each pair of the Glauber creation and annihilation operators α^+_k, α_k is replaced with the help of the Bogolubov transformation (5.6) by the correlated creation and annihilation operators

$$\begin{pmatrix} b_k \\ b_k^+ \end{pmatrix} = \begin{Vmatrix} u_k & v_k \\ v_k^* & u_k^* \end{Vmatrix} \begin{pmatrix} a_k \\ a_k^+ \end{pmatrix} \equiv V_k \begin{pmatrix} a_k \\ a_k^+ \end{pmatrix}, \tag{5.14}$$

where $\det V_k = |u_k|^2 - |v_k|^2 = 1$, k=1,2,...,N.

Since the correlated coherent states have the same properties as the Glauber coherent states, it is possible to use for describing the quadratic quantum systems in the correlated state representation the well developed scheme for studying these systems in the coherent state representation. In fact, transforming the Hamiltonian (5.13) with the help of Eq. (5.14) into the Hamiltonian $\hat{H}$ written down in the same form (5.13) in which the operators a, a^+ are replaced by the operators b, b^+ and the matrix Δ_a and the vector f are replaced by the new matrix Δ_b and the new vector h we obtain for the quantum systems with the Hamiltonian $\hat{H}$ in the correlated state representation in the same formulae that are used for the quadratic system in the coherent state representation.

Let us rewrite the Hamiltonian (5.13) in terms of the correlated operators b and b^+. We introduce the notations for the vector-rows $(\tilde{a}_1,\tilde{a}_2,\ldots,\tilde{a}_N)\equiv(\alpha_1, \alpha^+{}_1, \alpha_2, \alpha_2{}^+,\ldots\alpha_N, \alpha^+{}_N)$, $(\tilde{b}_1,\tilde{b}_2,\ldots,\tilde{b}_N)\equiv(b_1, b^+{}_1, b_2, b^+{}_2,\ldots,b_N, b^+{}_N)$ and the analogous notations for the vector-columns. Then the transformation (5.14) may be written in the form

$$(\tilde{\mathbf{b}}_1, \ldots, \tilde{\mathbf{b}}_N) = (\tilde{\mathbf{a}}_1, \ldots, \tilde{\mathbf{a}}_N)\, V^T, \quad \begin{pmatrix} \tilde{\mathbf{b}}_1 \\ \vdots \\ \tilde{\mathbf{b}}_N \end{pmatrix} = V \begin{pmatrix} \tilde{\mathbf{a}}_1 \\ \vdots \\ \tilde{\mathbf{a}}_N \end{pmatrix}, \tag{5.15}$$

where the 2N-dimensional matrix V is

$$V = \oplus \sum_{k=1}^{N} V_k. \tag{5.16}$$

In the expression for the Hamiltonian (5.13) the vector-rows (a, a^+) and the vector-columns $\begin{pmatrix} a \\ a^+ \end{pmatrix}$ are used. In these vectors the operators α_k, $\alpha^+{}_k$ are ordered in a different way. The transformation Φ_N which relates these differently ordered sets of the same operators may be found from the expression

$$\begin{pmatrix} \tilde{\mathbf{a}}_1 \\ \vdots \\ \tilde{\mathbf{a}}_N \end{pmatrix} = \Phi_N \begin{pmatrix} \mathbf{a} \\ \mathbf{a}^+ \end{pmatrix}. \tag{5.17}$$

All the nonzero elements of the 2N-dimensional matrix Φ_N are equal to unity, namely, we have $(\Phi_N)_{2k,\, N+k}=(\Phi_N)_{2k-1,\, k=1}$, k=1,2,...,N. The determinant of the matrix Φ_N is equal to unity and the inverse matrix is equal to the transposed matrix $\Phi_N{}^{-1}=\Phi^T{}_N$. It is obvious that the vector rows are transformed according to the rule $(\tilde{a}_1,\tilde{a}_2,\ldots,\tilde{a}_N)=(a, a^+)\Phi^T{}_N$. The analogous formulae hold for the correlated operators b, b^+.

Substituting the formula (5.17) into the right side of Eq. (5.15) and the analogous formula for the operators b, b^+ into the left side of Eq. (5.15) we find the transformation rule of the coherent operators into the correlated creation and annihilation operators

$$\begin{pmatrix} \mathbf{b} \\ \mathbf{b}^+ \end{pmatrix} = \Phi_N^T V \Phi_N \begin{pmatrix} \mathbf{a} \\ \mathbf{a}^+ \end{pmatrix}, \qquad (\mathbf{b}, \mathbf{b}^+) = (\mathbf{a}, \mathbf{a}^+)\, \Phi_N^T V^T \Phi_N \tag{5.18}$$

and the inverse transformation of the correlated operators into the coherent ones

$$\begin{pmatrix} \mathbf{a} \\ \mathbf{a}^+ \end{pmatrix} = \Phi_N^T V^{-1} \Phi_N \begin{pmatrix} \mathbf{b} \\ \mathbf{b}^+ \end{pmatrix}, \qquad (\mathbf{a}, \mathbf{a}^+) = (\mathbf{b}, \mathbf{b}^+)\, \Phi_N^T (V^{-1})^T \Phi_N \tag{5.19}$$

To obtain the expression for the Hamiltonian of the quantum system (5.13) in terms of the creation and annihilation operators we have to replace the operators a, a^+ in Eq. (5.13) by the correlated operators, b, b^+ according to the formula (5.19). We have

$$H_{\mathbf{b}} = \frac{1}{2}(\mathbf{b}, \mathbf{b}^+)\,\Phi_N^T (V^{-1})^T\,\Phi_N \Delta_{\mathbf{a}} \Phi_N^T V^{-1} \Phi_N \begin{pmatrix}\mathbf{b}\\ \mathbf{b}^+\end{pmatrix} + (\mathbf{f}, \mathbf{f}^*)\,\Phi_N^T V^{-1} \Phi_N \begin{pmatrix}\mathbf{b}\\ \mathbf{b}^+\end{pmatrix} \equiv$$
$$\equiv \frac{1}{2}(\mathbf{b}, \mathbf{b}^+)\,\Delta_{\mathbf{b}} \begin{pmatrix}\mathbf{b}\\ \mathbf{b}^+\end{pmatrix} + (\mathbf{h}, \mathbf{h}^*) \begin{pmatrix}\mathbf{b}\\ \mathbf{b}^+\end{pmatrix}. \tag{5.20}$$

Hence, when one replaces the coherent creation and annihilation operators by the correlated operators, the matrix Δ_a and the vector (f,f*) in the quadratic Hamiltonian of the general form are transformed in accordance with the formulae

$$\Delta_{\mathbf{b}} = \Phi_N^T (V^{-1})^T\,\Phi_N \Delta_{\mathbf{a}} \Phi_N^T V^{-1} \Phi_N, \qquad (\mathbf{h}, \mathbf{h}^*) = (\mathbf{f}, \mathbf{f}^*)\,\Phi_N^T V^{-1} \Phi_N. \tag{5.21}$$

It may be verified that the matrix Δ_b is a symmetrical one $\Delta^T{}_b = \Delta_b$, and the vector components are equal to the following numbers $h_k = u^*{}_k f_k - v^*{}_k f^*{}_k$, $h^*{}_k = u_k f^*{}_k - v_k f_k$. It means that the transformations (5.20)-(5.21) do not change the quadratic structure of the Hamiltonian H_b.

5.3 Propagators of Quantum Systems in Correlated State Representation

The propagators (the kernels of evolution operator) of the systems described by the Schrodinger type equation (we take $\hbar$=1)

$$i\,\frac{\partial \psi}{\partial t} = H\psi \tag{5.22}$$

with the quadratic Hamiltonians (5.13) or (5.20) may be found with the help of the simple and effective method of the integrals of the motion [8]. The linear integrals of the motion for the Hamiltonian (5.13) have the form

$$\hat{Q}_{\mathbf{a}}(t) = \Lambda_{\mathbf{a}}(t)\,\hat{\mathbf{q}}_{\mathbf{a}} + \mathbf{L}_{\mathbf{a}}, \quad \hat{\mathbf{q}}_{\mathbf{a}} = \begin{pmatrix}\mathbf{a}\\ \mathbf{a}^+\end{pmatrix}, \quad \mathbf{L}_{\mathbf{a}} = \begin{pmatrix} l \\ l_1 \end{pmatrix},$$
$$\hat{Q}_{\mathbf{a}}(0) = \hat{\mathbf{q}}_{\mathbf{a}}, \quad \Lambda_{\mathbf{a}} = \begin{Vmatrix} \xi & \eta \\ \eta_1 & \xi_1 \end{Vmatrix}, \tag{5.23}$$

where the 2N-dimensional matrix Λ_a and the 2N-dimensional vector L_a may be found as the solution to the following matrix equations

$$-i\dot{\Lambda}_{\mathbf{a}} = \Lambda_{\mathbf{a}} \Sigma \Delta_{\mathbf{a}}, \quad \Lambda_{\mathbf{a}}(0) = E_{2N}, \quad -i\dot{\mathbf{L}}_{\mathbf{a}} = \Lambda_{\mathbf{a}} \Sigma \mathbf{F}, \quad \mathbf{L}_{\mathbf{a}}(0) = 0. \tag{5.24}$$

Here E_{2N} is the 2N-dimensional unity matrix, the matrix $\Sigma = \begin{Vmatrix} 0 & E_N \\ -E_N & 0 \end{Vmatrix}$, the matrix Δ_a and the vector $F = \begin{pmatrix} f \\ f^* \end{pmatrix}$ are determined by the Hamiltonian (5.13).

In the case of the boson quantum systems, the coherent state representation by Glauber is the most convenient one, and the propagator of the system is given by the expression [8]

$$G(\alpha^*, \beta, t) = [\det \xi(t)]^{-1/2} \exp\Big\{-{}^1\!/_2\alpha^*\xi^{-1}\eta\alpha^* + \alpha^*\xi^{-1}\beta + {}^1\!/_2\beta\eta_1\xi^{-1}\beta -$$
$$-\alpha^*\xi^{-1}l + \beta(l_1 - \eta_1\xi^{-1}l) + {}^1\!/_2 l\eta_1\xi^{-1}l - \int_0^t l(\tau)\,\dot{l}_1(\tau)\,d\tau\Big\}. \tag{5.25}$$

For the Hamiltonian (5.20) expressed in terms of the correlated creation and annihilation operators we will write down the linear integrals of the motion in the form which is analogous to Eq. (5.23)

$$\hat{Q}_b(t) = \Lambda_b(t)\,\hat{\mathbf{q}}_b + \mathbf{L}_b, \qquad \hat{\mathbf{q}}_b = \begin{pmatrix} b \\ b^+ \end{pmatrix}, \qquad \mathbf{L}_b = \begin{pmatrix} \tilde{l} \\ \tilde{l}_1 \end{pmatrix},$$

$$\hat{Q}_b(0) = \hat{\mathbf{q}}_b, \qquad \Lambda_b = \begin{Vmatrix} \tilde{\xi} & \tilde{\eta} \\ \tilde{\eta}_1 & \tilde{\xi}_1 \end{Vmatrix}. \tag{5.26}$$

The matrix Λ_b and the vector L_b satisfy the equations

$$-i\dot{\Lambda}_b = \Lambda_b \Sigma \Delta_b, \qquad \Lambda_b(0) = E_{2N}, \qquad -i\dot{\mathbf{L}}_b = \Lambda_b \Sigma \mathscr{H}, \qquad \mathbf{L}_b(0) = 0, \tag{5.27}$$

where the matrix Δ_b and the vector $H=\begin{pmatrix} h \\ h^* \end{pmatrix}$ are determined by formulae (5.21).

Since the correlated coherent states have the same properties as the Glauber coherent states, the propagator of the system with the Hamiltonian (5.20) in the correlated state representation is described by the formula which is analogous to Eq. (5.25), namely,

$$G(\gamma^*, \delta, t) = [\det \tilde{\xi}(t)]^{-1/2} \exp\Big\{ -{}^1\!/_2\gamma^* \tilde{\xi}^{-1} \tilde{\eta} \gamma^* + \gamma^* \tilde{\xi}^{-1} \delta + {}^1\!/_2 \delta \tilde{\eta}_1 \tilde{\xi}^{-1} \delta -$$

$$- \gamma^* \tilde{\xi}^{-1} \tilde{l} + \delta(\tilde{l}_1 - \tilde{\eta}_1 \tilde{\xi}^{-1} \tilde{l}) + {}^1\!/_2 \tilde{l}\, \tilde{\eta}_1 \tilde{\xi}^{-1} \tilde{l} - \int_0^t \dot{\tilde{l}}(\tau)\, \tilde{l}_1(\tau)\, d\tau \Big\}, \tag{5.28}$$

where γ^* and δ label the correlated coherent states in accordance with the relations $b_k|\delta\rangle=\delta_k|\delta\rangle$ and $\langle\gamma|b^+_k=\gamma^*_k\langle\gamma|$.

To express the correlated integrals of the motion $\hat{Q}_b$ and, consequently, the correlated propagator (5.28) in terms of the integrals of the motion Q_a we replace the matrix Δ_b in the first equation (5.27) by its expression though the matrix Δ_a according to Eq. (5.21). Then we multiply this equation from the right side by the matrix $\Phi^T_N V \Phi_N$ and insert between the matrices Λ_b and Σ the unity matrix $\Phi^T_N V \Phi_N \bullet \Phi^T_N V^{-1} \Phi_N$. Finally we have the equation

$$-i\,\dot{\Lambda}_b \Phi_N^T V \Phi_N = \Lambda_b \Phi_N^T V \Phi_N \Phi_N^T V^{-1} \Phi_N \Sigma \Phi_N^T (V^{-1})^T \Phi_N \Delta_a. \tag{5.29}$$

Using the form of the matrices Φ_N, Σ and V we find the relations

$$\Sigma' = \Phi_N \Sigma \Phi_N^T = \oplus \sum_{k=1}^{N} \sigma_k, \qquad \sigma_k = \begin{Vmatrix} 0 & 1 \\ -1 & 0 \end{Vmatrix}, \tag{5.30}$$

$$V^{-1} \Sigma' (V^{-1})^T = \oplus \sum_{k=1}^{N} V_k^{-1} \sigma_k (V_k^{-1})^T = \Sigma', \tag{5.31}$$

since there is the relation $V^{-1}{}_k\sigma_k(V^{-1}{}_k)^T=\sigma_k$. Finally, $\Phi_N^T \Sigma' \Phi_N = \Phi_N^T \Phi_N \Sigma \Phi_N^T \Phi_N = \Sigma$, since we have the formula $\Phi_N^T \Phi_N = \Phi_N \Phi_N^T = 1$. Using these two formulae we rewrite the equation (5.29) in the form

$$-i\dot{\Lambda}_b \Phi_N^T V \Phi_N = \Lambda_b \Phi_N^T V \Phi_N \Sigma \Delta_a. \tag{5.32}$$

Multiplying the last equation from the left side by the matrix $\Phi_N^T V^{-1} \Phi_N$ and comparing the equation obtained

$$-i\,(\Phi_N^T V^{-1} \Phi_N \dot{\Lambda}_b \Phi_N^T V \Phi_N) = (\Phi_N^T V^{-1} \Phi_N \Lambda_b \Phi_N^T V \Phi_N)\, \Sigma \Delta_a \tag{5.33}$$

with the equation (5.24) for the matrix Λ_a, we find the connection of this matrix with the correlated matrix

$$\Lambda_a(t) = \Phi_N^T V^{-1} \Phi_N \Lambda_b(t) \Phi_N^T V \Phi_N, \quad \Lambda_b(t) = \Phi_N^T V \Phi_N \Lambda_a(t) \Phi_N^T V^{-1} \Phi_N. \quad (5.34)$$

The initial conditions for the matrix Λ_b directly follow from the second equation (5.34) taken at t=0. Analogously we find the relation of the vectors L_a and L_b using the equations (5.24), (5.27) and (5.34)

$$L_a = \Phi_N^T V^{-1} \Phi_N L_b, \quad L_b = \Phi_N^T V \Phi_N L_a. \quad (5.35)$$

Hence, the formula (5.34), (5.35) demonstrate one-to-one correspondence of the integrals of the motion $\hat{Q}_b$ to integrals of the motion $\hat{Q}_a$ creating the coherent representation of the quantum systems. The matrices $\tilde{\xi}, \tilde{\eta}, \tilde{\xi}_1, \tilde{\eta}_1$ are expressed in terms of the matrices ξ, η, ξ_1, η_1 by the formula

$$\begin{aligned}
(\tilde{\xi})_{km} &= \overset{*}{u}_k u_m \xi_{mk} + \overset{*}{u}_k v_m \eta^{(1)}_{mk} - \overset{*}{v}_k u_m \eta_{mk} - \overset{*}{v}_k v_m \xi^{(1)}_{mk},\\
(\tilde{\eta})_{km} &= - v_k u_m \xi_{mk} - v_k v_m \eta^{(1)}_{mk} + u_k u_m \eta_{mk} + u_k v_m \tilde{\xi}^{(1)}_{mk},\\
(\tilde{\eta}_1)_{km} &= \overset{*}{u}_k \overset{*}{v}_m \tilde{\xi}_{mk} + \overset{*}{u}_k \overset{*}{u}_m \eta^{(1)}_{mk} - \overset{*}{v}_k \overset{*}{v}_m \eta_{mk} - \overset{*}{v}_k \overset{*}{u}_m \tilde{\xi}^{(1)}_{mk},\\
(\tilde{\xi}_1)_{km} &= - v_k \overset{*}{v}_m \tilde{\xi}_{mk} - v_k \overset{*}{u}_m \eta^{(1)}_{mk} + u_k \overset{*}{v}_m \eta_{mk} + u_k \overset{*}{u}_m \tilde{\xi}^{(1)}_{mk},
\end{aligned} \quad (5.36)$$

where k, m=1,2,…,N. For the term L_b using the Eq. (5.35) we have

$$\begin{pmatrix} \tilde{l}_k \\ \tilde{l}_k^{(1)} \end{pmatrix} = V_k \begin{pmatrix} l_k \\ l_k^{(1)} \end{pmatrix}, \quad (5.37)$$

i.e., the components of this term transform like the creation and annihilation operators (5.14)

5.4. Stationary Quantum Systems Corresponding to Rotation Groups in Cayley-Klein Spaces in Correlated State Representation

In Section 4 we described the stationary quantum systems corresponding to the groups SO_{n+1}(j) in Glauber coherent state representation. In order to obtain the description of these systems in another basis (representation) which is the basis of the correlated coherent states we use the results of Sections 5.2 and 5.3. The Hamiltonian H_b(r, j) of the system (4.3) expressed through the correlated creation and annihilation operators b^+, b may be found using the formula (5.20) with the help of the matrix B_b(r,j). According to equations (4.5) and (5.21) this matrix is given by the formula

$$\mathcal{B}_b(\mathbf{r}, \mathrm{j}) = \Phi_N^T (V^{-1})^T \Phi_N \Psi(\mathrm{j}) \Delta_a(\mathbf{r}, \mathrm{j}) \Psi^{-1}(\mathrm{j}) \Phi_N^T V^{-1} \Phi_N. \quad (5.38)$$

The linear integrals of the motion Λ_b(r, j, t) may be obtained from the integrals of the motion (4.9) with help of the formula (5.34). They have the form

$$\Lambda_b(\mathbf{r}, \mathrm{j}, t) = \Phi_N^T V \Phi_N \Psi(\mathrm{j}) \begin{Vmatrix} \Xi(it\mathbf{r}, \mathrm{j}) & 0 \\ 0 & \Xi^T(-it\mathbf{r}, \mathrm{j}) \end{Vmatrix} \Psi^{-1}(j) \Phi_N^T V^{-1} \Phi_N. \quad (5.39)$$

The block-matrices $\tilde{\xi}, \tilde{\eta}, \tilde{\xi}_1, \tilde{\eta}_1$ of the correlated integrals of the motion Λ_b(r, j, t) are described by the formulae (5.36) where the matrices ξ, η, ξ_1, η_1 are given by Eq. (4.133) at ε=1. Finally, the propagator G(γ^*, δ, t, r, j) of the stationary quantum system (4.3) in the correlated state representation is expressed by the formula (5.28) at $\tilde{l} = \tilde{l}_1 = 0$. To find the propagator we have to fulfill the nonlinear operation calculating the inverse matrix ξ^{-1} that differs from calculating the Hamiltonian and the integrals of the motion. Because of this, the propagator is not described by the simple formulae like Eqs. (5.38), (5.39).

The case of the systems corresponding to the one-parametric groups $SO_2(j_1)$ is the simplest one. The matrix generators has the form $X(r,j_1)=r\begin{Vmatrix} 0 & -j_1^2 \\ 1 & 0 \end{Vmatrix}$, then the Hamiltonian (4.34) of the quantum system may be found with the help of the matrix $\Delta_a=\begin{Vmatrix} 0 & X^T(r,j) \\ X(r,j) & 0 \end{Vmatrix}$ (we omit the parameter index j_1). We consider two groups $SO_2(1)$ and $SO_2(\iota)$, i.e., $j=1,\ \iota$, then we have $\Psi(j)=\Psi^{-1}(j)=E$. According to Eqs. (5.38) the Hamiltonian of the system is expressed in terms of the correlated creation and annihilation operators b^+, b by the formula

$$\mathcal{B}_b(r,j) = (\tilde{V}^{-1})^T\,\Delta_a\tilde{V}^{-1}, \tag{5.40}$$

where

$$\tilde{V}^{-1} = \Phi_2^T V^{-1}\Phi_2 = \begin{Vmatrix} u_0^* & 0 & -v_0 & 0 \\ 0 & u_1^* & 0 & -v_1 \\ -v_0^* & 0 & u_0 & 0 \\ 0 & -v_1^* & 0 & u_1 \end{Vmatrix} \tag{5.41}$$

and it has the form

$$\mathcal{B}_b(r,j) = r\begin{Vmatrix} 0 & -u_0^*v_1^* + j^2u_1^*v_0^* & 0 & u_0^*u_1 - j^2v_0^*v_1 \\ -u_0^*v_1^* + j^2u_1^*v_0^* & 0 & v_0v_1^* - j^2u_0u_1^* & 0 \\ 0 & v_0v_1^* - j^2u_0u_1^* & 0 & -u_1v_0 + j^2u_0v_1 \\ u_0^*u_1 - j^2v_0^*v_1^* & 0 & -u_1v_0 + j^2u_0v_1 & 0 \end{Vmatrix}. \tag{5.42}$$

The matrix $B_b(r,j)$ is a symmetrical one that must be for the systems with Hamiltonians which are quadratic forms in creation and annihilation operators of bosons.

The finite rotation $\Xi(r,j)$ of the group $SO_2(j)$ is given by the formula (2.17). The propagator of the group quantum system is calculated with the help of the matrix

$$\Xi(itr, j) = \begin{Vmatrix} \mathrm{ch}\, jtr & -ij\,\mathrm{sh}\, jtr \\ ij^{-1}\,\mathrm{sh}\, jtr & \mathrm{ch}\, jtr \end{Vmatrix}. \tag{5.43}$$

The linear correlated integrals of the motion are given by the relations

$$\Lambda_b(r, j, t) = \tilde{V}\begin{Vmatrix} \Xi(itr, j) & 0 \\ 0 & \Xi^T(-itr, j) \end{Vmatrix}\tilde{V}^{-1} = \begin{Vmatrix} \tilde{\xi} & \tilde{\eta} \\ \tilde{\eta}_1 & \tilde{\xi}_1 \end{Vmatrix}, \tag{5.44}$$

where

$$\tilde{V} = \Phi_2^T V\Phi_2 = \begin{Vmatrix} u_0 & 0 & v_0 & 0 \\ 0 & u_1 & 0 & v_1 \\ v_0^* & 0 & u_0^* & 0 \\ 0 & v_1^* & 0 & u_1^* \end{Vmatrix}, \tag{5.45}$$

and the matrices $\tilde{\xi}, \tilde{\eta}, \tilde{\xi}_1, \tilde{\eta}_1$ have the form

$$\tilde{\xi} = \left\| \begin{matrix} \text{ch}\, jtr & (v_0 v_1^* - j^2 u_0 u_1^*) \dfrac{i}{j} \text{sh}\, jtr \\ (u_0^* u_1 - j^2 v_0^* v_1) \dfrac{i}{j} \text{sh}\, jtr & \text{ch}\, jtr \end{matrix} \right\|,$$

$$\tilde{\eta} = \left\| \begin{matrix} 0 & 1 \\ 1 & 0 \end{matrix} \right\| (-u_1 v_0 + j^2 u_0 v_1) \frac{i}{j} \text{sh}\, jtr,$$

$$\tilde{\eta}_1 = \left\| \begin{matrix} 0 & 1 \\ 1 & 0 \end{matrix} \right\| (u_0^* v_1^* - j^2 u_1^* v_0^*) \frac{i}{j} \text{sh}\, jtr,$$

$$\tilde{\xi}_1 = \left\| \begin{matrix} \text{ch}\, jtr & (-u_0^* u_1 + j^2 v_0^* v_1) \dfrac{i}{j} \text{sh}\, jtr \\ (-v_0 v_1^* + j^2 u_0 u_1^*) \dfrac{i}{j} \text{sh}\, jtr & \text{ch}\, jtr \end{matrix} \right\|. \tag{5.46}$$

The determinant of the matrix ξ is given by the expression

$$\det \tilde{\xi} = \text{ch}^2\, jtr + (u_0^* u_1 - j^2 v_0^* v_1)(v_0 v_1^* - j^2 u_0 u_1^*)\, j^{-2}\, \text{sh}^2\, jtr \tag{5.47}$$

and we obtain the explicit form of the matrices $\tilde{\xi}^{-1}, \tilde{\xi}^{-1}\tilde{\eta}, \tilde{\eta}_1\tilde{\xi}^{-1}$ which determine the time-dependence of the propagator given by the formula (5.28)

$$\tilde{\xi}^{-1} = (\det \tilde{\xi})^{-1} \left\| \begin{matrix} \text{ch}\, jtr & -(v_0 v_1^* - j^2 u_0 u_1^*) \dfrac{i}{j} \text{sh}\, jtr \\ -(u_0^* u_1 - j^2 v_0^* v_1) \dfrac{i}{j} \text{sh}\, jtr & \text{ch}\, jtr \end{matrix} \right\|,$$

$$\tilde{\xi}^{-1}\tilde{\eta} = \frac{(-u_1 v_0 + j^2 u_0 v_1)(i/j)\, \text{sh}\, jtr}{\det \tilde{\xi}} \times$$

$$\times \left\| \begin{matrix} -(v_0 v_1^* - j^2 u_0 u_1^*) \dfrac{i}{j} \text{sh}\, jtr & \text{ch}\, jtr \\ \text{ch}\, jtr & -(u_0^* u_1 - j^2 v_0^* v_1) \dfrac{i}{j} \text{sh}\, jtr \end{matrix} \right\|,$$

$$\tilde{\eta}_1\tilde{\xi}^{-1} = \frac{(u_0^* v_1^* - j^2 u_1^* v_0^*)(i/j)\, \text{sh}\, jtr}{\det \tilde{\xi}} \times$$

$$\times \left\| \begin{matrix} -(u_0^* u_1 - j^2 v_0^* v_1) \dfrac{i}{j} \text{sh}\, jtr & \text{ch}\, jtr \\ \text{ch}\, jtr & -(v_0 v_1^* - j^2 u_0 u_1^*) \dfrac{i}{j} \text{sh}\, jtr \end{matrix} \right\|. \tag{5.48}$$

Using Eqs. (5.47), (5.48) and the formula (5.28) we find the propagator of the quantum system in the correlated state representation corresponding to the group $SO_2(j)$, j=1, ?. We have the following expression

$$\begin{aligned} G(\gamma^*, \delta, t, r, j) &= (\det \tilde{\xi})^{1/2} \exp \{(\det \tilde{\xi})^{-1} [(\gamma_0^* \delta_0 + \gamma_1^* \delta_1)\, \text{ch}\, jtr - \\ &- (\gamma_1^* \delta_0 (u_0^* u_1 - j^2 v_0^* v_1) + \gamma_0^* \delta_1 (v_0 v_1^* - j^2 u_0 u_1^*))(i/j)\, \text{sh}\, jtr + \\ &+ (\delta_0 \delta_1 (u_0^* v_1^* - j^2 u_1^* v_0^*) + \gamma_0^* \gamma_1^* (u_1 v_0 - j^2 u_0 v_1))(i/j)\, \text{sh}\, jtr \cdot \text{ch}\, jtr + \\ &+ {}^1/_2 ((u_0^* v_1^* - j^2 u_1^* v_0^*)(\delta_0^2 (u_0^* u_1 - j^2 v_0^* v_1) + \delta_1^2 (v_0 v_1^* - j^2 u_0 u_1^*)) + \\ &+ (u_1 v_0 - j^2 u_0 v_1)(\gamma_0^{*2} (v_0 v_1^* - j^2 u_0 u_1^*) + \gamma_1^{*2} (u_0^* u_1 - j^2 v_0^* v_1)))\, j^{-2}\, \text{sh}^2\, jtr]\}. \end{aligned} \tag{5.49}$$

The formula (5.49) at j=ι, i.e., for one-dimensional Galilean group, transforms into the relation

$$\begin{aligned} G(\gamma^*, \delta, t, r, \iota) &= (1 + u_0^* u_1 v_0^* v_1 t^2 r^2)^{-1/2} \exp \{(1 + u_0^* u_1 v_0^* v_1 t^2 r^2)^{-1} [\gamma_0^* \delta_0 + \gamma_1^* \delta_1 + \\ &+ {}^1/_2 t^2 r^2 (u_0^* v_1^* (\delta_0^2 u_0^* u_1 + \delta_1^2 v_0 v_1^*) + u_1 v_0 (\gamma_0^{*2} v_0 v_1^* + \gamma_1^{*2} u_0^* u_1)) + \\ &+ itr (\delta_0 \delta_1 u_0^* v_1^* + \gamma_0^* \gamma_1^* u_1 v_0 - \gamma_1^* \delta_0 u_0^* u_1 - \gamma_0^* \delta_1 v_0 v_1^*)]\}. \end{aligned} \tag{5.50}$$

For comparison, let us write down the propagator of the same quantum system obtained from the kernel (4.36) of the finite rotation of the group $SO_2(\iota)$ in the coherent state basis with the help of the replacement of the group parameter r by itr

$$G(\alpha^*, \beta, t, r, \iota) = \exp\{\alpha_0^* \beta_0 + \alpha_1^* \beta_1 - itr\alpha_1^* \beta_0\}. \tag{5.51}$$

In the transformation V=1, i.e., $u_0=u_1=1$, $v_0=v_1=0$ the propagator (5.50) coincides with the propagator (5.51)

The propagator (5.51) describes the behavior of the quantum system in the state in which the quantum-mechanical uncertainties of two pairs of the canonically conjugate operators of the coordinate and momentum $\hat{q}_0$, $\hat{p}_0$, $\hat{q}_1$, $\hat{p}_1$ are minimal $\sigma_{q_0}\sigma_{p_0}=1/4$, $\sigma_{q_1}\sigma_{p_1}=1/4$ and these observables are statistically independent. The propagator (5.50) describes the behavior of the system for which the observables $\hat{q}_0$ and $\hat{p}_0$ as well as the observables $\hat{q}_1$ and $\hat{p}_1$ are statistically dependent ones (the correlation coefficient of the position q_0 and the momentum p_0 is equal to R_0, and the correlation coefficient of the position $\hat{q}_1$ and the momentum $\hat{p}_1$ is equal to R_1, where the numbers R_k, k=0,1 are connected with the parameters u_k, v_k of the transformation (5.14) by the relation (5.8)), that corresponds to the states in which the quantum-mechanical uncertainties of these observables are minimized $\sigma_{q_0}\sigma_{p_0}(1-R^2_0)=1/4$, $\sigma_{q_1}\sigma_{p_1}(1-R^2_1)=1/4$.

Let us point out that the operators of the coordinates and momenta with different indices are always independent in the scheme under consideration that follows from the special choice of the matrix V transforming the coherent creation and annihilation operators a^+, a given by the formula (5.14), (5.16). Generally speaking, the formulae (5.34), (5.35) for the linear integrals of the motion are valid for all such transformations V that satisfy the relation $V^{-1}\Sigma(V^{-1})^T=\Sigma$, i.e., for all transformations belonging to the symplectic group Sp(N,R,). For the general transformation $V \in$ Sp(N,R) all the observables $\hat{q}_k$, $\hat{p}_k$, k=0,1,2,...,N-1 turn out to be statistically dependent. The creation and annihilation operators given by the formula (5.18) create the states which may be called "symplectic" coherent states.

Accepted for publication in December 1988.

REFERENCES

1. Schwinger, J. "On angular momentum" In: "*Quantum theory of angular momentum,*" New York: Acad, Press. 1965.
2. Glauber, R.J. "Coherent and incoherent states of the radiation field" *Phys. Rev.*, 1963, vol. 131, no. 6, pp. 2766-2788.
3. Malkin, I.A., Man'ko, V.I., "Invariants and coherent states of arbitrary quantum systems" Preprint, Lebedev Phys. Inst. no. 15, Moscow, 1971.
4. Malkin, I.A., Man'ko, V.I., Trifonov, D.A. "Linear adiabatic invariants and coherent states" *J. Math. Phys.*, 1973, vol. 14, no. 5, pp. 576-582.
5. Malkin, I.A., Man'ko, V.I., Trifonov, D.A. "Dynamical symmetries of nonstationary quantum systems" *Nuovo Cim. A*, 1971, Vol. 4, pp. 773-782.
6. Dodonov, V.V., Malkin, I.A., Man'ko, V.I. "Integrals of the motion, Green functions and coherent states of dynamical systems" *Intern. J. Theor. Phys.*, 1975, vol. 14, no. 1, pp. 37-54.

7. Man'ko, V.I., Trifonov, D.A. "Matrix elements of finite transformations of Lie groups in the basis of coherent and Fock states" In: *"Group theoretical methods in physics,"* London, New York: Harwood Acad. Publ., 1987, vol. 3, pp. 795-810.
8. Dodonov, V.V., Man'ko, V.I. "Evolution of multidimensional systems. Magnetic properties of ideal gases of charged particles" Proc. Lebedev Phys. Inst., Moscow: Nauka, 1987, vol. 183, pp. 182-286.
9. Dodonov, V.V., Kurmyshev, E.V., Man'ko, V.I. "Generalized uncertainty relation and correlated coherent states" *Phys. Lett. A*, 1980, vol. 79, no. 2/3, pp. 150-152.
10. Dodonov, V.V., Kurmyshev, E.V., Man'ko, V.I. "The more exact uncertainty relation and correlated coherent states" Proc. First Intern. Seminar on Group Theoretical Methods in Physics, Zvenigorod (Russia), 1972, Moscow: Nauka, 1980, vol. 1, pp. 227-232.
11. Dodonov, V.V., Kurmyshev, E.V., Man'ko, V.I. "Correlated wave packets of minimal uncertainty" Preprint, Lebedev. Phys. Inst., no. 259, Moscow, 1984, p. 29.
12. Dodonov, V.V., Man'ko, V.I. "Invariants and correlated states of nonstationary quantum systems" Proc. Lebedev Phys. Inst., Moscow: Nauka, 1987, vol. 183, pp. 71-181.
13. Klauder, J.R. "Continuous-representation theory. I: Postulates of continuous-representation theory" *J. Math. Phys.*, 1963, vol. 4, no. 8, pp. 1055-1058.
14. Klauder, J.R. "Continuous-representation theory. II: Generalized relation between quantum and classical dynamics" *J. Math. Phys.*, 1963, vol. 4, no. 8, pp. 1058-1073.
15. Klauder, J.R. "Continuous-representation theory. III: On functional quantization of classical systems" *J. Math. Phys.* 1964, vol. 5, no. 2, pp. 177-187.
16. McKenna, J., Klauder, J.R. "Continuous-representation theory. IV: Structure of a class of function spaces arising from quantum mechanics" *J. Math. Phys.*, 1964, vol. 5, pp. 878-896.
17. Klauder, J.R., Skaugerstam, B.S. "Coherent states: Applications in physics and mathematical physics," Singapore: World. Sci. Publ., 1985, p. 327.
18. Leznov, A.N., Savel'yev, M.V. "Non-linear equations and graded Lie algebras" In: "Itogi nauki i tekhniki. Matematicheskiy analiz" [*Achievements in science and technology*. Mathematic analysis], Moscow: VINITI, 1984, vol. 22, pp. 101-136.
19. Leznov, A.N., Savel'yev, M.V. "Gruppovye metody integrirovaniya nelineynykh dinamicheskikh sistem" [*Group methods of integrating non-near dynamical systems*], Moscow: Nauka, 1985, p. 279.
20. Leznov, A.N., Man'ko, V.I., Savel'yev, M.V. "Soliton solutions to non-linear equations and group representation theory" Proc. Lebedev Phys. Inst, Moscow: Nauka, 1986, vol. 165, pp. 65-206.
21. Leznov, A.N., Man'ko, V.I., Chumakov, S.M. "Dynamical symmetries of nonlinear equations "Proc. Lebedev Phys. Inst., Moscow: Nauka, 1986, vol. 167, pp. 232-277.
22. Malkin, I.A., Man'ko, V.I. "The symmetry of hydrogen atom," *Sov. Phys. JETP Lett.*, 1965, vol. 2, pp. 146-149.
23. Malkin, I.A., Man'ko, V.I. "Symmetry of nitrogen atom," *Yadernaya fizika*, 1966, vol. 3, pp. 372-382.
24. Zaitzev, G.A., "Algebraicheskie problemy matematicheskoy i teoreticheskoy fiziki" [*Algebraic problems of mathematical and theoretical physics*], Moscow: Nauka, 1974, p. 191.
25. Gromov, N.A. "Limit transitions in spaces of constant curvature" Preprint series *"Scientific Reports"*, 1978, no. 37, Komi Branch Acad. Sci. Russia, Syktyvkar, p. 25.
26. Gromov, N.A. "On limit transitions in set of motion groups and Lie algebra of spaces of constant curvature" Matematicheskie zametki [*Mathematical notes*], 1982, vol. 32, no. 3, pp. 355-363.

27. Gromov, N.A. "Casimir operators of motion groups of spaces of constant curvature" *Teor. i Matem. Fizika,* 1981, vol. 49, no. 2, pp. 210-213.
28. Gromov, N.A. "Special unitary groups in fiber bundles" Preprint series *"Scientific Reports"*, 1984, no. 95, Komi Branch Acad. Sci. Russia, Syktyvkar, p. 20.
29. Gromov, N.A. "Symplectic groups in fiber bundles" Preprint series *"Scientific Reports"*, 1985, no. 126, Komi Branch Acad. Sci. Russia, Syktyvkar, p. 12.
30. Gromov, N.A. "Classical groups in Cayley-Klein spaces" Proc. Third Intern. Seminar on Group Theoretical Methods in Physics. Yurmala (Russia), 1985, Moscow: Nauka, 1986, vol. 2, pp. 183-190.
31. Gromov, N.A. "Orthogonal Cayley-Klein group representations in spaces of zero curvature" Preprint Series *"Scientific Reports,"* 1988, no. 180, Komi Branch Acad. Sci. Russia, Syktyvkar, p. 20.
32. Inonu, E., Wigner, E.P. *"On the contraction of groups and their representations"* Proc. Natl. Acad. Sci. USA, 1953, vol. 39, pp. 510-524.
33. Gromov, N.A., Man'ko, V.I. *"Matrix elements of representations in basis of coherent states and contractions"* Preprint, Lebedev Phys. Inst., no. 155, Moscow, 1988, p. 49.
34. Gromov, N.A., Man'ko, V.I. *"Matrix elements of representations and quantum Hamiltonians"* Preprint, Lebedev Phys. Inst., no. 196, Moscow, 1988, p. 41.
35. Gromov, N.A., Man'ko, V.I. *"Group quantum system in coherent states representation"* Preprint, Lebedev Phys. Inst., no. 21, Moscow, 1988, p. 27.
36. Yaglom, I.M., Rozenfel'd, B.A., Yasinskaya, E.U. *"Projective metrics"* UMN, 1964, vol. 19, no. 5, pp. 51-113.
37. Pimenov, R.I. "Unified axiomatics of space with maximal motion group" *Lithuanian Math. Proc."*, Vilnus, 1965, vol. 5, no. 3, pp. 457-486.
38. Gromov, N.A., Yakushevich, L.V. *"Kinematics as spaces of constant curvature"* Proc. Third Intern. Seminar on Group Theoretical Methods in Physics, Yurmala (USSR), 1985, Moscow: Nauka, 1986, vol. 2, pp. 191-198.
39. Bacry, H., Levy-Leblond, J.M. "Possible kinematics" *J. Math. Phys.*, 1968, vol. 9, no. 10, pp. 1605-1614.
40. Blokh, A. Sh. "Chislovye sistemy" *[Number systems]*, Minsk: Vysshaya shkola [High School], 1982, p. 158.
41. Gel'fand, I.M. "Center of infinitesimal group ring" *Math. Proc.*, Moscow, 1950, vol. 25, no. 1, pp. 103-112.
42. Barut, A., Raczka, R. "Teoriya predstavleniy grup i ee prilozheniya" *[Theory of group representations and applications]* Moscow: Mir, 1980, vol. 1, p. 456.
43. Gromov, N.A. "Limit transitions and Casimir operators of motion groups of spaces of constant curvature" Preprint series *"Scientific Reports"*, 1981, no. 75, Komi Branch Acad. Sci., USSR, Syktyvkar, p. 11.
44. Korn, G.A., Korn, T.M. "Spravochnik po matematike dlya nauchnykh rabotnikov i inzhenerov: Opredeleniya, formuly, teoremy" *[Mathematical handbook for scientists and engineers: Definitions, formulas, theorems]* Moscow: Nauka, 1984, p. 831.
45. Gromov, N.A. "Analogs of Fedorov parametrization of groups in fiber bundles" Vestsi AN BSSR, seriya fizicheskikh i matematicheskikh nauk *[Notes of Belorussian Acad. Sci. Physical and mathematical series]*, Minsk, 1984, no. 2, pp.108-114.
46. Fedorov, F.I. "Gruppa Lorentsa" *[Lorentze group]* Moscow: Nauka, 1979, p. 384.
47. Bourbaki, N. "Diferentsiruemye i analiticheskie mnogoobraziya: svodka rezultatov" *[Differentiable and analytical manyfolds: Review of results]* Moscow: Mir, 1978, p. 220.

48. Perroud, M. "The fundamental invariants of inhomogeneous classical groups" *J. Math. Phys.*, 1983, vol. 24, no. 6, pp. 1381-1391.

49. Jordan, P. "Der Zusammenhang der Symmetrischen und Linearen Grupen und das Mehrkorperproblem" *Ztschr. Phys.*, 1935, Bd. 94, S. 531-535.

50. Kim, S.K. "Theorems on the Jordan-Schwinger representations of Lie algebras" *J. Math. Phys.*, 1984, vol. 28, no. 11, pp.2540-2545.

51. Berezin, F.A. "Metod vtorichnogo kvantovaniya" *[Second quantization method]*, Moscow: Nauka, 1965, p. 236.

52. Bateman, H., Erdey, A. "Vysshie transtsendentnye funktsii" *[Higher transcendental functions]* Moscow: Nauka 1974, vol. 2, p. 295.

53. Gromov, N.A., Man'ko, V.I. "The Jordan-Schwinger representation of Cayley-Klein groups. I. The orthogonal groups" *J. Math. Phys.*, 1990, vol. 31, no. 5, pp. 1047-1053.

54. Gromov, N.A., Man'ko, V.I. "The Jordan-Schwinger representation of Cayley-Klein groups. II. The unitary groups" *J. Math. Phys.*, 1990, vol. 31, no. 5, pp. 1054-1059.

55. Gromov, N.A., Man'ko, V.I. "The Jordan-Schwinger representation of Cayley-Klein groups. III. The symplectic groups" *J. Math. Phys.*, 1990, vol. 31, no. 5, pp. 1060-1064.

PHYSICAL EFFECTS IN CORRELATED QUANTUM STATES

V.V. Dodonov, A.B. Klimov, V.I. Man'ko

Abstract: The properties of correlated coherent states (CCS) are studied. The physical significance of generalized coordinates and momenta of photon and phonon fields is elucidated. Methods of generation and calculation of correlated states are considered. Distribution functions of quanta in CCS are analyzed. An oscillating behavior of these functions is shown. The average value of the phase operator and its covariance in CCS are found. Spreading and tunnelling of a correlated packet are considered. The influence of correlation and squeezing coefficients on these effects is shown. The connection between correlated states and Berry's phase is considered. The orthogonality of multidimensional Hermite polynomials and the completeness of the system of quadratic exponentials are proved.

1. INTRODUCTION

The concept of correlated coherent states (CCS) of a harmonic oscillator was introduced in studies [1,2]. These states were defined as states minimizing the left-hand side of the Schrodinger-Robertson [3,4] uncertainty relation for the coordinate and momentum. They are described by means of wave functions which are Gaussian exponentials of the most general type. In the past, such generalizations of Glauber's coherent states [5] were subjected to intensive investigations both by theoreticians and experimenters. In literature, new states are met under different names. Each name emphasizes some characteristic feature of these states. For example, the following terms are known: "minimal uncertainty states" [6], "two-photon states" [7], "squeezed states" [8,9]. In studies [10,11] the concepts of correlated light and sound were introduced as states of field modes corresponding to the correlated coherent states of the related harmonic oscillators.

The aim of the present paper is to study in detail the properties of CCS. Special attention will be paid to those of them which are related to the existence of the correlation between a generalized coordinate and the corresponding momentum and to the minimization of the generalized uncertainty relation, i.e. to those features that distinguish CCS from other states which are unitary equivalent to CCS from the point of view of mathematics. In particular, we shall analyze the distribution functions of photons (phonons) for correlated light (sound) with different values of squeezing and correlation coefficients. The asymptotical values of this distribution function in terms of the Hermite polynomials will also be obtained. We give the physically visual interpretation of correlated and squeezed light as a field mode with different quantum fluctuations (noise) of electric and magnetic vectors of an electromagnetic wave and with a statistical correlation between these vectors (or between fluctuations of pressure and velocity in the case of sound).

Besides, we shall discuss the physical meaning of the measurement procedure for canonical variables – the coordinate and the momentum of the electromagnetic field oscillator corresponding to the given mode in terms of the measurement of electric and magnetic field strengths at points shifted for a quarter of a wavelength in space (at the same instant of time) or at the same space point but with the shift for a quarter of period in time. The analysis of the photons distribution function in a beam of a correlated light (or sound) will result in some specific effects distinguishing correlated fields from usual ones. Namely, for some values of squeezing and correlation coefficients, the probability to observe in an experiment the average number of photons may be equal to zero, in contradiction to the case of Glauber's coherent state, when photons statistics is determined by Poisson's distribution, for which the average value is the most probable in the measurement. This difference in photon counting experiments can be used as the characteristic feature of correlated light. We shall also discuss the relations between the known experiments on photons statistics and measurements of the intensity correlation function.

We shall also analyze other physical phenomena in which the statistical correlation between the coordinate and momentum can manifest itself, e.g. the effect of quantum tunnelling through a potential barrier, and elucidate the conditions under which quantum effects can be amplified due to the nonzero correlation coefficient. The relation between correlated states and the so called Berry's phase arising in adiabatic evolution of quantum systems will be considered as well.

We shall give some new formulae for the Hermite polynomials of several variables as additional mathematical results, in particular, the formula relating the two-dimensional Hermite polynomial at zero values of its arguments with the Legendre polynomial, and some formulae of the type of sum rules.

2. ELECTROMAGNETIC FIELD OSCILLATORS

Let us discuss the properties of oscillators of electromagnetic field making a start from the usual scheme of quantization of a free field (see, e.g. [12]). Thus we assume the field to be contained in a cube of volume V with periodic boundary conditions. Then we expand the electric field operator $\hat{E}$ over plane waves.

$$\hat{E} = \sum_k \hat{E}_k e^{ikx} + \sum_k \hat{E}_k^+ e^{-ikx}, \qquad \hat{E}_k \sim e^{-i\omega_k t}, \qquad \omega_k = kc,$$

and introduce the annihilation and creation operators [13]

$$\hat{b}_k = -i\left(\frac{V}{2\pi\hbar\omega_k}\right)^{1/2} \hat{E}_k, \qquad \hat{b}_k^+ = i\left(\frac{V}{2\pi\hbar\omega_k}\right)^{1/2} \hat{E}_k^+. \tag{2.1}$$

in order to represent the Hamiltonian in the oscillator form

$$H = \sum_k \hbar\omega_k \left(\hat{b}_k^+\hat{b}_k + \frac{1}{2}\right), \qquad [\hat{b}_k, \hat{b}_k^+] = 1,$$

all operators being given in the Heisenberg picture. The field Hamiltonian can also be expressed through generalized canonical coordinates and momenta

$$H = \frac{1}{2}\sum_k (\omega_k^2 Q_k^2 + P_k^2),$$

provided Q_k and P_k are related to E_k as follows,

$$P_k = -(V/\pi)^{1/2}\,\mathrm{Re}\,E_k, \qquad Q_k = (V/\pi\omega_k^2)^{1/2}\,\mathrm{Im}\,E_k, \qquad [Q_k, P_k] = i\hbar. \tag{2.2}$$

The symbols Re$\hat{E}$ and Im$\hat{E}$ mean the operators

$$\mathrm{Re}\,\hat{E} = \frac{1}{2}(\hat{E} + \hat{E}^+), \qquad \mathrm{Im}\,\hat{E} = \frac{1}{2i}(\hat{E} - \hat{E}^+).$$

Our first goal is to elucidate the physical meaning of canonical coordinates and momenta in connection with the question on the possibility of measuring Q_k and P_k. We suppose that we can measure the components of electric and magnetic fields at the given point of space and at the given instant of time. The problem of possibility of such measurements was studied by Bohr and Rosenfeld [14] who showed that this possibility does exist in principle, and the limiting precision of the joint measurements of fields E and H is determined by the uncertainty relation resulting from the commutation relations [12,14]

$$[\hat{E}_j(\mathbf{x}), \hat{H}_l(\mathbf{x}')] = i\varepsilon_{jlk}\frac{\partial^2}{\partial t\,\partial x_k} D_0(\mathbf{x} - \mathbf{x}'), \qquad j, l, k = 1, 2, 3,$$

where

$$D_0(\mathbf{x}) = \frac{1}{(2\pi)^3}\int d^3k\, e^{ikx}\,\frac{\sin(\omega t)}{\omega}.$$

Let us consider a single plane wave (field mode) described by the operator

$$E = E_k e^{ikx} + E_k^+ e^{-ikx}, \qquad E_k(t) = E_k(0)\,e^{-i\omega t}.$$

Suppose that the coordinates of the point at which the electric field is measured are equal to zero. Then measuring the electric field strength is equivalent to measuring the canonical momentum of the corresponding field oscillator

$$P_k = -{}^1/_2\,(V/\pi)^{1/2}\,E. \tag{2.3}$$

We suppose the measuring device is tuned to the distinguished field mode and does not react to other modes. Now let us assume that we can measure the rate of change of the field E at the given point of space and at the given instant of time, i.e. the quantity $\dot{E} = aE / at$, which can be represented also as

$$\dot{E} = 2\,\mathrm{Re}\,(\dot{E}_k e^{ikx}) = 2\,\mathrm{Re}\,(-i\omega_k E_k e^{ikx}).$$

At the point x=0 we get

$$\dot{E} = 2\,\mathrm{Re}\,(-i\omega_k E_k) = 2\omega_k\,\mathrm{Im}\,E_k. \tag{2.4}$$

Consequently, we can measure the canonical coordinate of the field oscillator

$$Q_k = (V/\pi)^{1/2}(2\omega_k^2)^{-1}\,\mathrm{E}. \tag{2.5}$$

The measurement of $\dot{E}$ can be performed in different ways. For example, let us compare two formulae:

$$E(t) = E_\alpha \sin(\omega t),\ \dot{E}(t) = \omega E_\alpha \cos(\omega t) = \omega E_\alpha \sin(\omega t_*),\ t_* = t + T/4,$$

where T is the period of oscillations. Consequently, measuring the field at the instant of time shifted "forward" in time for a quarter period we measure Q_k:

$$Q_k(t = t_0) = \left(\frac{V}{\pi}\right)^{1/2}(2\omega_k)^{-1}\,E\,(t_0 + T/4). \tag{2.6}$$

The following commutation relations hold:

$$[E\,(t + T/4),\ E\,(t)] = -i\hbar\,(4\pi\omega/V). \tag{2.7}$$

Taking into account Maxwell's equation $\dot{E} = crotH$ we can write

$$Q_k = \left(\frac{V}{\pi}\right)^{1/2}(c/2\omega_k^2)\,\mathrm{rot}\,\mathbf{H}. \tag{2.8}$$

To measure rotH one can meaure, e.g., the magnetic field circulation along a contour perpendicular to the field, according to the Stokes theorem

$$\int_S d\mathbf{S}\,\mathrm{rot}\,\mathbf{H} = \int_l d\mathbf{l}\mathbf{H}.$$

For the plane wave E=[H,n], where n=k/|k|. Then

$$Q_k|_{t_0} = (V/\pi)^{1/2}(2\omega_k)^{-1}[\mathbf{H}\,(t_0 + T/4),\,\mathbf{n}]. \tag{2.9}$$

Note that for a plane wave $E = E_0 \sin(kx - \omega t)$ the shift in time for T/4 is equivalent to the shift in space "backward" for a quarter of wavelength. The field commutators

$$\begin{aligned} &[E\,(t + T/4),\ E\,(t)] = -i\hbar\,(4\pi\omega/V),\\ &[E\,(r = -\lambda/4,\ t),\ E\,(r = 0,\ t)] = -i\hbar\,(4\pi\omega/V) \end{aligned} \tag{2.10}$$

lead to the uncertainty relations

$$\begin{aligned} &\Delta E\,(t + T/4)\,\Delta E\,(t) \gtrsim h\,(2\pi\omega/V),\\ &\Delta E\,(r - \lambda/4,\ t)\,\Delta E\,(r,\ t) \gtrsim \hbar\,(2\pi\omega/V). \end{aligned} \tag{2.11}$$

If the shape of a cavity containing the field differs from a cube, then the relations

$$\begin{aligned} &P_k \sim E\,(t), \quad Q_k \sim E\,(t + T/4),\\ &P_k \sim E\,(t), \quad Q_k \sim (\mathrm{rot}\,\mathbf{H})_k. \end{aligned}$$

still hold due to the harmonic time dependences of fields: E, $H \propto e^{-1\omega t}$, while other relations are specific for the field quantization inside a cubic cavity.

Let us note an analogy between field oscillators and a mechanical oscillator. Indeed, the commutation relation

$$[x,\ p] = i\hbar$$

can be rewritten as $[x, \dot{x}] = i\hbar/m$, where m is a mass of a mechanical oscillator. Then due to the relations

$$x = x_\alpha \sin(\omega t),\ \dot{x} = \omega x_\alpha \cos(\omega t) = \omega x_\alpha \sin(\omega t_*),\ t_* = T/4,$$

this commutator can be rewritten as follows,

$$[x\ (t),\ x\ (t + T/4)] = i\hbar/m\omega.$$

The corresponding uncertainty relation is

$$\Delta x\,(t)\,\Delta x\,(t + T/4) \geqslant \hbar/2m\omega.$$

The fluctuations of the electric field in Glauber's coherent state $|\beta\rangle$ are given by the relations

$$\langle\beta\,|\,E\,|\,\beta\rangle = i\,(2\pi\hbar\omega/V)^{1/2}[\beta e^{i(kx-\omega t)} - \beta^* e^{-i(kx-\omega t)}],$$
$$\langle\beta\,|\,E^2\,|\,\beta\rangle = -(2\pi\hbar\omega/V)[(\beta e^{i(kx-\omega t)} - \beta^* e^{-i(kx-\omega t)})^2 - 1],$$
$$\sigma\,(E) = \langle\beta\ |\ E^2\ |\ \beta\rangle - (\langle\beta\ |\ E\ |\ \beta\rangle)^2 = 2\pi\hbar\,\omega/V.$$

Analogously, for a mechanical oscillator

$$\sigma\,(x) = \langle\beta\ |\ x^2\ |\ \beta\rangle - (\langle\beta\ |\ x\ |\ \beta\rangle)^2 = \hbar/2\omega m.$$

Note that in the field case the role of mass is played by the quantization volume V. In the oscillator case the fluctuation $\sigma(x)$ tends to zero when $m\to\infty$. Analogously, for a field $\sigma(E)\to 0$ when $V\to\infty$. This result seems paradoxical at first glance. Moreover, it evokes some spontaneous protest, because changing the volume of a box (which in fact plays only a subsidiary role making it possible to consider a discrete wavenumbers spectrum *for the sake of convenience of quantization*) we change the intensity of the electric field strength fluctuations, i.e. the real physical quantity. The solution of this paradox is connected with the assumption that the device distinguishes the single field mode and measures not the whole strength but only its part relating to the given mode (which certainly depends on the volume).

If the field is expanded over plane waves, then in every mode the strength of the electric field equals the strength of the magnetic field. Thus only one of these quantities can be chosen as a canonical variable. But if we use standing waves as the basic field modes, then we can choose as canonically conjugated variables the amplitudes of electric and magnetic fields in definite points of space. For example, let us consider an electromagnetic field inside a rectangular parallelepiped with ideal walls. The standing waves satisfying Helmholtz's equation inside the resonator, boundary conditions at the walls, and the transversality condition divE=0 are characterized by the following distributions of electric and magnetic fields in each mode,

$$\mathbf{E}\,(\mathbf{r},t) = E\,(t)\begin{pmatrix}\tau_x \cos(k_x x)\sin(k_y y)\sin(k_z z)\\ \tau_y \sin(k_x x)\cos(k_y y)\sin(k_z z)\\ \tau_z \sin(k_x x)\sin(k_y y)\cos(k_z z)\end{pmatrix},$$

$$\mathbf{B}\,(\mathbf{r},t) = B\,(t)\begin{pmatrix}(n_y\tau_z - n_z\tau_y)\sin(k_x x)\cos(k_y y)\cos(k_z z)\\ (n_z\tau_x - n_x\tau_z)\cos(k_x x)\sin(k_y y)\cos(k_z z)\\ (n_x\tau_y - n_y\tau_x)\cos(k_x x)\cos(k_y y)\sin(k_z z)\end{pmatrix}. \qquad (2.12)$$

Here $\tau=(\tau_x, \tau_y, \tau_z)$ is the unity polarization vector relating to the unity vector $n=(k_x, k_y, k_z)/|k|^{-1}$ by the orthogonality relation $\tau\cdot n=0$. Nonnegative coefficients k_x, k_y, k_z satisfy the relation $k^2{}_x+k^2{}_y+k^2{}_z=\omega^2/c^2$. They are quantized in a usual way. The amplitude E(t) varies in time as follows,

$$E(t) = a\cos(\omega t) + b\sin(\omega t).$$

Then due to the equation $\mathrm{rot}E=-c^{-1}\partial B/\partial t$ the amplitude B(t) equals

$$B(t) = b\cos(\omega t) - a\sin(\omega t).$$

Consequently,

$$\dot{E} = \omega B, \quad \dot{B} = -\omega E,$$

so that one can quantize the field chosing the quantity E/ω as a generalized coordinate and the quantity B(t) as a generalized momentum. But the relations of the corresponding operators $\hat{E}$ and $\hat{B}$ to the electric and magnetic field components operators are not direct due to (2.12). Only in exceptional cases can they be simple and visual. For example, in the case of a cubic resonator the mode with the electric field vector polarized along the z-axis is characterized by the following distributions of fields,

$$\mathbf{E} = E(t)\begin{pmatrix} 0 \\ 0 \\ \sin(k_x x)\sin(k_y y) \end{pmatrix}, \quad \mathbf{B} = B(t)\begin{pmatrix} n_y \sin(k_x x)\cos(k_y y) \\ -n_x\cos(k_x x)\sin(k_y y) \\ 0 \end{pmatrix}.$$

If one considers the lowest mode with $k_x=k_y=\pi/L$, then canonical variables E(t) and B(t) coincide with the strengths of electric and magnetic fields at the points lying in the line segment (L/4, L/4, z).

3. SOUND QUANTIZATION IN SOLIDS AND LIQUIDS

The quantization procedure for the sound field in solids is similar to that for the electromagnetic field. The physical quantities now are the displacement vector u(x) and the corresponding canonical momentum $P(x)=\rho\dot{u}$ which can be expanded over plane waves as follows (for periodic boundary conditions) [15]

$$\mathbf{u}(x) = \left(\frac{\hbar}{2\rho V}\right)^{1/2}\sum_{k,\alpha}(\omega_\alpha(k))^{-1/2}[c_{k\alpha}e^\alpha e^{ikr} + c^+_{k\alpha}e^{*\alpha}e^{-ikr}],$$

$$\mathbf{P}(x) = i\left(\frac{\hbar\rho}{2V}\right)^{1/2}\sum_{k,\alpha}(\omega_\alpha(k))^{1/2}[c^+_{k\alpha}e^{*\alpha}e^{-ikr} - c_{k\alpha}e^\alpha e^{ikr}],$$

$$c_k \sim e^{-i\omega_k t}, \tag{3.1}$$

where ρ is the density of the medium, e^α is the polarization vector, and the normalization coefficient is chosen in such a way that the energy of the field assumes the usual oscillator form

$$H = \sum_{k,\alpha}\hbar\omega_\alpha(k)\left(c^+_{k\alpha}c_{k\alpha} + \frac{1}{2}\right) = \frac{1}{2}\sum_{k,\alpha}(P^2_{k\alpha} + \omega^2_\alpha(k)\,Q^2_{k\alpha}),$$

with

$$c_{k\alpha} = (2\hbar\omega_\alpha(k))^{-1/2}[\omega_\alpha(k)\,Q_{\alpha k} + iP_{\alpha k}], \quad [Q_{\alpha k}, P_{\alpha' k'}] = i\hbar\delta_{\alpha\alpha'}\delta_{kk'}.$$

Our goal is to express the coefficients $Q_{\alpha k}$ and $P_{\alpha k}$ through physical variables. For this purpose let us recall the relation between the pressure p and the displacement u [15,16]

$$p = E\,\mathrm{div}\,\mathbf{u}.$$

Here E is Young's modulus. The corresponding Fourier components are related as follows,

$$p_k = iEku_k, \qquad p = \sum_k p_k e^{ikr} + p_k^+ e^{-ikr}.$$

Then for the single mode with ω_k=ck one has (c is the sound velocity)

$$p_k = iEku_k = iE\,(\hbar k/2\rho Vc)^{1/2}\, c_k = \frac{iE}{2c}\,(\rho V)^{-1/2}(\omega Q_k + iP_k),$$
$$p_k^+ = -\frac{iE}{2c}\,(\rho V)^{-1/2}\,(\omega Q_k - iP_k),$$

hence

$$Q_k = (2c/E\omega)\sqrt{\rho V}\ \mathrm{Im}\, p_k, \qquad P_k = -(2c/E)\sqrt{\rho V}\ \mathrm{Re}\, p_k. \tag{3.2}$$

Further, we repeat the reasonings of the previous section concerning measurements at time instants separated in a quarter of period. The measurable quantity now is the pressure p. Then at the point r=0 we have

$$p = 2\ \mathrm{Re}\, p_k, \qquad p_k = p_k\,(0)\, e^{-i\omega t}. \tag{3.3}$$

On the other hand,

$$\mathrm{Im}\, p_k \sim \dot p\,(t_0) = \omega p\left(t_0 + \frac{T}{4}\right). \tag{3.4}$$

In the plane wave case

$$\dot p\,(t = t_0,\ r = 0) = \omega p\ (t = t_0,\ r = -\lambda/4). \tag{3.5}$$

Consequently, the generalized coordinate Q_k and the generalized momentum P_k can be expressed as follows:

$$Q_k\,(r = 0, t = t_0) = c\omega E^{-1}\sqrt{\rho V}\ p\,(r = 0, t_0 + T/4) =$$
$$= c\omega E^{-1}\sqrt{\rho V}\ p\,(r = -\lambda/4),$$
$$P_k\,(r = 0, t = t_0) = -cE^{-1}\sqrt{\rho V}\ p\,(r = 0, t = t_0). \tag{3.6}$$

The consequence of the commutation relations between P_k and Q_k is (for a single mode)

$$[p\ (t + T/4),\ p\ \ (t)] = -i\hbar\ E^2/c^2\rho V\omega.$$

The corresponding uncertainty relations are

$$\Delta p\ (r = 0,\ t_0 + T/4)\ \Delta p\ (r = 0,\ t_0) \geqslant \hbar\ E^2/2c^2\rho V,$$
$$\Delta p\ (r = -\lambda/4,\ t_0)\ \Delta p\ (r = 0,\ t_0) \geqslant \hbar\ E^2/2c^2\rho V$$

Quite similar relations hold for the quantized acoustic fields in liquids. Using the corresponding equations given, e.g., in [16] one can get the following relations,

$$H = \sum_k \left(c_k^+ c_k + \frac{1}{2}\right)\hbar\omega_k = \frac{1}{2}\sum_k (P_k^2 + \omega_k^2 Q_k^2),$$
$$P_k = \frac{2}{c}\left(\frac{V}{\rho_0}\right)^{1/2} \mathrm{Re}\, p_k, \qquad Q_k = -\frac{2}{\omega c}\left(\frac{V}{\rho_0}\right)^{1/2} \mathrm{Im}\, p_k,$$
$$c_k = \frac{i}{c}\left(\frac{2V}{\rho_0\omega\hbar}\right)^{1/2} p_k, \qquad \omega = ck. \tag{3.7}$$

$$P_k\ (r = 0,\ t = t_0) = c^{-1}\,(V/\rho_0)^{1/2} p\ (t = t_0,\ r = 0),$$
$$Q_k\ (r = 0,\ \ t = t_0) = -c^{-1}\,(V/\rho_0)^{1/2}\ p\ (t_0 + T/4,\ \ r = 0) =$$
$$= -c^{-1}\,(V/\rho_0)^{1/2} p\ (t_0,\ r = -\lambda/4),$$
$$[p\ (t + T/4),\ p\ (t)] = -ihc^2\rho_0/V. \tag{3.8}$$

4. CORRELATED COHERENT STATES OF A HARMONIC OSCILLATOR

Correlated coherent states (CCS) were introduced in studies [1,2] (see also [11,17] as states for which the left-hand side of the precise Schrodinger-Robertson uncertainty relation

$$\sigma_p^2\sigma_q^2 - \sigma_{pq}^2 \equiv \sigma_p^2\sigma_q^2\,(1 - r^2) \geqslant \hbar^2/4 \tag{4.1}$$

reaches its minimum value for the fixed correlation coefficient r defined as follows (we have changed notations in comparison with [1,2,11,17]: in that paper symbols σ_p and σ_q were used instead of $\sigma^2{}_p$ and $\sigma^2{}_q$),

$$\begin{gathered}\sigma_p^2 = \langle\hat{p}^2\rangle - \langle\hat{p}\rangle^2, \quad \sigma_q^2 = \langle\hat{q}^2\rangle - \langle\hat{q}\rangle^2,\\ \sigma_{pq} = {}^1\!/_2\,\langle\hat{p}\hat{q} - \hat{q}\hat{p}\rangle - \langle\hat{p}\rangle\langle\hat{q}\rangle,\\ r = \sigma_{pq}/\sigma_p\sigma_q.\end{gathered} \tag{4.2}$$

We assume hereafter that coordinates and momenta have been made dimensionless in a proper manner, so that Planck's constant formally equals unity, whereas $\hat{q}$ and $\hat{p}$ operators form the operators of annihilation and creation of quanta registered in experiments according to formulas

$$\hat{a} = \frac{\hat{q} + i\hat{p}}{\sqrt{2}}, \quad \hat{a}^+ = \frac{\hat{q} - i\hat{p}}{\sqrt{2}}, \quad \hat{q} = \frac{\hat{a} + \hat{a}^+}{\sqrt{2}}, \quad \hat{p} = \frac{i}{\sqrt{2}}(\hat{a}^+ - \hat{a}). \tag{4.3}$$

The states for which relation (4.1) becomes the equality turn out to be eigenstates of the operator [1,2]

$$\hat{b}\,(r, \eta) = \frac{1}{2\eta}\left[1 - \frac{ir}{\sqrt{1 - r^2}}\right]\hat{q} + i\eta\hat{p}, \tag{4.4}$$

where

$$\sigma_q = \eta, \quad \sigma_p = \frac{1}{2\eta\sqrt{1 - r^2}}, \quad \sigma_{pq} = \frac{r}{2\sqrt{1 - r^2}}, \tag{4.5}$$

so that r and η are real parameters, $0<\eta<\infty$, $-1<\eta<1$.

The explicit expression for a correlated coherent state in the coordinate representation is as follows:

$$\begin{gathered}\Psi_\beta\,(x\,|\,r, \eta) = (2\pi\eta^2)^{-1/4}\exp\left[-\frac{x^2}{4\eta^2}\left(1 - \frac{ir}{\sqrt{1 - r^2}}\right) + \right.\\ \left. + \frac{\beta x}{\eta} - \frac{1}{2}(\beta^2 + |\,\beta\,|^2)\right], \quad \hat{b}\Psi_\beta = \beta\Psi_\beta.\end{gathered} \tag{4.6}$$

Operators $\hat{a}$ and $\hat{a}^+$ are connected with operators $\hat{b}$ and $\hat{b}^+$ by means of the Bogoliubov canonical transformation

$$\hat{b} = u\hat{a} + v\hat{a}^+, \quad \hat{b}^+ = v^*\hat{a} + u^*\hat{a}^+, \tag{4.7}$$

with

$$\begin{gathered}u = \frac{1}{2\sqrt{2}\,\eta}\left[1 - \frac{ir}{\sqrt{1 - r^2}}\right] + \frac{\eta}{\sqrt{2}},\\ v = \frac{1}{2\sqrt{2}\,\eta}\left[1 - \frac{ir}{\sqrt{1 - r^2}}\right] - \frac{\eta}{\sqrt{2}}.\end{gathered} \tag{4.8}$$

As a consequence we have the identity

$$|\,u\,|^2 - |\,v\,|^2 = 1, \tag{4.9}$$

ensuring the validity of the commutation relations

$$[\hat{a}, \hat{a}^+] = [\hat{b}, \hat{b}^+] = 1.$$

Average values of the coordinate and momentum in state (4.6) are as follows:

$$\langle q \rangle = \sqrt{2}\,\mathrm{Re}\,[(u - v)\,\beta^*] = 2\eta\,\mathrm{Re}\,\beta,$$
$$\langle p \rangle = \sqrt{2}\,\mathrm{Im}\,[(u^* + v^*)\beta] = \eta^{-1}\left[\,\mathrm{Im}\,\beta + \frac{r}{\sqrt{1-r^2}}\,\mathrm{Re}\,\beta\right]. \quad (4.10)$$

Variances (4.2) can be expressed in terms of the Bogoliubov transformation coefficients u, v according to formulas

$$\sigma_q^2 = {}^1/_2 + |\,v\,|^2 - \mathrm{Re}\,(uv^*), \qquad \sigma_p^2 = {}^1/_2 + |\,v\,|^2 + \mathrm{Re}\,(uv^*),$$
$$\sigma_{pq} = \mathrm{Im}\,(uv^*). \quad (4.11)$$

Linear canonical transformation (4.7) is defined by two complex parameters satisfying constraing (4.9). Consequently, there are only three independent real parameters. However, one of them is trivial, since it corresponds to an admissible multiplication of operator $\hat{b}$ by a phase factor exp(iϕ), ϕ being an arbitrary phase. Therefore only two parameters are significant. This means that correlated coherent states can be considered as the most general pure quantum states described by Gaussian wave functions. In other words, any Gaussian pure state minimizes the generalized uncertainty relation (4.1), i.e. it is a correlated coherent state. At the same time, eigenstates of operators like (4.7) can be met in literature under different names mentioned in the Introduction (see also studies [6-9, 18-23]). Therefore it is expedient to show the place of CCS in the set of numerous generalizations of Glauber's coherent states. Last years eigenstates of operators like (4.7) with u≠0 are usually called "squeezed" states. This name is explained by the fact that if r=0 and $\sigma_p\sigma_q=1/2$, then either σ_p or σ_q inevitably occurs less than $2^{-1/2}$ for any non-Glauber's state. In other words, fluctuations of one of two quadrature components are "squeezed" by amplifying fluctuations of another component. Therefore term "squeezed states," to our opinion, well describes the essence of the phenomenon in the case of zero correlation coefficient, when parameters u, v can be assumed real (more precisely, their phases coincide). In this case a consequence of (4.9) is the natural parametrization

$$u = \mathrm{ch}\,\tau, \qquad v = \mathrm{sh}\,\tau, \quad (4.12)$$

resulting in the following variant of eqs. (4.10) and (4.11),

$$\langle q \rangle = \sqrt{2}\,e^{-\tau}\,\mathrm{Re}\,\beta, \qquad \langle p \rangle = \sqrt{2}\,e^{\tau}\,\mathrm{Im}\,\beta,$$
$$\sigma_q^2 = {}^1/_2 e^{-2\tau}, \qquad \sigma_p^2 = {}^1/_2 e^{2\tau}. \quad (4.13)$$

A more general parametrization is chosen frequently in the form

$$u = \mathrm{ch}\,\tau, \qquad v = \mathrm{sh}\,\tau \cdot e^{-i\vartheta} \quad (4.14)$$

and corresponding states are still called "squeezed" [21-24]. But for $\vartheta \neq 0$ this term does not correspond to the point properly, since instead of (4.13) the following formulas hold,

$$\sigma_q^2 = {}^1/_2\,(\mathrm{ch}\,2\tau - \mathrm{sh}\,2\tau \cdot \cos\vartheta),$$
$$\sigma_p^2 = {}^1/_2\,(\mathrm{ch}\,2\tau + \mathrm{sh}\,2\tau \cdot \cos\vartheta), \quad (4.15)$$

so that for values of phase ϑ close to $\pi/2$ both variances can be much greater than fluctuations in Glauber's coherent states. For $\vartheta \neq 0$ the correlation coefficient differs from zero:

$$\sigma_{pq} = {}^1/_2 \operatorname{sh} 2\tau \cdot \sin \vartheta, \qquad r = \frac{\operatorname{sh} 2\tau \cdot \sin \vartheta}{\sqrt{1 + \operatorname{sh}^2 2\tau \cdot \sin^2 \vartheta}}. \tag{4.16}$$

For this reason, in our opinion, just term "correlated state" emphasizing the statistical dependence of quadrature components, i.e. their nonzero correlation coefficient, is the most adequate in the case of complex parameters u and v (with different phases). Different parametrizations of generalized coherent states and their unitary equivalence were considered in detail in surveys [17,24].

Instead of parameter η characterizing the variance of coordinate, i.e. the "absolute" squeezing of its fluctuations, a new parameter

$$k = \sigma_q/\sigma_p, \tag{4.17}$$

characterizing the "relative" squeezing of quadrature components with respect to each other can be introduced. Since this parameter is frequently used in papers on squeezed states, we rewrite formulas obtained above in terms of quantities k and r:

$$\eta = \left(\frac{k}{2\sqrt{1-r^2}}\right)^{1/2},$$

$$u = \frac{1}{2}\left(\frac{\sqrt{1-r^2}}{k}\right)^{1/2}\left[1 - \frac{ir}{\sqrt{1-r^2}}\right] + \frac{1}{2}\left(\frac{k}{\sqrt{1-r^2}}\right)^{1/2},$$

$$v = \frac{1}{2}\left(\frac{\sqrt{1-r^2}}{k}\right)^{1/2}\left[1 - \frac{ir}{\sqrt{1-r^2}}\right] - \frac{1}{2}\left(\frac{k}{\sqrt{1-r^2}}\right)^{1/2}. \tag{4.18}$$

Remember once more that even with small relative squeezing, i.e. with $k \to 0$, the absolute squeezing of coordinate can be large provided $|r| \to 1$.

Higher moments of the coordinate can be calculated with the aid of formula [11]

$$\langle x^n \rangle = \left(-\frac{\sigma^2}{2}\right)^{n/2} H_n\left[\frac{\bar{x}}{2}\left(-\frac{\sigma^2}{2}\right)^{-1/2}\right], \tag{4.19}$$

valid for Gaussian distribution functions of the type

$$W(x) = |\Psi(x)|^2 = (2\pi\sigma^2)^{-1/2} \exp[-(x - \bar{x})^2/2\sigma^2].$$

Here $H_n(z)$ is the Hermite polynomial. In terms of k, r variables we get from (4.19)

$$\langle q \rangle = 2\left(\frac{k}{2\sqrt{1-r^2}}\right)^{1/2} \operatorname{Re}\beta, \qquad \sigma_q^2 = \frac{k}{2\sqrt{1-r^2}},$$

$$\langle q^n \rangle = \left(\frac{i}{2}\left(\frac{k}{\sqrt{1-r^2}}\right)^{1/2}\right)^n H_n(-i\sqrt{2}\operatorname{Re}\beta). \tag{4.20}$$

Let us discuss in principle the problem of generating correlated coherent states. For this purpose consider a general problem of solving nonstationary Schrodinger equation for an oscillator with time-dependent frequency $\Omega(t)$ (remember that in the stationary case the frequency is assumed to be equal to unity). The solution of this problem is well known. It is given in great detail in review [11], where it is shown that there exists a complete normalized system of solutions (more precisely, an overcomplete system in the same sense as the usual system of Glauber's coherent states) in the form of Gaussian exponentials

$$\Psi_\alpha(x, t) = \pi^{-1/4}\varepsilon^{-1/2} \exp\left[\frac{i\dot{\varepsilon}}{2\varepsilon}x^2 + \frac{\sqrt{2}\alpha x}{\varepsilon} - \frac{\alpha^2}{2}\frac{\varepsilon^*}{\varepsilon} - \frac{|\alpha|^2}{2}\right], \tag{4.21}$$

determined by an arbitrary complex number α and an arbitrary complex solution of the classical equation of motion for a parametrically excited oscillator $\varepsilon(t)$:

$$\ddot{\varepsilon} + \Omega^2(t)\,\varepsilon = 0. \tag{4.22}$$

Since complex conjugated function $\varepsilon^*(t)$ satisfies the same equation as well, and Wronskian does not depend on time, it is convenient to impose on solutions a normalization constraint preserved in time

$$\dot{\varepsilon}\varepsilon^* - \dot{\varepsilon}^*\varepsilon = 2i. \tag{4.23}$$

The choice of constant in the right hand side corresponds in the case of $\Omega=1$ to the solution $\varepsilon(t)=e^{it}$. Then function (4.21) is the Glauber coherent state. Any other solution $\varepsilon(t)$ yields, in general, a correlated state due to formulas [11]

$$k = |\,\varepsilon/\dot{\varepsilon}\,|, \qquad r = \mathrm{Re}\,(\dot{\varepsilon}\varepsilon^*)/\,|\,\varepsilon\dot{\varepsilon}\,|, \qquad r^2 = 1 - |\,\varepsilon\dot{\varepsilon}\,|^{-2}, \tag{4.24}$$

which can be obtained if one compares eqs. (4.6) and (4.21) (in formula for r^2 identity (4.23) is taken into account). Thus the most natural way to create a correlated coherent state is to subject the oscillator to the parametric swinging. For the first time a possibility of obtaining the states with variances less than in the ground state by means of parametric swinging was investigated in detail apparently in studies [25-27]. Although the problem of parametric excitation of a quantum oscillator was investigated (from a different point of view) even earlier [28] (see also monograph [29] where rather large list of references was given).

All experimental methods of generating squeezed states correspond in fact to one or the other way of parametric excitation. We confine ourselves to the references to surveys [30,31] and the special issue [32] devoted to squeezed states.

In general, both correlation coefficient and squeezing coefficient vary in time. The constant frequency case will be analyzed in detail in Section 8. Meanwhile, here we want to find such dependences of frequency on time which allow us to construct states with time-independent parameters r and k. Let us begin with correlation coefficient. Due to (4.23) and (4.24) condition r=const is equivalent to $\mathrm{Re}(\dot{\varepsilon}\varepsilon^*)$=const. Let us single out the modulus and phase of the complex function $\varepsilon(t)$

$$\varepsilon(t) = \rho(t)\exp[i\varphi(t)]. \tag{4.25}$$

Then $\mathrm{Re}(\dot{\varepsilon}\varepsilon^*)=\rho\dot{\rho}=1/2\,d\rho^2/dt$. Consequently,

$$\rho^2(t) = 2bt + c, \qquad b,\ c = \mathrm{const}. \tag{4.26}$$

Equation for $\rho(t)$ is obtained from Eq. (4.22), if one takes into account the identity

$$\dot{\varphi} = \rho^{-2}. \tag{4.27}$$

following from (4.23). As a result we have the equation

$$\ddot{\rho} - \rho^{-3} + \Omega^2\rho = 0, \tag{4.28}$$

allowing to find the dependence $\Omega(t)$ for the given function $\rho(t)$. For function (4.26) we get

$$\Omega(t) = \sqrt{1+b^2}/(2bt+c). \tag{4.29}$$

Then correlation and squeezing coefficients can be expressed as follows,

$$k = \frac{\rho^2}{\sqrt{\rho^2\dot{\rho}^2+1}} = \frac{2bt+c}{\sqrt{1+b^2}} = \frac{1}{\Omega(t)}, \tag{4.30}$$

$$r = \frac{\rho\dot{\rho}}{\sqrt{\rho^2\dot{\rho}^2+1}} = \frac{b}{\sqrt{1+b^2}}. \tag{4.31}$$

Consequently, to preserve the correlation coefficient the frequency has to vary according to the law

$$\Omega(t) = [2rt + c\sqrt{1 - r^2}]^{-1}. \tag{4.32}$$

It should be mentioned that the arising CCS described by formulas (4.21), (4.25)-(4.27) is in fact *unsqueezed*, in spite of relation (4.30). The point is that the squeezing coefficient in (4.30) was defined with respect to vacuum fluctuations of the oscillator with unit frequency. For the oscillator with frequency ω the vacuum state fluctuations $\sigma^{(0)}_q$ and $\sigma^{(0)}_p$ are related by formula $\sigma^{(0)}_p=\omega\sigma^{(0)}_q$ (for unit mass). Therefore it is more natural to define the relative squeezing coefficient for the oscillator with frequency ω as

$$k_\omega = \omega\sigma_q/\sigma_p = \omega k. \tag{4.33}$$

In the example considered above $k_\omega \equiv 1$ for all instants of time. Moreover, the absolute squeezing is at all absent, since

$$\begin{aligned} \sigma_q &= |\varepsilon|/\sqrt{2} = \rho/\sqrt{2} = (2\Omega)^{-1/2}(1 - r^2)^{-1/4}, \\ \sigma_p &= (\Omega/2)^{1/2}(1 - r^2)^{-1/4} \end{aligned} \tag{4.34}$$

and both these values are $(1-r^2)^{-1/4}$ times higher than zero fluctuations of coordinate and momentum for the oscillator with frequency Ω. Thus, this example demonstrates distinctly the difference between squeezed and correlated states.

The condition of independence on time of coefficient k (i.e. the relative squeezing normalized by the initial unit frequency) leads due to-2 the first formula from (4.30) to the equation

$$\rho\dot{\rho} = \sqrt{k^{-2}\rho^4 - 1}. \tag{4.35}$$

Its trivial solution $\dot{\rho}\equiv 0$, $\rho^2=k$ corresponds to the constant frequency $\omega=k^{-1}$, i.e. to unsqueezed ($k_\omega=1$) and noncorrelated states. If $\dot{\rho}\neq 0$, then the integration of (4.35) results in

$$\rho^2 = k\,\mathrm{ch}\,(2t/k) \tag{4.36}$$

(the initial instant of time is assumed to be zero). From (4.28) we get $\Omega^2=-k^{-2}$, which corresponds to the motion in the centrifugal potential $U(x)=-1/2k^{-2}x^2$. In this case absolute values of both variances and the correlation coefficient grow in time monotonously:

$$\sigma_q^2 = {}^1/_2 k\,\mathrm{ch}\,(2t/k), \quad \sigma_p^2 = {}^1/_2 k^{-1}\,\mathrm{ch}\,(2t/k), \quad r = \mathrm{th}\,(2t/k). \tag{4.37}$$

It is not difficult to derive a general dependence $\Omega(t)$ resulting in any given function r(t). Indeed, due to Equations (4.5), (4.23), (4.24) function $g=\rho^2$ satisfies the equation

$$\dot{g} = 2\rho\dot{\rho} = 2r\,(1 - r^2)^{-1/2} = 4\sigma_{pq}\,(t). \tag{4.38}$$

Therefore with the given dependence r(t) or $\sigma_{pq}(t)$ function g(t) can be obtained by means of a quadrature. After this function $\Omega(t)$ can be found from Equation (4.28), if one takes into account the relation

$$\ddot{g} = 4\dot{\sigma}_{pq} = 2\rho\ddot{\rho} + 2\dot{\rho}^2,$$

$$\Omega^2(t) = \frac{1 + 4\sigma_{pq}^2 - 2g\dot{\sigma}_{pq}}{g^2} = \frac{1 - 4g\dot{r}\,(1 - r^2)^{-1/2}}{g^2\,(1 - r^2)}. \tag{4.39}$$

Suppose that frequency Ω was equal to unity for $t \leq 0$, then began to change in time and finally assumed an asymptotic value Ωf. Let for $t \leq 0$ the state of oscillator was unsqueezed and noncorrelated, i.e. the initial function $\varepsilon(t)$ was $\varepsilon_{in}(t)=e^{it}$. Then solution of Equation (4.22) with $t \to \infty$, when the frequency has moved up into the new stationary regime, can be expressed as follows:

$$\varepsilon(t) = \Omega_f^{-1/2}(\xi e^{it\Omega_f} + \eta e^{-it\Omega_f}), \tag{4.40}$$

where ξ and η are constant complex coefficients satisfying relation $|\xi|^2 - |\eta|^2 = 1$ which is equivalent to (4.23). But Equation (4.22) is equivalent to the equation describing the propagation of a wave through some potential barrier characterized by a varying in space wave number $k(x)$, if one makes redesignations $t \to x$, $\Omega \to k$. Then the quantity η^*/ξ^* is nothing but the amplitude reflection coefficient from this barrier for the wave going from right to left. On the other hand, Equation (4.40) leads to the relation

$$\frac{\eta}{\xi} = \exp(2i\Omega_f)\frac{i\Omega_f\varepsilon - \dot{\varepsilon}}{i\Omega_f\varepsilon + \dot{\varepsilon}}.$$

Consequently, the energy reflection coefficient $R = |\eta/\xi|^2$ (which does not depend on what side the wave goes from) equals

$$R = \frac{\Omega_f^2|\varepsilon|^2 + |\dot{\varepsilon}|^2 - 2\Omega_f}{\Omega_f^2|\varepsilon|^2 + |\dot{\varepsilon}|^2 + 2\Omega_f}.$$

Substituting into this expression formulas

$$|\varepsilon|^2 = k/\sqrt{1-r^2}, \qquad |\dot{\varepsilon}|^2 = 1/k\sqrt{1-r^2}, \tag{4.41}$$

resulting from (4.24) we arrive at the equation

$$k_*^2(1-R) - 2k_*(1+R)\sqrt{1-r^2} + 1 - R = 0, \qquad k_* = \Omega_f k. \tag{4.42}$$

Solving it with respect to k_* we get two values

$$k_* = [(1+R)\sqrt{1-r^2} \pm [(1+R)^2(1-r^2) - (1-R)^2]^{1/2}](1-R)^{-1}. \tag{4.43}$$

The requirement that k_* must be a real number leads to the restriction on possible values of the correlation coefficient with the given value of the energy reflection coefficient from the barrier

$$|r| \leqslant 2\sqrt{R}/(1+R). \tag{4.44}$$

The state with the maximum correlation coefficient turns out to be unsqueezed: $k_*=1$. Analyzing expression (4.43) as the function of parameter r one can see that with the given value of R the squeezing coefficient is confined within the limits

$$(1-\sqrt{R})/(1+\sqrt{R}) \leqslant k_* \leqslant (1+\sqrt{R})/(1-\sqrt{R}). \tag{4.45}$$

The extremal values are reached for noncorrelated states.

In the subsequent calculations we shall need the overlap integral between correlated coherent state (4.6) and Fock's state

$$\Psi_n(x) = \pi^{-1/4}(2^n n!)^{-1/2}\exp[-x^2/2]H_n(x), \tag{4.46}$$

which is the eigenstate of the number of quanta operator $\hat{a}^+\hat{a}$. It was calculated for the u, v-parametrization of the Bogoliubov transformation (4.7) in [8,17]:

$$\langle n|\beta uv\rangle = \frac{1}{\sqrt{u}\sqrt{n!}}\left(\frac{u-v}{u^*-v^*}\right)^{1/4}\left(\frac{v}{2u}\right)^{n/2}\exp\left(-\frac{|\beta|^2}{2} + \frac{\beta^2 v^*}{2u}\right)H_n\left(\frac{\beta}{\sqrt{2uv}}\right), \tag{4.47}$$

$H_n(x)$ is the Hermite polynomial. (More general overlap integrals like $\langle\alpha,u',v'|\beta,u,v\rangle$ and $\langle n,u',v'|\beta,u,v\rangle$ were calculated in [17] as well.)

5. DISTRIBUTION OF QUANTA IN CORRELATED COHERENT STATE

In this section we investigate the photons distribution function W_n in a correlated coherent state and apply the results to electromagnetic and sound fields.

Distribution function W_n is defined as the modulus squared coefficient of the CCS expansion over Fock's states which can be registered with a detector available. Then formula (4.7) yields

$$W_n = \frac{1}{|u|}\frac{1}{n!}\exp\left[-|\beta|^2 + \frac{\beta^2}{2}\frac{v^*}{u} + \frac{\beta^{*2}}{2}\frac{v}{u^*}\right]\left|\frac{v}{2u}\right|^n \left|H_n\left(\frac{\beta}{\sqrt{2uv}}\right)\right|^2. \quad (5.1)$$

The most interesting cases in investigating W_n seem the limits corresponding to strong squeezing $k\to 0$, $k\to\infty$ and strong correlation $|r|\to 1$. One can expect that the behavior of W_n in these cases will be substantially different from that in usual CS ($k=1$, $r=0$), when it is represented by Poisson's distribution.

Taking into account the relations between u, v and k, r coefficients (4.18) one can obtain the following limit expressions for W_n in different cases,

1) $k\to 0,\ r=0$:

$u\to 1/2\sqrt{k},\quad v\to 1/2\sqrt{k},$

$$W_n \simeq 2^{1-n}k^{1/2}(n!)^{-1}\exp[-2(\mathrm{Im}\,\beta)^2]\,|H_n(\beta\sqrt{2k})|^2; \quad (5.2)$$

2) $k\to\infty,\ r=0$:

$u\to\sqrt{k}/2,\quad v\to -\sqrt{k}/2,$

$$W_n \simeq 2^{1-n}k^{-1/2}(n!)^{-1}\exp[-2(\mathrm{Re}\,\beta)^2]\,|H_n(-i\beta\sqrt{2/k})|^2; \quad (5.3)$$

3) $k=1,\quad 1-r^2=\xi^2\to 0$:

$u\to(1-i)/2\sqrt{\xi},\quad v\to-(1+i)/2\sqrt{\xi},$

$$W_n \simeq 2^{1/2-n}\xi^{1/2}(n!)^{-1}\exp[-2(\mathrm{Re}\,\beta)^2]\,|H_n(-i\beta\sqrt{\xi})|^2; \quad (5.4)$$

4) $k\to 0,\quad 1-r^2=\xi^2\to 0$:

$u\to -i/2\sqrt{\xi k},\quad v\to -i/2\sqrt{\xi k},$

$$W_n \simeq 2^{1-n}(\xi k)^{1/2}(n!)^{-1}\exp[-2(\mathrm{Re}\,\beta)^2]\,|H_n(-i\beta\sqrt{2\xi k})|^2; \quad (5.5)$$

5) $k\to\infty,\quad 1-r^2=\xi^2\to 0$:

$u\to \tfrac{1}{2}\sqrt{k/\xi},\quad v\to -\tfrac{1}{2}\sqrt{k/\xi},$

$$W_n \simeq 2^{1-n}(\xi/k)^{1/2}(n!)^{-1}\exp[-2(\mathrm{Re}\,\beta)^2]\,|H_n(-i\beta\sqrt{2\xi/k})|^2. \quad (5.6)$$

All these asymptotics contain the Hermite polynomials $H_n(x)$ with $x\to 0$. The correspoding values can be taken from [33]:

$$H_{2n}(x)\sim\frac{(2n)!}{n!}(-1)^n, \qquad nx^2\ll 1,$$

$$H_{2n+1}(x)\sim\frac{(2n+1)!}{n!}(-1)^n\,2x, \qquad nx^2\ll 1,$$

$$H_{2n}\sim 2^{(2n+1)/2}(2n)^n e^{-n}e^{x^2/2}(-1)^n\cos(\sqrt{4n+1}\,x), \qquad n\gg 1.$$

$$H_{2n+1}\sim 2^{n+1}(2n+1)^{(2n+1)/2}e^{x^2/2}e^{-(2n+1)/2}(-1)^n\sin(\sqrt{4n+3}\,x), \qquad n\gg 1.$$

For sufficiently small values of x both inequalities n>>1 and nx^2<<1 can be valid. Thus the asymptotics given above may overlap. The final results can be represented as follows.

If r=0, k→0, then

$$W_{2n} \simeq \frac{\sqrt{4k}}{2^{2n}} \frac{(2n)!}{(n!)^2} \exp[-2(\mathrm{Im}\,\beta)^2], \qquad |\beta|^2 kn \ll 1,$$

$$W_{2n+1} \simeq \frac{\sqrt{(4k)^3}}{2^{2n}} \frac{(2n+1)!}{(n!)^2} \exp[-2(\mathrm{Im}\,\beta)^2]|\beta|^2, \qquad |\beta|^2 kn \ll 1;$$

$$W_{2n} \simeq 2\sqrt{k/\pi n}\, \exp[-2(\mathrm{Im}\,\beta)^2]\,|\cos(\beta\sqrt{2k(4n+1)})|^2, \qquad n \gg 1,$$

$$W_{2n+1} \simeq 2\sqrt{k/\pi n}\, \exp[-2(\mathrm{Im}\,\beta)^2]\,|\sin(\beta\sqrt{2k(4n+3)})|^2, \qquad n \gg 1. \tag{5.7}$$

All other expressions (5.3)-(5.6) can be written in the unified form

$$W_{2n} \simeq \frac{\sqrt{4\varepsilon}}{2^{2n}} \frac{(2n)!}{(n!)^2} \exp[-2(\mathrm{Re}\,\beta)^2], \qquad |\beta|^2 \varepsilon n \ll 1,$$

$$W_{2n+1} \simeq \frac{\sqrt{(4\varepsilon)^3}}{2^{2n}} \frac{(2n+1)!}{(n!)^2} \exp[-2(\mathrm{Re}\,\beta)^2]|\beta|^2, \qquad |\beta|^2 \varepsilon n \ll 1;$$

$$W_{2n} \simeq 2\sqrt{\varepsilon/\pi n}\, \exp[-2(\mathrm{Re}\,\beta)^2]\,|\mathrm{ch}(\beta\sqrt{2\varepsilon(4n+1)})|^2, \qquad n \gg 1,$$

$$W_{2n+1} \simeq 2\sqrt{\varepsilon/\pi n}\, \exp[-2(\mathrm{Re}\,\beta)^2]\,|\mathrm{sh}(\beta\sqrt{2\varepsilon(4n+3)}|^2, \qquad n \gg 1, \tag{5.8}$$

The parameter ε assumes different values depending on the relations between parameters r and k:

2) ε(r=0,k→∞)=1/k;
3) ε(r→1, k=1)=$2^{-1/2}\xi$, $\xi=(1-r^2)^{1/2}$;
4) ε(r→1, k→0)=ξk;
5) ε(r→1, k→∞)=ξ/k.

It is interesting to analyze the dependence of the distribution function on the "conditional squeezing"

$$\tilde{k} = k|\langle p\rangle/\langle q\rangle| = |(\sigma_q/\langle q\rangle)\cdot(\langle p\rangle/\sigma_p)|, \tag{5.9}$$

because this parameter can be varied through the change of average values of coordinate and momentum <q>, <p>, given by Equations (4.10). Introducing an auxiliary parameter φ= | Imβ/Reβ | we can write

$$k = 2\xi r^{-2}(\tilde{k} - \xi\varphi)^2$$

and

$$u(\tilde{k}r) = \frac{1}{2\sqrt{2}} \frac{r}{|\tilde{k} - \xi\varphi|}\left[1 - \frac{ir}{\xi}\right] + \frac{1}{\sqrt{2}} \frac{|\tilde{k} - \xi\varphi|}{r},$$

$$v(\tilde{k}r) = \frac{1}{2\sqrt{2}} \frac{r}{|\tilde{k} - \xi\varphi|}\left[1 - \frac{ir}{\xi}\right] - \frac{1}{\sqrt{2}} \frac{|\tilde{k} - \xi\varphi|}{r}.$$

The parameter x equals as above

$$\xi=(1-r^2)^{1/2}$$

Instead of (5.8) we have now asymptotics

$$W_{2n} \simeq 2\sqrt{2}\,\varepsilon\,(2n)!\,(n!)^{-2} \exp[-2(\mathrm{Re}\,\beta)^2], \qquad |\beta|^2 \varepsilon n \ll 1,$$

$$W_{2n} \simeq 2\sqrt{2}\,(2\varepsilon)^3(2n+1)!\,(n!)^{-2} \exp[-2(\mathrm{Re}\,\beta)^2]|\beta|^2, \qquad |\beta|^2 \varepsilon n \ll 1;$$

$$W_{2n} \simeq \sqrt{2/\pi n}\, 2\varepsilon \exp[-2(\mathrm{Re}\,\beta)^2]\,|\mathrm{ch}(2\varepsilon\beta\sqrt{4n+1})|^2,$$

$$W_{2n+1} \simeq \sqrt{2/\pi n}\, 2\varepsilon \exp[-2(\mathrm{Re}\,\beta)^2]\,|\mathrm{sh}(2\varepsilon\beta\sqrt{4n+3})|^2, \tag{5.10}$$

with $\varepsilon \ll 1$ assuming the following values,

$$\varepsilon(\tilde{k}=1,\xi\to 0)=\xi;$$
$$\varepsilon(\tilde{k}\to 0,\xi\to 0,\xi\varphi \ll \tilde{k})=\tilde{k}\xi;$$
$$\varepsilon(\tilde{k}\to\infty,\xi\to 0,\tilde{k}^2\xi \gg 1)=1/2\tilde{k}^2;$$
$$\varepsilon(\tilde{k}\to\infty,\xi\to 0,\tilde{k}^2\xi \ll 1)=\tilde{k}\xi.$$

Varying <q> and <p> one can change the "conditional squeezing" and obtain essentially non-Poissonian photons distribution functions W_n, provided $r\neq 0$. The most interesting case in this sense is $k\to\infty$, $\xi\to 0$.

It is not difficult to calculate the mean number of photons, absolute and relative variances in CCS:

$$\bar{n}=\sum_n nW_n=|v|^2(1+|\beta|^2)+|u|^2|\beta|^2-\beta^2 v^* u^*-\beta^{*2}uv,$$
$$\bar{n^2}=\sum_n n^2W_n=2\bar{n}\,[2|u|^2-1]+(\bar{n})^2-2|v|^4-|\beta|^2,$$
$$\sigma_n=\bar{n^2}-(\bar{n})^2,\qquad \Delta_n=\sigma_n/(\bar{n})^2. \tag{5.11}$$

We get two types of formulas corresponding to the asymptotics considered above. If $r=0$, $k\to 0$, $k|\beta|^2\to 0$, then

$$\bar{n}=\frac{1}{4k}[1+4(\mathrm{Im}\,\beta)^2],\qquad \sigma_n=\frac{1}{8k^2}[1+8(\mathrm{Im}\,\beta)^2],$$
$$\Delta_n=2\frac{1+8(\mathrm{Im}\,\beta)^2}{[1+4(\mathrm{Im}\,\beta)^2]^2}=\begin{cases}2, & |\mathrm{Im}\,\beta|\ll 1,\\ (\mathrm{Im}\,\beta)^{-2}, & |\mathrm{Im}\,\beta|\gg 1.\end{cases} \tag{5.12}$$

Note that the values $\bar{n}$, σ_n, Δ_n in CCS differ essentially from the analogous values in usual CS. The mean number of photons (and proportional to its mean energy) in CS is determined by the quantity $|\beta|^2$ only. For CCS in the case of $k\to 0$ the mean number of photons is determined by two quite different parameters k and $(\mathrm{Im}\beta)^2$. Therefore even if $\beta=i\mathrm{Im}\beta$, when the centers of both wave packets coincide, the energy of CCS can be much greater due to increased quantum fluctuations. On the other hand, the value of Δ_n in CCS is determined by $\mathrm{Im}\beta$ only, whereas in usual CS it is equal to $(\bar{n})^{-1}=|\beta|^{-2}$. Thus for $|\beta|\gg 1$ the parameter Δ_n tends to zero in the case of Poissonian statistics, while for non-Poissonian CCS this parameter can remain finite provided $|\mathrm{Im}\beta|\ll 1$.

The second set of asymptotical formulas can be written in a unique manner as follows:

$$\bar{n}={}^1/_4\varepsilon^{-1}[1+4(\mathrm{Re}\,\beta)^2],\qquad \sigma_n={}^1/_8\varepsilon^{-2}[1+8(\mathrm{Re}\,\beta)^2],$$
$$\Delta_n=2\frac{1+8(\mathrm{Re}\,\beta)^2}{[1+4(\mathrm{Re}\,\beta)^2]^2}=\begin{cases}2, & |\mathrm{Re}\,\beta|\ll 1,\\ (\mathrm{Re}\,\beta)^{-2}, & |\mathrm{Re}\,\beta|\gg 1,\end{cases}$$

where parameter $\varepsilon \ll 1$ assumes the same values as in Equations (5.8).

We shall continue the analysis of formulas (5.11) in Section 8.

It is not difficult to generalize the results obtained above to oscillators of electromagnetic and phonon fields. For example, in the electromagnetic field case we have established earlier the following relation between canonical variables and the electric field strength:

$$P(t_0)=-\frac{1}{2}\left(\frac{V}{\pi}\right)^{1/2}E(t_0),\qquad Q(t_0)=\frac{1}{2\omega}\left(\frac{V}{\pi}\right)^{1/2}E\left(t_0+\frac{T}{4}\right).$$

Then we have a very interesting expression for the squeezing coefficient

$$k = \sigma_{E(t+T/4)}/\sigma_{E(t)}.$$

Other expressions for k can be derived from (2.3), (2.6), (2.8), (2.9). If E_β is the mean value of the electric field strength in CCS $|\beta\rangle$, then

$$\operatorname{Im}\tilde{\beta} = -\gamma E_\beta(t), \qquad \operatorname{Re}\tilde{\beta} = \gamma E_\beta(t + T/4),$$
$$\tilde{\beta} = \beta u^* - \beta^* v, \qquad \gamma = {}^1/_2\,(V/2\pi\hbar\omega)^{1/2}. \tag{5.13}$$

The correlation coefficient can be defined respectively as

$$r = \frac{\sigma_{E(t)E(t+T/4)}}{\sigma_{E(t)}\sigma_{E(t+T/4)}}, \qquad \sigma_{E(t)E(t+T/4)} = \frac{1}{2}\left\langle\left[E(t), E\left(t + \frac{T}{4}\right)\right]_+\right\rangle -$$
$$-\langle E(t)\rangle\left\langle E\left(t + \frac{T}{4}\right)\right\rangle,$$

Note that the quantum correlation coefficient is expressed through the quantity which can really be measured in experiments: the value of the temporal electric field correlation function corresponding to the quarter of period shift.

For the phonon field we have from (3.2)-(3.6) quite analogous relations

$$P(t_0) = \frac{c}{E}\sqrt{\rho V}\,p(t_0), \qquad Q(t_0) = \frac{c\omega}{E}\sqrt{\rho V}\,p\left(t_0 + \frac{T}{4}\right),$$

$$k = \frac{\sigma_{p(t+T/4)}}{\sigma_{p(t)}}, \qquad \operatorname{Im}\beta = -\chi p(t), \qquad \operatorname{Re}\beta = \chi p\left(t + \frac{T}{4}\right),$$
$$\chi = \frac{c}{2E}\left(\frac{2\rho V}{\hbar\omega}\right)^{1/2}. \tag{5.14}$$

For the lowest electromagnetic mode of a cubic resonator the parameter r can be defined also as the correlation coefficient between fluctuations of electric and magnetic fields in the points belonging to the line segment (L/4, L/4, z). Analogously, the squeezing coefficient can be defined as the ratio of variances of electric and magnetic strengths in the same points (see end of Section 2).

Note, in conclusion of this part that variances of electric field and vector potential operators in a "two-photon" state were calculated also in studies [34,35].

6. OSCILLATIONS OF PHOTONS DISTRIBUTION FUNCTION AND EXPERIMENTS ON PHOTONS CORRELATIONS

Asymptotics obtained above give certain notion on behavior of the distribution function W_n under the strong correlation and squeezing, when it differs essentially from the Poisson distribution corresponding to the usual coherent state

$$W_n^0 = |\beta|^2 (n!)^{-1} e^{-|\beta|^2}.$$

1. Consider the correlated vacuum, when $\beta=0$, but $r\neq 0$ or $k\neq 1$. Then we have

$$W_{2n+1} = 0,$$
$$W_{2n} = \left(\frac{k\sqrt{1-r^2}}{2k\sqrt{1-r^2}+1+k^2}\right)^{1/2}\left|\frac{1+k^2-2k\sqrt{1-r^2}}{1+k^2+2k\sqrt{1-r^2}}\right|^{n/2}\frac{(2n)!}{2^{2n+1}(n!)^2}.$$

Therefore, in the correlated vacuum we can register with non-zero probabilities only even numbers of photons. In a sense, this fact explains the term "two-photon states" [7].

In the limit of strong correlation and squeezing we get

$$W_{2n+1} = 0, \qquad W_{2n} \simeq \frac{\sqrt{\varepsilon}}{2^{2n-1}}\frac{(2n)!}{(n!)^2},$$

with the parameter $\varepsilon<<1$ assuming the following values,

$k\to 0, r=0: \varepsilon=k; k\to\infty, r=0: \varepsilon=k^{-1};$

$k=1, \xi\to 0: \varepsilon=2^{-1/2}\xi; k\to 0, \xi\to 0: \varepsilon=\xi k; k\to\infty, \xi\to 0: \varepsilon=\xi/k,$

where $\xi=[1-r^2]^{1/2}$.

2. The function $W^0{}_n$ depends only on $|\beta|^2$, whereas a CCS at fixed $|\beta|^2$ behaves substantially different, depending on the ratio of the real and imaginary parts of β. For example, for $|\mathrm{Im}\beta|<<|\mathrm{Re}\beta|$ the distribution function W_n oscillates strongly when $r=0$ andk$\to$0, i.e., when the variance of the coordinate is much less than the variance of the momentum. If $|\mathrm{Im}\beta|>>|\mathrm{Re}\beta|$, these oscillations will be exponentially suppressed. For certain other combinations of parameters (r,k) the function W_n can also strongly oscillate with $|\mathrm{Im}\beta|>>|\mathrm{Re}\beta|$.

3. As noted above, the properties of $W^0{}_n$ are determined only by the value of $|\beta|^2$. Thus, the average number of photons in the usual CS is $n=|\beta|^2$. At the same values of $|\beta|^2$ the average number of photons in a CCS is determined by squeezing and correlation. Figure 1 shows the functions W_n and $W^0{}_n$ for specified $\beta=2.5+2.5i$, $r=0$, $k=0,1$. The average number of photons is $\bar{n}_0\cong 12$ in the CS and $\bar{n}\cong 60>>n_0$ in the CCS. It can be stated that the difference between CS and CCS for one and the same classical trajectory is determined by quantum fluctuations

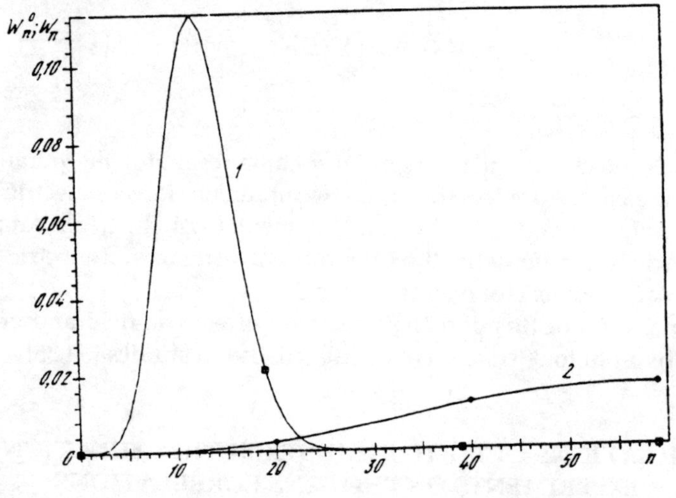

Figure 1. Photons distribution functions $W^0{}_n(1)$ and W_n (2) in the case of $\beta=2.5+2.5i$, $r=0$, $k=0.1$.

4. Let us consider the fluctuations of the distribution function W_n. Figure 2a shows W_n with $\beta=0.5+2i$, $r=0$, $k=0.1$, whereas Figure 2b corresponds to the same values of β and r, but $k=10$. We see strong oscillations when $k=10$. The period of fluctuations, i.e., the distance between peaks, is $T_n=2$. It is seen from the plot that even though the average number of photons in the given CCS is $\bar{n}\cong 5$, the probability of recording 5 photons in experiments is much lower than that of recording 2 and 4 photons. Thus, in registering CCS we obtain a certain effective quantization of a photon beam over the number of photons, in the sense of the probability of recording them. Figures 2b, 2d, 2e show certain (β, r, k) combinations in which such an "effective quantization" is strongly manifested.

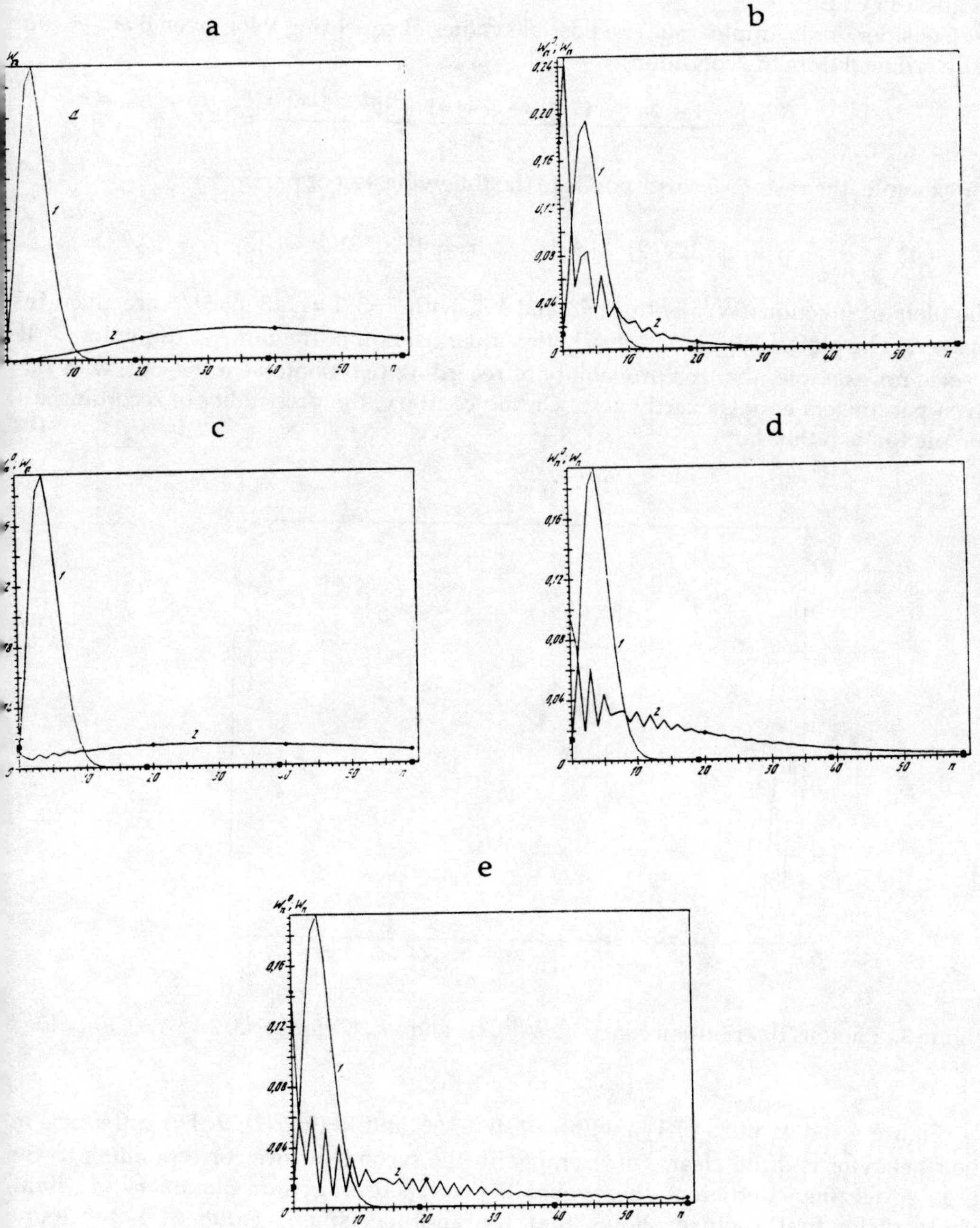

Figure 2. Photons distribution functions W^0_n (1) and W_n (2) in the case of β=2.5 + 2.5i: a) r=0, k=0.1; b) r=0, k=10; c) r=0.9, k=0.1; d) r=0.9, k=1; e) r=0, k=10.

5. From the expression for the average number of photons (5.11) it is clear that a number of sets of parameters (β, r, k) can lead to one and the same average number of photos $\bar{n}$ in CCS.

Consider, for example, r=0. The possible values of squeezing k for given β and $\bar{n}$ can be determined from the condition

$$k = \frac{1 + 2\bar{n} \pm [4\bar{n} + 4\bar{n}^2 - 4|\beta|^2 + (\beta^2 - \beta^{*2})]^{1/2}}{1 + (\beta + \beta^*)^2}. \tag{6.1}$$

For example, the case $\bar{n}=3$ corresponds to the following sets of parameters:

$$1)\ k = 1,\ \beta = \sqrt{3};\quad 2)\ k = 1/13,\quad \beta = \sqrt{3};\ 3)\ k = 13,\ \beta = i\sqrt{3}.$$

The plots of functions W^0_n with $\bar{n}_0=3$ and W_n with $\bar{n}=3$, $k=1/13$, $\beta=3^{1/2}$ are given in Figure 3. The plot of W_n shows clearly the "quantization in the number of photons." It is seen, for example, that the probability of recording two photons in the CCS with the given parameters equals exactly zero. On the contrary, the probability of recording no one photon is rather large.

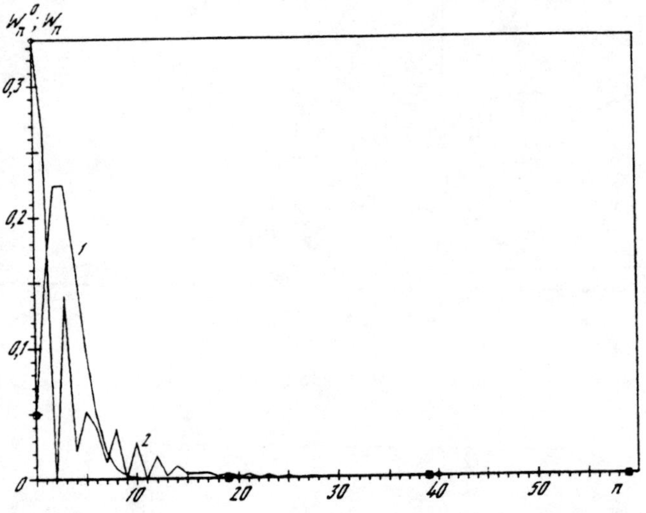

Figure 3. Photons distribution functions W^0_n (1) and W_n (2) for $\beta=\sqrt{3}$, $\bar{n}=3$, r=0, k=1/13.

Figure 4 shows plots of W_n with $\bar{n}=4$, $\beta=2^{1/2}$, and $k=(9\pm6\sqrt{2})/9$. The difference in their behavior is quite clear. For example, in the second picture corresponding to the large squeezing coefficient the probability of recording four quanta is maximal, whereas the first picture shows that for another (small) value of k the same probability is equal to zero. The possibility of oscillations of the photon distribution function for squeezed light was also indicated in [36, 37].

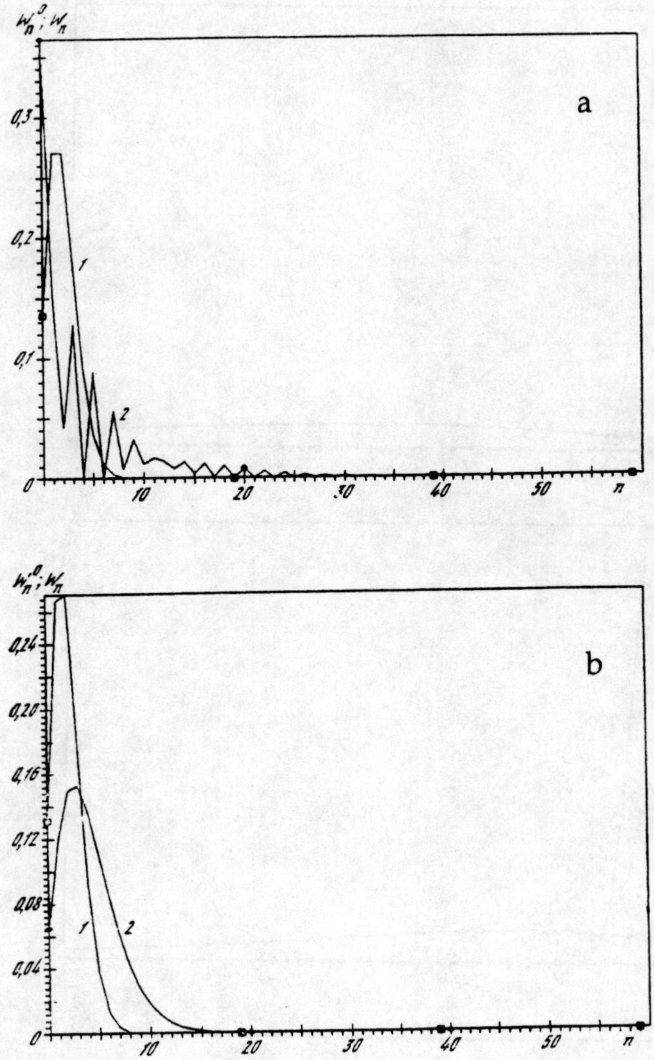

Figure 4. Photons distribution functions $W^0{}_n$ (1) and W_n (2) for $\beta=\sqrt{2}$, $\bar{n}=4$, r=0: a) $k=(9-6\sqrt{2})/9$; b) $k=(9+6\sqrt{2})/9$.

The plots of functions W_n and $W^0{}_n$ for some other parameters are given in Figure 5.

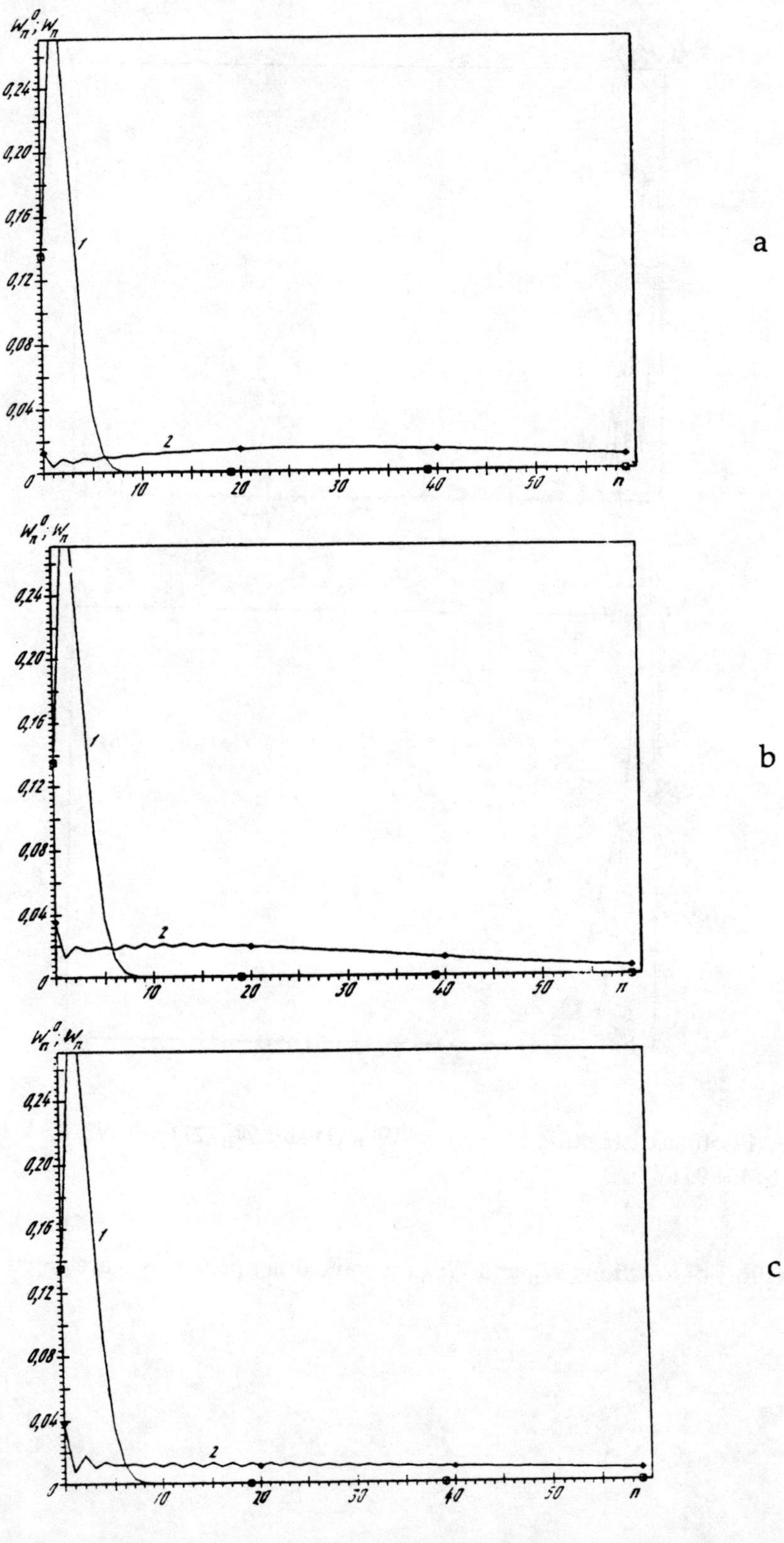

Figure 5. Photons distribution functions W^0_n (1) and W_n (2) for β=1+i, r=0.9: a) k=0.1; b) k=1; c) k=10.

Let us conclude by considering the connection between experiments on photon counting, i.e., measurements of the function W_n, with experiments on intensity correlations. A lot of measurements of the intensity correlation function were carried out [30-32] with "squeezed light," and revealed antibunching of photons at certain values of the squeezing. In the language of correlation functions, photon antibunching means that a dip at $\tau=0$ exists in the second order correlation functions

$$G^{(2)}(\tau)=\lim_{t\to\infty}\langle:I(t+\tau)I(t):\rangle,$$

where $\langle:\ldots:\rangle$ denotes averaging of normally ordered intensity operators. For the relative coherence function

$$g^{(2)}(\tau)=G^{(2)}(\tau)/\langle I\rangle^2$$

photon antibunching means $g^{(2)}(0)<1$. For sufficiently short measurement times $g^{(2)}(0)$ is expressed in terms of the fluctuation of the number of photons [23, 30, 38]

$$\xi = \frac{\overline{n^2} - \bar{n}^2 - \bar{n}}{\bar{n}} = g^{(2)}(0) - 1.$$

Antibunching corresponds to $\xi<0$. In [23,30] the funtion ξ for different squeezing coefficients without allowance for correlation was investigated. From (5.11) we obtain $\xi\bar{n}$ as a function of k, r, and $\alpha=\beta u^*=vb^*$.

$$\bar{n}\xi = |\,\alpha u^* - \alpha^* v\,|^2 - |\,\alpha\,|^2 + |\,v\,|^2\,(|\,u\,|^2 + |\,v\,|^2). \tag{6.2}$$

For real $\alpha>>1/k$ we have with $k\to 0$, $r=0$

$$\bar{n}\xi = (k-1)\,\alpha^2 < 0, \tag{6.3}$$

which corresponds to photon antibunching [23].

If $r\to 1$ and $k=1$, there is no antibunching, i.e., $\bar{n}\xi>0$. For $r\to 1$ and $k\to 0$ antibunching can be observed if $k<<(1-r^2)^{1/2}$. Thus we find that correlation can suppress squeezing, so that if the correlation is strong no antibunching effect can occur in correlated squeezed light even if the squeezing satisfies and is strong enough (6.3).

7. FLUCTUATIONS OF PHASE IN CORRELATED STATES

There exist different definitions of phase operator in quantum mechanics [38]. In this section we investigate the properties of the phase operator defined as follows [38, 39]

$$\exp[i\hat{\varphi}]\,|\,n\rangle = \begin{cases} |\,n-1\rangle, & n\neq 0,\\ 0, & n=0,\end{cases}$$

$$\exp[-i\hat{\varphi}]\,|\,n\rangle = |\,n+1\rangle. \tag{7.1}$$

However operator $\exp(i\hat{\varphi})$ is not hermitian. Therefore the hermitian combinations are introduced [38] to describe observables

$$\cos\hat{\varphi} = {}^1/_2\,(e^{i\hat{\varphi}} + e^{-i\hat{\varphi}}), \qquad \sin\hat{\varphi} = (1/2i)\,(e^{i\hat{\varphi}} - e^{-i\hat{\varphi}}). \tag{7.2}$$

The nonzero matrix elements of these operators are

$$\langle n-1\,|\cos\hat{\varphi}\,|\,n\rangle = \langle n\,|\cos\hat{\varphi}\,|\,n-1\rangle = 1/2,$$
$$\langle n-1\,|\sin\hat{\varphi}\,|\,n\rangle = -\langle n\,|\sin\hat{\varphi}\,|\,n-1\rangle = 1/2i.$$

Let us obtain the expressions for the average values of operators (7.2) and their variances in CCS in terms of the coefficients of the expansion of CCS over the Fock basis. If

$$|\beta uv\rangle = \sum_{n=0}^{\infty} c_n |n\rangle,$$

with $c_n = \langle n|\beta uv\rangle$ given by (4.47), then we get the following expressions for average values

$$\langle\beta uv|\cos\hat{\varphi}|\beta uv\rangle = \mathrm{Re}\sum_{n=0}^{\infty} c_n^* c_{n+1},$$

$$\langle\beta uv|\sin\hat{\varphi}|\beta uv\rangle = \mathrm{Im}\sum_{n=0}^{\infty} c_n^* c_{n+1} \tag{7.3}$$

and for variances

$$\langle\beta uv|\cos^2\hat{\varphi}|\beta uv\rangle = \frac{1}{2} - \frac{1}{4}|c_0|^2 + \frac{1}{2}\mathrm{Re}\sum_{n=0}^{\infty} c_n^* c_{n+2},$$

$$\langle\beta uv|\sin^2\hat{\varphi}|\beta uv\rangle = \frac{1}{2} - \frac{1}{4}|c_0|^2 - \frac{1}{2}\mathrm{Re}\sum_{n=0}^{\infty} c_n^* c_{n+2},$$

$$\langle\beta uv|[\cos\hat{\varphi}, \sin\hat{\varphi}]_+|\beta uv\rangle = \mathrm{Im}\sum_{n=0}^{\infty} c_n^* c_{n+2},$$

$$\langle\beta uv|[\cos\hat{\varphi}, \sin\hat{\varphi}]_-|\beta uv\rangle = -|c_0|^2/2i. \tag{7.4}$$

Note that all average values and variances are determined by the sums

$$\sum_{n=0}^{\infty} c_n^* c_{n+1} = \frac{1}{|u|}\exp\left[-|\beta|^2 + \frac{\beta^2}{2}\frac{v^*}{u} + \frac{\beta^{*2}}{2}\frac{v}{u^*}\right]\left(\frac{v}{2u}\right)^{1/2}\Psi_1,$$

$$\sum_{n=0}^{\infty} c_n^* c_{n+2} = \frac{1}{|u|}\exp\left[-|\beta|^2 + \frac{\beta^2}{2}\frac{v^*}{u} + \frac{\beta^{*2}}{2}\frac{v}{u^*}\right]\left(\frac{v}{2u}\right)\Psi_2. \tag{7.5}$$

with Ψ_1 and Ψ_2 given by

$$\Psi_1 = \sum_{n=0}^{\infty}\left|\frac{v}{2u}\right|^n \frac{1}{n!}\frac{H_n^*(x)\,H_{n+1}(x)}{\sqrt{n+1}},$$

$$\Psi_2 = \sum_{n=0}^{\infty}\left|\frac{v}{2u}\right|^n \frac{1}{n!}\frac{H_n^*(x)\,H_{n+2}(x)}{\sqrt{(n+1)(n+2)}}, \qquad x = \beta/\sqrt{2uv}. \tag{7.6}$$

Let us first investigate sums Ψ_1 and Ψ_2 with $\beta=0$ (i.e., for a correlated vacuum). Since $H_{2n+1}(0)=0$, $H_{2n}(0)=(2n)!(n!)^{-1}(-1)^n$, then $\Psi_1(0)=0$, and, consequently,

$$\langle 0uv|\cos\hat{\varphi}|0uv\rangle = \langle 0uv|\sin\hat{\varphi}|0uv\rangle = 0.$$

The expression for $\Psi_2(0)$ in the limit of strong correlation or squeezing can be written for $|u|>>1$ as follows:

$$\Psi_2(0) = \sum_{n=0}^{\infty}\left|\frac{v}{2u}\right|^{2n}\frac{(-1)}{n!}\frac{(2n+2)!}{(n+1)!}\frac{1}{\sqrt{(2n+1)(2n+2)}} \simeq$$

$$\simeq -2\sum_{n=0}^{\infty}\left|\frac{v}{2u}\right|^{2n}\frac{(2n)!}{(n!)^2}\left[1 - \frac{1}{2(2n+1)} + \ldots\right] \simeq -2|u| + 1 + O\left(\frac{1}{|u|}\right).$$

The summation was performed with the aid of formulas 5.2.13(1) and 5.2.13(10) from [40]. Taking into account (4.18) and neglecting the term $|c_0|^2 = |u|^{-1}$ in (7.4) we get the following expressions for variances in the correlated vacuum:

$$\sigma^2_{\cos\varphi} = <0uv|\cos^2\hat{\varphi}|0uv> = \frac{1}{2}[1-\mathrm{Re}(\frac{v}{u})] = \frac{k^2+k(1-r^2)^{1/2}}{1+k^2+2k(1-r^2)^{1/2}},$$

$$\sigma^2_{\sin\varphi} = <0uv|\sin^2\hat{\varphi}|0uv> = \frac{1}{2}[1+\mathrm{Re}(\frac{v}{u})] = \frac{1+k(1-r^2)^{1/2}}{1+k^2+2k(1-r^2)^{1/2}}. \quad (7.7)$$

However, in the vacuum case the physical sense of operators $\cos\hat{\varphi}$ and $\sin\hat{\varphi}$ is rather vague due to the relation $<0|\cos\hat{\varphi}|0>^2+<0|\sin\hat{\varphi}|0>^2=0$, which obviously contradicts our intuitive ideas on the properties of phase variable. Besides, from (7.4) we see that for any state $<\cos2\hat{\varphi}>+<\sin2\hat{\varphi}>=1-1/2|c_0|^2$. Consequently, one may expect a proper behavior of the operators discussed only for highly excited states, when $|c_0|^2<<1$. Therefore, let us find average values and variances of operators (7.2) for $|\beta|>>1$. First we investigate the sum $\Psi_1(\lambda)$, where $\lambda>>1$ will be defined further:

$$\Psi_1(\lambda) = \sum_{n=0}^{\infty} \frac{t^n}{n!\sqrt{n+1}} H_n^*(x)\, H_{n+1}(x), \qquad t = \left|\frac{v}{2u}\right|.$$

We take into account the integral representation

$$\frac{1}{\sqrt{n+1}} = \frac{1}{\Gamma(1/2)} \int_0^{\infty} \frac{dz}{\sqrt{z}} e^{-z(n+1)}.$$

Then $\Psi_1(\lambda)$ assumes the form

$$\Psi_1(\lambda) = \frac{1}{\Gamma(1/2)} \int_0^{\infty} \frac{dz}{\sqrt{z}} e^{-z} \sum_{n=0}^{\infty} (te^{-z})^n H_n^*(x)\, H_{n+1}(x).$$

The series in the integrand can be calculated exactly [33]. Making the change of variables $z=y^2$ we get

$$\Psi_1(\lambda) = \frac{2}{\sqrt{\pi}} \int_{-\infty}^{+\infty} dy e^{-y^2} \frac{x - 2tx^* e^{-y^2}}{[1-(2te^{-y^2})^2]^{3/2}} \exp\left[\lambda \frac{e^{y^2} - t(x^2+x^{*2})/|x|^2}{e^{2y^2}-(2t)^2}\right]. \quad (7.8)$$

We have defined $\lambda=4t|x|^2=|\beta/u|^2>>1$. Now we can calculate integral (7.8) by the steepest descent method with the accuracy up to $\lambda^{-3/2}$ in order to obtain the expressions for $<\cos\varphi>$ and $<\sin\varphi>$ with the accuracy $|\beta|^{-2}$:

$$\Psi_1(\lambda) = \frac{2}{\sqrt{\lambda}} e^{\lambda S(y_0)} \left(a_0 + \frac{a_1}{\lambda}\right),$$

where functions $f(y)$ and $S(y)$ are given by

$$f(y) = \frac{e^{-y^2}x - e^{-2y^2}2tx^*}{[1-(2te^{-y^2})^2]^{3/2}}, \qquad S(y) = \frac{e^{y^2} - t(x^2+x^{*2})/|x|^2}{e^{2y^2}-(2t)^2}.$$

The extremum point is $y_0=0$. The final expressions for $<\cos\varphi>_\beta$ and $<\sin\varphi>_\beta$ are as follows:

$$<\cos\varphi>_\beta = \Delta^{-1/2}\{\cos(\theta-\delta) - |v/u|\cos(\theta-\gamma)\}$$
$$-[8|u|^4|\beta|^2\Delta^{5/2}]^{-1}\{\cos(\theta-\delta)(3-2|u|^2\Delta) - |v/u|\cos(\theta-\gamma)(3+2|u|^2\Delta)\}, \quad (7.9)$$

$$<\sin\varphi>_\beta = \Delta^{-1/2}\{\sin(\theta-\delta) - |v/u|\sin(\theta-\gamma)\}$$
$$-[8|u|^4|\beta|^2\Delta^{5/2}]^{-1}\{\sin(\theta-\delta)(3-2|u|^2\Delta) + |v/u|\sin(\theta-\gamma)(3+2|u|^2\Delta)\}, \quad (7.10)$$

where

$$\Delta=1+|v/u|^2-2|v/u|\cos(2\theta-\delta-\gamma),$$
$$\theta=\arg(\beta),\ \delta=\mathrm{art}(u),\ \gamma=\arg(v). \tag{7.11}$$

From (7.9) and (7.10) in the limit v=0, u=1 one can easily obtain the average values of operators (7.2) in Glauber's coherent states [38]

$$\langle\cos\varphi\rangle_{KC}\simeq\cos\theta\left(1-\frac{1}{8|\beta|^2}\right),\qquad \langle\sin\varphi\rangle_{KC}\simeq\sin\theta\left(1-\frac{1}{8|\beta|^2}\right).$$

To find the variances of the same operators for $|\beta|>>1$ we shall look for the value of $\Psi_2(\lambda)$ with the accuracy up to $|\beta|^{-2}$ For this purpose we use the expansion

$$\frac{1}{[(n+1)(n+2)]^{1/2}}=\frac{1}{[(n+1)^2+(n+1)]^{1/2}}\simeq\frac{1}{n+1}-\frac{1}{2(n+1)^2}+\ldots$$

Then

$$\Psi_2(\lambda)\simeq\sum_{n=0}^{\infty}H_n^*(x)\,H_{n+2}(x)\frac{t^n}{n!}\left[\frac{1}{n+1}-\frac{1}{2(n+1)^2}\right].$$

Using integral representations for $(n+1)^{-1}$ and $(n+1)^{-2}$ we get

$$\Psi_2(\lambda)=\int_0^{\infty}dz\left[1-\frac{z}{2}\right]e^{-z}\sum_{n=0}^{\infty}\frac{1}{n!}H_n^*(x)\,H_{n+2}(x)\,(te^{-z})^n.$$

Performing the summation (see [33]) we get the integral representation

$$\Psi_2(\lambda)=\int_0^{\infty}dz\frac{e^{-z}[1-z/2]}{[1-(2te^{-z})^2]^{3/2}}\left[4\frac{(x-2tx^*e^{-z})^2}{1-(2te^{-z})^2}-2\right]\times$$
$$\times\exp\left[\lambda\frac{e^{z}-t(x^2+x^{*2})/|x|^2}{e^{2z}-(2t)^2}\right]=\int_0^{\infty}dzf(z)\,e^{\lambda S(z)}.$$

This integral can be calculated again with the steepest descent method. We do not give here, however, the corresponding explicit formulas for $<\cos^2\varphi>$ and $<\sin^2\varphi>$, since they are rather cumbersome. Instead we prefer to introduce a new parameter σ^2_φ characterizing the uncertainty of phase.

Let us suppose that φ is the classical random variable with average value φ_0 and fluctuating part $\delta\varphi$, so that $\varphi=\varphi_0+\delta\varphi$. Developing $\cos\varphi$ and $\sin\varphi$ into Taylor's series up to the second order terms and performing averaging, one gets

$$\overline{\delta\varphi^2}=\overline{\cos^2\varphi}-[\overline{\cos\varphi}]^2+\overline{\cos^2\varphi}-[\overline{\cos\varphi}]^2,$$

where bar means the classical averaging. Replacing classical averaging with the quantum one we can define the quantum phase variance as follows,

$$\sigma^2_\varphi=\sigma^2_{\cos\varphi}+\sigma^2_{\sin\varphi}=1-<\cos\hat\varphi>^2-<\sin\hat\varphi>^2-\frac{1}{2}|c_0|^2. \tag{7.12}$$

Neglecting the exponentially small contribution of the term $|c_0|^2$ which is proportional to $\exp(-|\beta|^2)$ we obtain from (7.9)-(7.12) (provided $|\beta|>>1$)

$$\sigma^2_\varphi=\{4|u|^4|\beta|^2\Delta^2\}^{-1} \tag{7.13}$$

In the case of v=0 (for Glauber's states) we have $\sigma^2_\varphi=1/4|\beta|^2$. For strongly squeezed and correlated states, when $|u|>>1$, (7.13) reads

$$\sigma_\varphi^2 = 4|\beta|^{-2}|u|^4\{8|u|^2(2|u|^2-1)\sin^2(\theta-\frac{1}{2}(\delta+\gamma))+\cos(2\theta-\delta-\gamma)\}^{-2}$$

Depending on the relations between parameters θ, δ, γ this phase variance varies from $16|u|^4$ to $(16|u|^4)^{-1}$ of its value in the Glauber state with the same $|\beta|^2$. The comparison of the phase fluctuations with the photon number fluctuations is given in the next section.

8. CCS EVOLUTION IN A STATIONARY QUADRATIC POTENTIAL

In this section we analyze in detail the special case of time-independent quadratic Hamiltonians (in one dimension). The evolution of CCS in the x-representation is determined by the overlap integral

$$\langle x|\beta uvt\rangle = \langle x|U(t)|\beta uv\rangle = \sum_n \langle x|U(t)|n\rangle\langle n|\beta uv\rangle. \quad (8.1)$$

The time dependence of the n-quantum state is well known:

$$\langle x|U(t)|n\rangle = \Psi_n(x,t) = \frac{1}{\pi^{1/4}}\frac{e^{-it(n+1/2)}}{\sqrt{2^n n!}}e^{-x^2/2}H_n(x). \quad (8.2)$$

The overlap integral $\langle n|\beta uv\rangle$ is given by (4.47). (We assume that the oscillator has unity mass and frequency.) From (8.1) and (8.2) we get

$$\langle x|\beta uvt\rangle = \frac{1}{\pi^{1/4}}\exp\left[-\frac{it}{2}-\frac{x^2}{2}-\frac{|\beta|^2}{2}+\frac{\beta^2}{2}\frac{v^*}{u}\right]\left[\frac{1}{u}\left(\frac{u-v}{u^*-v^*}\right)^{1/2}\right]^{1/2}\times$$
$$\times\sum_{n=0}^{\infty}\frac{e^{-itn}}{2^n n!}\left(\frac{v}{u}\right)^{n/2}H_n(x)H_n\left(\frac{\beta}{\sqrt{2uv}}\right).$$

Performing the summation with the aid of known formulas for the Hermite polynomials (see, e.g., [33] we obtain the final result

$$\langle x|\beta uvt\rangle = \frac{1}{\pi^{1/4}}\left[\frac{1}{u}\left(\frac{u-v}{u^*-v^*}\right)^{1/2}\right]^{1/2}\exp\left[-\frac{it}{2}-\frac{x^2}{2}-\frac{|\beta|^2}{2}+\frac{\beta^2}{2}\frac{v^*}{u}\right]\times$$
$$\times\left(1-\frac{v}{u}e^{-2it}\right)^{-1/2}\exp\left[\left(\frac{\sqrt{2}\,x\beta e^{it}}{u}-\left(\frac{v}{u}x^2+\frac{\beta^2}{2u^2}\right)e^{-2it}\right)\left(1-e^{-2it}\frac{v}{u}\right)^{-1}\right]. \quad (8.3)$$

With $v=0$, $|u|=1$ we have the standard expression [11]

$$\langle x|\beta\rangle = \pi^{-1/4}\exp\left[-\frac{it}{2}-\frac{x^2}{2}-\frac{|\beta|^2}{2}+\sqrt{2}\,x\beta e^{-it}-\frac{\beta^2}{2}e^{-2it}\right]. \quad (8.4)$$

As was shown in [17], the evolution of CCS as an arbitrary quadratic Hamiltonian can be represented by a simple formula

$$|\beta uv\rangle \xrightarrow{U(t)} |\beta(t)\,u(t)\,v(t)\rangle.$$

To find the explicit dependences of parameters β, u, v on time one should compare (8.3) with the same expression at the initial instant

$$\Psi_{\beta uv}(x) = \langle x|\beta uvt=0\rangle = \frac{1}{\pi^{1/4}}\left[\frac{u-v}{u^*-v^*}\right]^{1/4}\frac{1}{\sqrt{u-v}}\times$$
$$\times\exp\left[-\frac{u+v}{u-v}\frac{x^2}{2}+\frac{\sqrt{2}\,\beta x}{u-v}-\frac{\beta^2}{2}\frac{u^*-v^*}{u-v}-\frac{|\beta|^2}{2}\right]. \quad (8.5)$$

Then one may write

$$\langle x|\beta uvt\rangle = e^{i\varphi}\langle x|\beta, u(t), v(t)\rangle, \quad (8.6a)$$

where

$$u(t) = ue^{it} \qquad v(t) = ve^{-it} \qquad e^{i\varphi} = \left(\frac{u-v}{\ldots}\right)^{1/2} = \mathrm{const}\,(t).$$

In this notation

$$\langle x\,|\,\beta uvt\rangle = \frac{e^{i\varphi}}{\pi^{1/4}}\frac{1}{\sqrt{u(t)-v(t)}}\exp\left[-\frac{x^2}{2}-\frac{|\beta|^2}{2}+\frac{\beta^2}{2}\frac{v^*(t)}{u(t)}\right]\times$$
$$\times\exp\left[\left(\sqrt{2}\,x\beta - v(t)\,x^2-\frac{\beta^2}{2u(t)}\right)(u(t)-v(t))^{-1}\right]. \qquad (8.6b)$$

Analogously, using the time-dependent wave function of the n-quantum state in the momentum representation

$$\Psi_n(p,t) = \frac{1}{\pi^{1/4}}\frac{e^{-p^2/2}}{\sqrt{2^n n!}}\exp\left[-i\left(t+\frac{\pi}{2}\right)\left(n+\frac{1}{2}\right)\right]H_n(p)$$

one can obtain the expression for CCS in the p-representation

$$\langle p\,|\,\beta uvt\rangle = \frac{e^{i\varphi}}{\pi^{1/4}}\exp\left[-\frac{p^2}{2}-\frac{|\beta|^2}{2}+\frac{\beta^2}{2}\frac{v^*(t)}{u(t)}\right]\times$$
$$\times\exp\left[\left(-i\sqrt{2}\,p\beta + v(t)\,p^2+\frac{\beta^2}{2u(t)}\right)(u(t)+v(t))^{-1}\right]\frac{e^{-i\pi/4}}{\sqrt{u(t)+v(t)}}.$$

In the energy representation one has

$$\langle n\,|\,\beta uvt\rangle = \langle n\,|\,U(t)\,|\,\beta uv\rangle = \int d^2\alpha\,\langle n\,|\,U(t)\,|\,\alpha\rangle\langle\alpha\,|\,\beta uv\rangle. \qquad (8.7)$$

Then the formula for usual CS in the Fock basis [5.11]

$$\langle n\,|\,U(t)\,|\,\alpha\rangle = e^{-it/2-int}e^{-|\alpha|^2/2}\alpha^n\,(n!)^{-1/2}. \qquad (8.8)$$

leads to the relation

$$\langle n\,|\,\beta uvt\rangle = e^{-it/2-int}\langle n\,|\,\beta uv\rangle = \langle n\ |\beta u(t)\,v(t)\rangle. \qquad (8.9)$$

In the Wigner picture a time-dependent CCS can be expressed as follows [17],

$$W_{uv\beta} = \frac{1}{2\pi}\int_{-\infty}^{+\infty} d\xi\left\langle q+\frac{\xi}{2}\,|\,uv\beta t\right\rangle\left\langle uv\beta t\,|\,q-\frac{\xi}{2}\right\rangle e^{-ip\xi} =$$
$$= \frac{1}{2\pi\Delta^{1/2}}\exp\left[-\frac{\sigma_p^2(t)}{2\Delta}\tilde{q}^2(t)-\frac{\sigma_q^2(t)}{2\Delta}\tilde{p}^2(t)+\frac{\sigma_{qp}(t)}{\Delta}\tilde{q}(t)\,\tilde{p}(t)\right], \qquad (8.10)$$

where

$$\Delta = \sigma_q^2\sigma_p^2\,(1-r^2) = {}^1/_4,\quad \tilde{q}(t) = q(t)-\langle q(t)\rangle,\quad \tilde{p}(t) = p(t)-\langle p(t)\rangle,$$

all variances and averages being given by Equations (4.10), (4.11) with functions u(t), v(t) defined above.

The Gaussian exponential in the right-hand side of (8.10) is in fact the most genera two-dimensional normal distribution determined by five independent parameters three variances and two averages. The restriction $\Delta=1/4$ singles out the subclass o Gaussian exponentials corresponding to *pure quantum* states [1]. Every norma distribution can be clearly represented in the phase plane (p,q) by means of the *constan probability ellipses* W(p,q)=const. defined by the equation [17]

$$(\sigma_p^2\tilde{q}^2+\sigma_q^2\tilde{p}^2-2\sigma_{qp}\tilde{q}\tilde{p})\,\Delta^{-1} = \lambda^2 = \text{const}. \qquad (8.11)$$

The center of the ellipse moves along the trajectory determined by the relations

$$\langle q(t)\rangle = \sqrt{2}\,\mathrm{Re}\,[e^{-it}(u^*\beta - v\beta^*)],\quad \langle p(t)\rangle = \sqrt{2}\,\mathrm{Im}\,[e^{-it}(u^*\beta - v\beta^*)]. \qquad (8.12)$$

The characteristic equation of ellipse (8.11) and its solutions are as follows,

$$\Lambda^2 - \Lambda(\sigma_q^2 + \sigma_p^2) + \sigma_q^2\sigma_p^2 - \sigma_{qp} = 0$$

$$\Lambda_{1,2} = {}^1/_2\{\sigma_q^2 + \sigma_p^2 \pm [(\sigma_q^2 + \sigma_p^2)^2 - 4(\sigma_q^2\sigma_p^2 - \sigma_{qp}^2)]^{1/2}\}.$$

In the case of the CCS evolution in a stationary quadratic potential two time-independent invariants exist:

$$I_1 = \Lambda_1 + \Lambda_2 = \sigma_q^2 + \sigma_p^2 = 1 + 2|v|^2, \qquad I_2 = \sigma_q^2\sigma_p^2 - \sigma_{qp}^2 = \Lambda_1\Lambda_2 = {}^1/_4.$$

The first invariant I_1 is in fact the energy of fluctuations. It is conserved only for time-independent potentials. The second invariant I_2 has a much more general nature, since its value does not depend on time for quite arbitrary nonstationary quadratic Hamiltonians [41] (it is the simplest example of the so called *universal quantum invariants*). The roots of the characteristic equation are expressed in terms of the invariants as follows,

$$\Lambda_{1,2} = {}^1/_2(I_1 \pm \sqrt{I_1^2 - 1}) = \text{const}\,(t).$$

The principal semiaxes are given by the formulas

$$a^2 = \lambda\Delta/\Lambda_2, \qquad b^2 = \lambda\Delta/\Lambda_1. \tag{8.13}$$

In terms of the correlation and squeezing coefficients

$$I_1 = 1 + \frac{1 + k^2 + 2k\sqrt{1 - r^2}}{2k\sqrt{1 - r^2}}.$$

The angle φ between the principal axis and q-axis satisfies the equation

$$\operatorname{tg} 2\varphi = \frac{2\sigma_{qp}}{\sigma_q^2 - \sigma_p^2} = -\frac{\operatorname{Im}(uv^* e^{2it})}{\operatorname{Re}(uv^* e^{2it})}. \tag{8.14a}$$

In terms of correlation and squeezing:

$$\operatorname{tg} 2\varphi = kr/(k^2 - 1). \tag{8.14b}$$

Thus the correlation coefficient is proportional to tg(2φ), if $k \neq 1$. If one circumscribes a rectangle around the ellipse with sides x_1 and y_1 parallel to p- and q-axes: $x_1 = 2\lambda\sigma_q$, $y_1 = 2\lambda\sigma_p$, then the ratio

$$x_1/y_1 = \sqrt{\sigma_q^2/\sigma_p^2} = k \tag{8.15}$$

is nothing but the squeezing coefficient

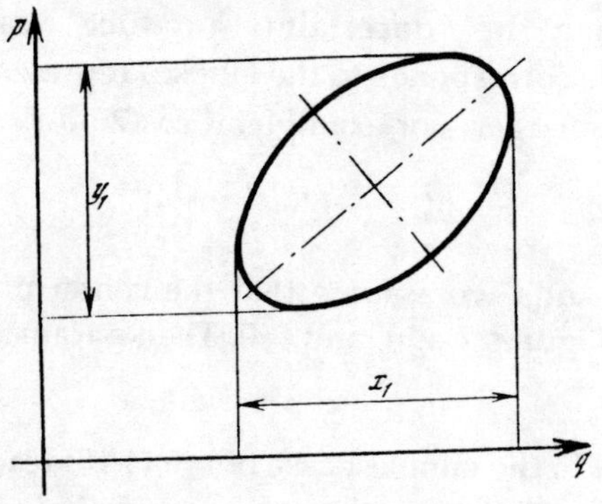

Figure 6. The constant probability ellipse W(p,q)=*const* in the phase plane of (p,q)-variables; $k = x_1/y_1$.

We see that the constant probability ellipse preserves its shape during the evolution in a stationary potential, but its principal axes rotate in accordance with (8.14a).

Now let us analyze the time dependences of the correlation and squeezing coefficients:

$$r(t) = \frac{\mathrm{Im}\,(uv^*e^{2it})}{[(1/2 + |v|^2)^2 - (\mathrm{Re}\,(uv^*e^{2it}))^2]^{1/2}}, \tag{8.16a}$$

$$k(t) = \left[\frac{1/2 + |v|^2 - \mathrm{Re}\,(uv^*e^{2it})}{1/2 + |v|^2 + \mathrm{Re}\,(uv^*e^{2it})}\right]^{1/2}. \tag{8.16b}$$

Let r(t=0)=0 and k(0)=0. Then

$$r(t) = \frac{(1-k^2)\sin 2t}{[(1+k^2) - (1-k^2)\cos^2 2t]^{1/2}},$$

$$r_{\max} = r(t = \pi/4 + \pi n/2) = (1-k^2)(-1)^n/(1+k^2). \tag{8.17}$$

Under the same initial conditions

$$k(t) = \left[\frac{(1+k^2) - (1-k^2)\cos 2t}{(1+k^2) + (1-k^2)\cos 2t}\right]^{1/2}$$

so that the squeezing coefficient periodically changes with period π:if $k(0)=k_0$, then $k(\pi n+\pi/2)=1/k_0$. Thus we see that a pure squeezed state becomes correlated during the time evolution.

Now let us consider the relations between the shape of the constant probability ellipse, photons number fluctuations, and phase fluctuations in CCS. The phase variable is well defined provided $|\beta|>>1$. Since asymptotic formulas (5.12) for moments of the photons distribution function were derived in the opposite limit case $k|\beta|<<1$, they cannot be used now. Therefore it is convenient to rewrite exact formulas (5.11) as follows:

$$\bar{n} = |v|^2 + |u|^2 |\beta|^2 \Delta,$$

$$\sigma_n = 2|uv|^2 + |\beta|^2 \{2|u|^2 (2|u|^2 - 1)\Delta - 1\},$$

parameter Δ being defined by (7.11). Taking into account (7.13) we get the following relation for the "phase-number of quanta uncertainty product" in the limit of $|\beta|>>1$:

$$\sigma_n \sigma_\varphi^2 = \frac{2|u|^2 (2|u|^2 - 1)\Delta - 1}{4|u|^4 \Delta^2}$$

If $\cos(2\theta-\delta-\gamma)=\pm 1$, then the "uncertainty product" reaches its minimal value $(\sigma_n\sigma^2_\varphi)\min=1/4$. This case corresponds to the ellipses represented in Figure 7. For other orientations the product is larger. For example, if $\cos(2\theta-\delta-\gamma)=0$, then

$$\sigma_n \sigma_\varphi^2 = \frac{1}{2} - [2(|u|^2 + |v|^2)]^{-2}$$

To illustrate these results we assume that the center of the constant probability ellipse lies in x-axis (see Figures 7 a,b), and t=0. Then parameter

$$\alpha = u^*\beta - v\beta^*.$$

must be real due to (8.12). In the simplest case of r=0 (4.18) reads

$$u = 1/2\,(1/\sqrt{k} + \sqrt{k}), \qquad v = 1/2\,(1/\sqrt{k} - \sqrt{k}),$$

so that

$$\alpha = i \operatorname{Im} \beta/\sqrt{k} + \sqrt{k} \operatorname{Re} \beta,$$

and β must be real. Figure 7a corresponds to the case k>1. Then δ=0, γ=π (see previous section), and for k>>1 the phase variance σ^2_φ assumes its minimal value $(2\beta k)^{-2}$ which coincides with $(\sigma_p/\langle x\rangle)^2$. In terms of the average number of quanta $\bar{n}=k\beta^2$ we have $\sigma^2_\varphi=(4\bar{n}k)^{-1}$, $\sigma_n=k/\bar{n}>\bar{n}$: this is the case of the super-Poissonian statistics of quanta [30].

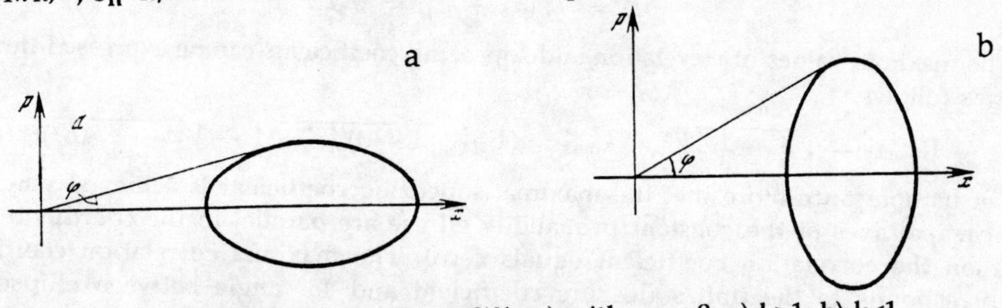

Figure 7. The constant probability ellipses W(p,x) with <p>=0: a) k>1; b) k<1.

For the ellipse of the same shape but turned to 90 degrees (Figure 7b) we have δ=γ=0, and the same formulas correspond to the maximal phase variance and minimal fluctuations of the photon number, since in this case k<1. The statistics is sub-Poissonian.

The explicit expressions for the moments of the quantum distribution function for unsqueezed states (r=0) are as follows:

$$\bar{n} = (4k)^{-1}\{(1-k)^2 + 4(\operatorname{Im}\beta)^2 + 4k^2(\operatorname{Re}\beta)^2\},$$
$$\sigma_n = (8k^2)^{-1}\{(1-k^2)^2 + 8(\operatorname{Im}\beta)^2 + 8k^4(\operatorname{Re}\beta)^2\}.$$

Fluctuations of the photon distribution function in squeezed light were investigated also in Reference [36] on the basis of the analysis of the overlap integrals between the Wigner functions describing squeezed and Fock's states.

We would like to draw attention to an important difference between oscillators with time-independent frequencies and nonstationary oscillators discussed in Section 4. It was shown that any linear canonical transformation of quantum creation and annihilation operators is determined by two nontrivial parameters, for example, by squeezing and correlation coefficients. In a general case these coefficients are independent, because, as was demonstrated, there exist unsqueezed states with nonzero correlation coefficients for all instants of time. But in the stationary case, correlation and squeezing coefficients periodically change in time according to (8.16) and can be expressed through each other. Their maximal values are connected with a simple formula resulting from (8.16) and (8.17):

$$k^2_{\max} = (1 + r_{\max})/(1 - r_{\max}), \qquad r = (k^2_{\max} - 1)/(k^2_{\max} + 1). \tag{8.18}$$

Thus in a stationary case we have a single nontrivial parameter: either k_{max} or r_{max}. In other words, in a specific case of a stationary Hamiltonian one may believe that correlated and squeezed states are identical. This is ultimately related to the additional symmetry of stationary Hamiltonians with respect to translations in time. As a consequence of this symmetry, an arbitrary phase in the definition of creation and annihilation operators (see (4.7)) can be eliminated by means of a shift of the initial time instant due to the simple temporal dependence of the annihilation operator in the

Heisenberg picture: $\hat{a}(t)=\hat{a}(0)\exp(-i\omega t)$. Thus one of two independent parameters can be made trivial. On the contrary, in the nonstationary case the initial time instant is fixed, and both transformation parameters appear truly independent.

Since correlation and squeezing coefficients continuously change in the stationary case, the most convenient parameter characterizing the quantum state is the conserved energy of fluctuations [11]

$$E = {}^1/_2\,(\sigma_p^2 + \sigma_q^2) = {}^1/_2 I_1.$$

The maximal values of correlation and squeezing coefficients can be expressed through it as follows,

$$r_{\max} = \sqrt{1 - 1/4E^2} \qquad k_{\max}^2 = (1 + \sqrt{1 - 1/4E^2})/(1 - \sqrt{1 - 1/4E^2}). \quad (8.19)$$

Let us note once more that the maximal squeezing coefficient is achieved when the principal axes of the constant probability ellipse are parallel to the coordinate axes (then the correlation coefficient equals zero). The maximal correlation coefficient corresponds to the unit squeezing coefficient and 45° angle between ellipse and coordinate axes. As to the examples given in Section 4, in these cases constant probability ellipses changed both their shapes and orientation with respect to the coordinate axes in accordance with Equations (8.13), (8.14). Nonetheless the area of the ellipse is not changed in time for quite arbitrary time-dependent quadratic Hamiltonians due to the quantum generalization to the Liouville theorem [41].

9. QUANTUM EFFECTS IN CORRELATED STATES

One of the main peculiarities of correlated states is the increased level of quantum fluctuations determined by three parameters β, u, v. Moreover, the representation of the generalized Schrodinger-Robertson uncertainty relation (4.1) in the form

$$\sigma_p \sigma_q \geq \frac{1}{2}\hbar(1-r^2)^{-1/2}$$

naturally leads to the conjecture [2] that the correlation coefficient may exercise a significant influence on the behavior of a system in the cases when quantum mechanical effects are determinative. In this section we analyze the possibility of such an effect in two examples: spreading of free correlated wavepacket, and tunnelling of a correlated packet through a potential barrier.

9.1 Spreading of a Free Packet

Consider the time evolution of a correlated packet in a free space. Suppose the initial packet to be Gaussian:

$$\Psi(x, t = 0) = \frac{1}{(2\pi\sigma^2)^{1/4}} \exp\left[-\frac{x^2(1 - i\nu)}{4\sigma^2} + ikx\right], \quad (9.1)$$

where $v=r(1-r^2)^{-1/2}$. Then one has for the subsequent instants of time

$$\Psi(x, t) = \int dx' G(x, x', t)\,\psi(x', 0), \quad (9.2)$$

where

$$G(x, x', t) = \frac{1}{\sqrt{2\pi i t}} \exp\left[i\,\frac{x^2 + x'^2 - 2xx'}{2t}\right]$$

is the free particle Green function. Using (9.1) and (9.2) one can easily calculate the probability density $\rho(x,t)=|\psi(x,t)|^2$:

$$\rho(x,t)=\left[\frac{2\sigma^2}{\pi[t^2+(vt+2\sigma^2)^2]}\right]^{1/2}\exp\left\{-\frac{x^2}{[(1/4\sigma^2)^2+(v/4\sigma^2+1/2t)^2]}\frac{1}{8t^2\sigma^2}+\right.$$
$$\left.+\frac{2kx-k^2t}{(1/4\sigma^2)+(v/4\sigma^2+1/2t)^2}\frac{1}{8t\sigma^2}\right\}.$$

Restoring Planck's constant and designating particle's velocity V=$\hbar$k (we suppose the unit mass) we get for large time $t\gg 2\sigma^2/v\hbar$

$$\rho(x,t)=\left(\frac{2\sigma^2}{t^2\hbar_*^2}\right)^{1/2}\exp\left[-\frac{2\sigma^2(x-Vt)^2}{\hbar_*^2t^2}\right]. \tag{9.3}$$

We recognize in this expression the standard formula describing spreading of the usual (uncorrelated) Gaussian wave packet in a free space, provided Planck's constant $\hbar$ is replaced by the "effective Planck's constant" $\hbar_*=\hbar(1-r^2)^{-1/2}$. This result is in agreement with the intuitive idea concerning the influence of correlation on quantum effects following from the uncertainty relation

$$\sigma_x\sigma_p\geqslant\hbar_*/2.$$

9.2 Transmission of Packet Through a Potential Barrier

We shall consider a correlated packet localized initially in a potential well of the form

$$V(x)={}^1/_2\omega^2(x^2-\delta x^3). \tag{9.4}$$

The state of the packet inside the well is quasistationary. Such kind of states were discussed in detail in [42]. However, here we shall apply another method, which is based essentially on the concrete form of potential (9.4). The underbarrier tunnelling at finite temperature, i.e., with Boltzmann's energy levels populations in the potential well, was considered in studies [43,44]. It was shown that the inverse tunnelling time can be calculated according to the formula

$$\tau^{-1}=\sum_{n=0}^{N}\tau_n^{-1}\rho_n, \tag{9.5}$$

$\tau^{-1}{}_n$ being the inverse tunnelling time from the n-th energy level, ρ_n-the population of this level. The value of N, i.e., the number of energy levels giving significant contribution to the tunnelling rate, is determined, as a rule, from physical considerations.

Specifically, it was shown in [45] that with Boltzmann's energy levels population and for low temperatures, when tunnelling prevails over the thermal activation, one may take into account only tunnelling from the ground state due to the strong exponential decrease of levels populations with the increase of the energy of the levels. This is not true, however, for coherent states, since the energy levels population differs from Boltzmann's one,

$$\rho_n(\alpha)=|\alpha|^{2n}(n!)^{-1}e^{-|\alpha|^2}, \tag{9.6}$$

and this fact will be taken into further account.

Let us analyze in more detail potential (9.4). For $\delta\ll(\omega/\hbar)^{1/2}$ and low energies of the packet, one may believe that the packet is in a harmonic oscillator potential with the corresponding levels populations. Tunnelling will be taken into account in the frame of the quasiclassical approach. Now it is time to use the specific features of potential

(9.4). A simple consequence of Equation (8.9) is the time independence of the energy levels population for CCS in the quadratic potential

$$\rho_n(\beta uv) = |\langle n \mid \beta uvt\rangle|^2 = |\langle n \mid \beta uv\rangle|^2. \tag{9.7}$$

If the height of the potential barrier is V_0, then the anharmonicity parameter in (9.4) reads

$$\delta = {}^1/_3\omega\sqrt{2/3V_0}.$$

The inverse tunnelling time from the n-th level in the quasiclassical approximation is given by

$$\tau^{-1}{}_n=(\omega/2\pi)\exp\,[\text{expression in square brackets from (9.8)}] \tag{9.8}$$

with $E_n=\hbar\omega(n+1/2)$. The exponential factor (the permeability of the barrier) can be written in the following compact form:

$$D=\exp[-27\frac{V_0}{\hbar\omega}F\left(\frac{4}{27}\frac{E_n}{V_0}\right)], \tag{9.9}$$

where

$$F(t)=\int_\alpha^b dx[x^2-x^3-t], \tag{9.10}$$

and $\alpha<b$ are positive solutions of the equation

$$x^2-x^3-t=0 \tag{9.11}$$

(turning points of the classical motion). Designating the third (negative) solution of the cubic equation by c and introducing the paremter ξ according to relation $(\alpha\text{-}c)=\xi^2(b\text{-}c)$ one can get the following exact formula for F(t) in terms of the elliptic integrals:

$$F(t)=\frac{2}{15}(b-c)^{5/2}\{2(1-\xi^2+\xi^4)E([1-\xi^2]^{1/2})-\xi^2(1+\xi^2)K([1-\xi^2]^{1/2})\}$$

We assume the energy of the packet to be much less than the height of the potential barrier. Then $t<<1$, and the following approximate relations hold:

$$\alpha=-c=t^{1/2},b=1-t,\xi^2=2t^{1/2},$$

$$F(t)=\frac{2}{15}[2+\frac{9}{4}\bullet t\ln t+0(t)] \tag{9.12}$$

Thus we arrive at the expression

$$\tau_n^{-1}=(\omega/2\pi)\exp[-\frac{36}{5}Q+\frac{3}{5}(2n+1)\ln(\frac{2Q}{2n+1})], \tag{9.13}$$

where the dimensionless parameter

$$Q=V_0/(\hbar\omega)>>1$$

is introduced. Since logarith is rather slowly varying function, we replace it by some constant parameter

$$\lambda=\frac{3}{5}\ln\left(\frac{2Q}{2n+1}\right), \tag{9.14}$$

where n corresponds approximately to the mean energy of the packet (further we shall find the self-consistent value of this parameter). Let us first evaluate the tunnelling time for the uncorrelated (Glauber's) coherent state. Substituting (9.6) and (9.13) into (9.5) we see that the series can be calculated exactly, so that choosing N=∞ we get

$$\tau_\alpha^{-1} = \tau_0^{-1} \sum_{n=0}^{\infty} \frac{|\alpha|^{2n}}{n!} \exp(2\lambda n - |\alpha|^2) = \tau_0^{-1} \exp[|\alpha|^2 (e^{2\lambda} - 1)], \tag{9.15}$$

where $\tau^{-1}{}_0$ is given by formula (9.13) with n=0. The main contribution to sum (9.15) is due to terms with numbers n belonging to the interval $|n-\bar{n}| \le n^{-1/2}$ (since we have in fact a sum over Poisson's distribution) with $\bar{n} = |\alpha|^2 e^{2\lambda}$. Our approach is self-consistent provided the variation of the parameter λ in this interval can be neglected, and the maximal significant energy level number is much less than the whole number of levels inside the potential well Q. Therefore, we must demand the fulfillment of the inequalities $Q \gg \bar{n} \gg 1$. Substituting into (9.14) the value $n=\bar{n}$ we can express λ through α and Q:

$$\lambda = (3/11)\ln(Q/|\alpha|^2) \tag{9.16}$$

Then the self-consistency conditions are

$$Q^{-6/5} \ll |\alpha|^2 \ll Q \tag{9.17}$$

The final expression for the inverse tunnelling time is as follows,

$$\tau_\alpha^{-1} = \tau_0^{-1} \exp\{|\alpha|^{10/11} \cdot Q^{6/11}\} \gg \tau_0^{-1} \tag{9.18}$$

The tunnelling time monotonously decreases with increasing mean energy of the coherent wave packet.

Now let us consider the tunnelling of the correlated packet with ρ_n given by (5.1). We confine ourselves to the case of small squeezing and correlation, when

$$z = |ve^{2\lambda}/u| < 1, \tag{9.19}$$

(otherwise the replacement of the upper limit of summation in (9.5) by infinity is impossible, and a delicate account of the behavior of energy levels in the vicinity of the top of the barrier is necessary). We have to calculate the sum

$$\tau_{\beta uv}^{-1} = \tau_0^{-1} |u|^{-1} \exp\{-|\beta|^2 + \mathrm{Re}(\beta^2 v^*/u)\} \sum_{n=0}^{\infty} \frac{2^n}{n! 2^n} |H_n(x)|^2, \tag{9.20}$$

where $x = \beta(2uv)^{-1/2}$. It is calculated exactly due to known formulas for the Hermite polynomials [33]. Thus we obtain

$$\tau_{\beta uv}^{-1} = \tau_0^{-1} |u|^{-1} (1-z^2)^{-1/2} \exp\{-|\beta|^2 + \mathrm{Re}(\beta^2 v^*/u) + |\beta|^2 \frac{e^{2\lambda}}{|u|^2} \frac{1 - z \cdot \cos(2\theta - \delta - \gamma)}{1 - z^2}\}. \tag{9.21}$$

The sense of parameters θ, δ, γ is the same as in Sections 7 and 8. To determine the value of λ, consider the case of relatively small squeezing or correlation and relatively large mean energy of the packet. Then $|\beta|$ is large, $|v|$ is small, so that $|x| \gg 1$, and $H_n(x) \approx (2x)^n$. Therefore the behavior of terms entering the series (9.20) is qualitatively the same as that in (9.15), if one replaces $|\alpha|^2$ by $|\beta|^2/|u|^2$. Consequently, we may choose

$$\lambda = (3/11)\ln(Q|u|^2/|\beta|^2), \quad z = |v \cdot u^{1/11} \cdot Q^{6/11} \cdot \beta^{-12/11}|, \tag{9.22}$$

provided the following restrictions are fulfilled:

$$Q^{-6/5} << |\beta|^2 / |u|^2 << Q \tag{9.23}$$

To compare (9.18) and (9.21), i.e., to analyze the influence of correlation and squeezing on the tunnelling time, we have to establish the relation between parameters α and β.

Suppose first (see Equation (4.7)) $\beta = u\alpha + v\alpha^*$ with $|v| << 1$. Keeping only the first-order terms with respect to $|\beta|$ in (9.21) and taking into account that $\varepsilon^{2\lambda} >> 1$ we get

$$\tau^{-1}_{\beta uv} / \tau^{-1}_{\alpha} = \exp[-|v|(Q^6/|\alpha|)^{2/11} \cos(2\theta - \delta - \gamma)] \tag{9.24}$$

We see that correlation and squeezing may both increase and decrease the tunnelling rate depending on phase relations between coefficients β, u, v. Since $Q^6/|\alpha| >> 1$, the effect may be significant. To elucidate the situation let us consider two special cases represented in Figure 7.

We have r=0, u≈1, v≈-k/2, where k=k-1, |k|<<1. Then θ=δ=0, γ=0 for k<0 and γ=π for k>0. Thus we obtain from (9.24)

$$\tau^{-1}_{\beta uv} / \tau^{-1}_{\alpha} = \exp[\frac{k}{2}(Q^6/|\alpha|)^{2/11}] \tag{9.25}$$

This result has a clear physical ground. The tunnelling rate is increased for k>0 (Figure 7a), when energy fluctuations are large, and it is decreased for k<0, (Figure 7b), when energy fluctuations are small.

Now let us suppose that the mean energies of Glauber's and squeezed packets coincide. For the sake of simplicity we assume all coefficients real, moreover, only coefficient v may be negative (this means that we consider again the situations represented in Figure 7). Due to (5.11) we have the relation

$$\alpha^2 = v^2 + \beta^2(u-v)^2 \approx \beta^2(u-v)^2 \tag{9.26}$$

Due to condition (9.19) we are to consider only the case of $|v| << 1$, since $e^{2\lambda} >> 1$. Then we again arrive at formula (9.25).

10. CORRELATED STATES, ADIABATIC INVARIANTS AND BERRY'S PHASE

In 1984 Berry [46] considered the problem of the evolution of a Hamiltonian eigenstate corresponding to the discrete art of the spectrum in the case when parameters of the Hamiltonian (combined into a single vector $R=(R_1, R_2, \ldots, R_N)$) vary slowly in time. Let us designate the Hamiltonian eigenfunctions as $|\Phi_n[R]>$. If at some time instant t_0 the wave function Ψ(t) coincided with $|\Phi_n[R(t_0)]>$, then at any subsequent time instant Ψ(t) will coincide with $|\Phi_n[R(t_0)]>$ up to a phase vector, provided vector R(t) changes adiabatically. This statement, known as the Born-Fock theorem, was proved as far back as the first years of the existence of quantum mechanics. Its physical significance is that quantum numbers are adiabatic invariants. Almost 60 years ago people thought that the above mentioned phase factor had quite dynamical origin, so that it was believed to be equal to

$$\exp\left[-\frac{i}{\hbar}\int_{t_0}^{t} d\tau E_n[R(\tau)]\right],$$

$E_n[R]$ being the corresponding energy level.

It was Berry's discovery that besides such a trivial factor there also exists a nontrivial one which is quite independent on the rate of parameters variation but is determined only by the state $|\Phi_n(R)]>$ and the contour C which vector R moves along in the parameter space. The correct formula is as follows,

$$|\Psi(t)\rangle = |\Phi_n[R(t)]\rangle \exp\Big[-\frac{i}{\hbar}\int_{t_0}^{t} d\tau E_n[R(\tau)] - \\ -\int_c \Big\langle \Phi_n[R]\Big| \sum_{i=1}^{N} \frac{\partial}{\partial R_i}\Phi_n[R]\Big\rangle dR_i\Big]. \tag{10.1}$$

In fact the existence of nontrivial phase factors (geometrical phases) was discovered in optics almost 45 years in an earlier Berry paper. The history of the development of the geometrical phase concept can be found, e.g., in Reference [47]. Further we follow our paper [48].

Formula (10.1), being valid for arbitrary quantum systems, is nonetheless rather complicated from the viewpoint of calculations, since it requires the knowledge of the explicit dependence of eigenfunctions on the Hamiltonian parameters and the skill of calculating the corresponding integrals. Our aim is to calculate Berry's phase for the generalized harmonic oscillator described by the Hamiltonian

$$\hat{H}(t) = {}^1/_2\,[\mu(t)\,\hat{p}^2 + \nu(t)\,\hat{x}^2 + \rho(t)\,(\hat{x}\hat{p} + \hat{p}\hat{x})], \tag{10.2}$$

not appealing to (10.1) but using the theory of linear adiabatic integrals of motion developed in studies [29, 49].

Note that Hamiltonian (10.2) is a linear form with respect to generators of the 0(2.1) group. Therefore the quantities μ, ν, ρ are the parameters determining elements of this group.

Although the final result in the case under study has been already obtained by Berry himself [50] and by means of another method in Reference [51], our approach is much simpler so that it deserves attention. Moreover, we will establish the relations between Berry's phase and the coordinate-momentum correlation coefficient in the state $|\Phi_n[R)]>$ and show that this phase is not equal to zero only for correlated quantum states.

We begin with constructing solutions of the Schrodinger equation corresponding to nonstationary Hamiltonian (10.2). The simplest method described in detail, e.g., in [11.29] is based on finding explicit time-dependent (in the Schrodinger picture) linear with respect to coordinate and momentum operators integrals of motion

$$\hat{A}(t) = \varepsilon(t)\,\hat{p} + \gamma(t)\,\hat{x}, \tag{10.3}$$

satisfying the equation

$$i\hbar\partial\hat{A}/\partial t = [\hat{H}, \hat{A}], \tag{10.4}$$

that leads to a system of ordinary differential equations for the coefficients ε and γ

$$\dot{\varepsilon} - \rho\varepsilon + \mu\gamma = 0, \tag{10.5}$$

$$\dot{\gamma} - \nu\varepsilon + \rho\gamma = 0. \tag{10.6}$$

It follows from (10.3) and (10.5) that any linear integral of motion is expressed through function $\varepsilon(t)$, its derivative $d\varepsilon/dt$ and the coefficients of the Hamiltonian as follows:

$$\hat{A}(t) = \varepsilon(t)\,\hat{p} + \mu^{-1}(\rho\varepsilon - \dot{\varepsilon})\,\hat{x}. \tag{10.7}$$

Function ε(t) is an arbitrary solution of the second-order differential equation

$$\ddot{\varepsilon} - \frac{\dot{\mu}}{\mu}\dot{\varepsilon} + \left(\mu\nu - \rho^2 - \dot{\rho} + \frac{\dot{\mu}}{\mu}\rho\right)\varepsilon = 0, \tag{10.8}$$

which follows from (10.5) and (10.6). It is convenient to make the substitution $\varepsilon=\mu^{1/2}x$ to expel the first-order derivative term

$$\ddot{\chi} + \left[\omega^2 - \dot{\rho} + \frac{\dot{\mu}}{\mu}\rho + \frac{1}{2}\frac{\ddot{\mu}}{\mu} - \frac{3}{4}\left(\frac{\dot{\mu}}{\mu}\right)^2\right]\chi = 0, \tag{10.9}$$

where we have introduced the notation

$$\omega^2 = \mu\nu - \rho^2. \tag{10.10}$$

Then formula (10.7) can be rewritten as follows,

$$\hat{A}(t) = \sqrt{\mu}\,[\chi\hat{p} + ((\rho - \dot{\mu}/2\mu)\chi - \dot{\chi})\,\hat{x}/\mu]. \tag{10.11}$$

Since Hamiltonian (10.2) is presumed hermitian, all its coefficients are real. Therefore the complex conjugate function x* is a solution to Equation (10.9) as well as function x(t). We shall impose on the solutions the normalization condition

$$\dot{\chi}\chi^* - \dot{\chi}^*\chi = i/\hbar. \tag{10.12}$$

(The left-hand side of this relation does not depend on time since it is nothing but the Wronskian of Equation (10.9)). Then operator $\hat{A}(t)$ and its hermitian conjugate counterpart $\hat{A}^+(t)$ satisfy the bosonic commutation relation $[\hat{A}(t),\ \hat{A}^+(t)]=1$. Summing up, we can say that the adiabatic evolution of quantum system (10.2) is determined completely by the adiabatic solution of Equation (10.9) (the same statement holds, certainly, for the exact evolution). To find this solution we will use the known method [52] (see also [29]). We introduce the large parameter T (overall time of evolution) and assume t=Tτ, the nondimensional "time" τ being changed from 0 to 1. The solution is supposed to have the form

$$\chi(T\tau) = K(\tau)\exp\left[iT\int_0^\tau dx S(x)\right], \tag{10.13}$$

with slowly changing real functions K(τ) and S(τ). Calculating derivatives according to the formula $d/dt=T^{-1}d/d\tau$ and designating with a prime the derivative over the "slow time" τ we rewrite Equation (10.9) as follows,

$$-S^2K + i\left(\frac{S'}{T}K + 2S\frac{K'}{T}\right) + \frac{K''}{T^2} + K\left(\omega^2 - \frac{\rho'}{T} + \frac{\mu'}{\mu}\frac{\rho}{T} + \frac{1}{2T^2}\frac{\mu''}{\mu} - \frac{3}{4T^2}\left(\frac{\mu'}{\mu}\right)^2\right) = 0. \tag{10.14}$$

The imaginary part of this equation gives rise to the relation

$$SK^2 = \text{const.} \tag{10.15}$$

This equality is exact. Excluding with its aid function K(τ) from Equation (10.14) we get the equation

$$-S^2 + \omega^2 + \frac{1}{T}\left(\rho\frac{\mu'}{\mu} - \rho'\right) + \frac{1}{T^2}\left(\frac{\mu''}{2\mu} - \frac{3}{4}\left(\frac{\mu'}{\mu}\right)^2 - \frac{S''}{2S} + \frac{3}{4}\left(\frac{S'}{S}\right)^2\right) = 0. \tag{10.16}$$

Further we look for function S(τ) in the form of the asymptotic series

$$S(\tau) = S_0(\tau) + \frac{1}{T} S_1(\tau) + \frac{1}{T^2} S_2(\tau) + \dots \tag{10.17}$$

Then we have in the zeroth order approximation the natural formula

$$S_0(\tau) = \omega(\tau), \tag{10.18}$$

while the first order correction is

$$S_1(\tau) = (2\omega)^{-1} (\rho\mu'/\mu - \rho'). \tag{10.19}$$

This correction differs from zero provided $\rho \neq 0$. When $T \to \infty$ it yields the finite value for the geometrical phase of function x(t):

$$\chi(t \mid T \to \infty) = (2\hbar\omega(t))^{-1/2} \exp\left[i \int_0^t d\xi \omega(\xi) + i\Omega\right] = (2\hbar\omega(t))^{-1/2} e^{i\Phi}, \tag{10.20}$$

$$\Omega = \int_0^\tau d\tau' \frac{1}{2\omega}\left(\rho\frac{\mu'}{\mu} - \rho'\right) = \int_C \frac{1}{2\omega}\left(\rho\frac{d\mu}{\mu} - d\rho\right), \tag{10.21}$$

C being the contour along which coefficients ρ, μ, ω vary in their parameter space (in the parameter space of the dynamical group 0(2.1)). The determinative for the possibility of transforming the integral over "slow time" τ' into the countour integral in the parameter space (thus independent on the rate of the variation of parameters) is the proportionality of function $S_1(\tau)$ to the first derivatives μ' and ρ'. Coefficient $(2\hbar)^{-1/2}$ in(10.20) is chosen in accordance with condition (10.12).

In the adiabatic limit, one should set $dx/dt \approx i\omega x$ in (10.11) and neglect the small quantity dμ/dt. Then integral of motion (10.11) can be expressed as follows:

$$\hat{A}(t) = (2\hbar\omega\pi)^{-1/2} e^{i\Phi(t)} [\mu\hat{p} - i(\omega + i\rho)\hat{x}]. \tag{10.22}$$

The phase Φ(t) was defined in (10.20). Then the following formula is valid:

$$\hat{A}^+\hat{A} = \hat{H}(t)/\hbar\omega - 1/2, \tag{10.23}$$

so that the Hamiltonian eigenfunctions $|\Phi_n[R]\rangle$ discussed in the beginning of the paper coincide with eigenfunctions of operator $\hat{A}^+\hat{A}$. But the method of constructing such eigenfunctions is well known: first the ground state is determined from the equation $\hat{A}|0\rangle=0$, then the excited states $|n\rangle$ are obtained by acting the creation operators $(A^+)^n$ on the vacuum. In the stationary case, when the ω frequency does not depend on time, phase Φ(t) equals ωt, and the n-quantum state acquires the phase factor

$$\exp[-i\omega t(n+\frac{1}{2})] = \exp[-i\Phi(t)(n+\frac{1}{2})]$$

The second expression obviously remains valid for adiabatic variations of the Hamiltonian due to (10.22) and (10.23). Thus for Berry's phase, i.e., for that part of the phase in eq. (10.1) which differs from the dynamical contribution $-\int d\tau \cdot \omega(\tau)(n+\frac{1}{2})$, we get immediately from (10.20) and (10.21) the value $-\Omega(n+\frac{1}{2})$. This result coincides with that obtained in [50, 51], but it is obtained without lengthy calculations of cumbersome integrals in formula (10.1) and, moreover, without any appeal to the explicit form of function $|\Phi_n[R]\rangle$ (a similar method based on solving in the adiabatic approximation Heisenberg's equations of motion for operators $\hat{\alpha}$ and $\hat{\alpha}^+$ factorizing Hamiltonian (10.2) was used in Reference [53]).

Now we would like to draw the reader's attention to the following important circumstance: the phase Ω differs from zero only provided the coefficient ρ standing at the mixed product $\hat{x}\hat{p}+\hat{p}\hat{x}$ in Hamiltonian (10.2) is nonzero. But as was shown in [1,2], with ρ≠0 the state $|\Phi_n\rangle$ obtained by acting operator $\hat{A}^+$ (10.22) on the vacuum state is the correlated state with a nonzero correlation coefficient between the coordinate and the momentum

$$r = -\rho/\sqrt{\omega^2+\rho^2} = -\rho/\sqrt{\mu\nu}. \tag{10.24}$$

This coefficient is the same for all states. The remarkable property of eigenstates of operator $\hat{A}$ (10.22) is that they minimize the generalized Schrodinger-Robertson uncertainty relation (4.1). Correspondingly the left-hand side of (4.1) equals $\hbar^2(n+1/2)^2$ for state $|\Phi_n\rangle$. Thus we have shown that for systems with quadratic Hamiltonians like (10.2) the nonzero Berry phase appears only for the correlated states. This explains why this phase was discovered so late, although the problem of the oscillator with adiabatically changing parameters was repeatedly investigated both in classical and quantum mechanics: simply the special case of Hamiltonian (10.2) with ρ=0 was studied earlier. But in such a case only the terms of even orders in expansion (10.17) differ from zero [29,52], and for T→∞ they give rise to an exponentially small contribution to phase Φ(t).

It is not difficult to generalize the results obtained to the case of inhomogeneous quadratic forms, when terms like $f(t)\hat{x}+\varphi(t)\hat{p}$ are added to Hamiltonian (10.2). Then the extra term

$$\delta(t) = \int(\varepsilon\varphi - \gamma f)\,dt', \tag{10.25}$$

ought to be added to the integral of motion (10.30). Having calculated this term in the adiabatic approximation one can obtain the extra phase factor in the state $|\Phi_n[R,t]\rangle$. This factor proves to be independent on the quantum number n.

APPENDIX A
On The Completeness of Multidimensional Hermite Polynomials

n-dimensional Hermite polynomials $H_{m_1\ldots m_n}(x_1\ldots x_n)$ are introduced in a standard manner [33] through the generating function

$$\exp\left[W\mathbf{x}\alpha - \frac{1}{2}\alpha W\alpha\right] = \sum \frac{\alpha_1^{m_1}}{m_1!}\cdots\frac{\alpha_n^{m_n}}{m_n!} H^W_{m_1 m_n}(\mathbf{x}), \tag{A.1}$$

where $x=\begin{pmatrix}x_1\\ \vdots\\ x_n\end{pmatrix}, \alpha=\begin{pmatrix}\alpha_1\\ \vdots\\ \alpha_n\end{pmatrix}$ are n-dimensional vectors, and W is a symmetric matrix determining $H^W_{m_1\ldots m_n}$. It is well known that the Hermite polynomials possess the biorthogonality property. Namely, if functions $G^W_{k_1\ldots k_n}(x)$ are defined through the generating function

$$\exp\left[\mathbf{x}\alpha - \frac{1}{2}\alpha W\alpha\right] = \sum \frac{\alpha_1^{k_1}}{k_1!}\cdots\frac{\alpha_n^{k_n}}{k_n!} G^W_{k_1\ldots k_n}(\mathbf{x}),$$

then the following relation holds [33]:

$$\int dx_1\ldots dx_n H^W_{m_1\ldots m_n}(x)\, G^W_{k_1\ldots k_n}(\mathbf{x}) \exp[-{}^1/_2\mathbf{x}W\mathbf{x}] = \\ = \delta_{m_1k_1}\ldots\delta_{m_nk_n} m_1!\ldots m_n!, \tag{A.2}$$

where W is any positively definite matrix which can be analytically continued to the complex plane so that $\exp(-\frac{1}{2}xWx)$ would decrease at infinity.

This relation shows that an arbitrary function $f(x)\in L^2$ can be expanded in a series over functions $H^W(x)$ or $G^W(x)$:

$$f(\mathbf{x}) = \sum a_{m_1\ldots m_n} H^W_{m_1\ldots m_n}(\mathbf{x}) = \sum b_{k_1\ldots k_n} G^W_{k_1\ldots k_n}(\mathbf{x}),$$

$$a_{m_1\ldots m_n} m_1!\ldots m_n! = \int d\mathbf{x} G^W_{m_1\ldots m_n}(\mathbf{x}) \exp[-{}^1/_2\mathbf{x}W\mathbf{x}]\tilde{f}(\mathbf{x}),$$

$$b_{k_1\ldots k_n} k_1!\ldots k_n! = \int d\mathbf{x} H^W_{k_1\ldots k_n}(\mathbf{x}) \exp[-{}^1/_2\mathbf{x}W\mathbf{x}]\tilde{f}(\mathbf{x}). \tag{A.3}$$

Here $\tilde{f}(x)=(2\pi)^{-n/2}(\det W)^{1/2} f(x)$.

Now it is natural to pose the question, is it possible to expand any function $f(x)\in L^2$ in a series over the Hermite polynomials defined by matrix W which may be not positively definite? The answer is contained in the following new orthogonality relation:

$$\int d\mathbf{x}\Psi(\mathbf{x})\, H^W_{m_1\ldots m_n}(A\mathbf{x})\, H^{*W}_{k_1\ldots k_n}(A\mathbf{x}) = \delta_{m_1k_1}\ldots\delta_{m_nk_n} m_1!\ldots m_n!, \tag{A.4}$$

where

$$A = -i(\lambda^+)^{-1}, \quad \Gamma = (\lambda^+\lambda)^{-1}, \quad W = -\lambda^*\lambda^{-1},$$
$$\Psi(\mathbf{x}) = (2\pi)^{-n/2}(\det(\lambda\lambda^*))^{-1/2}\exp[-{}^1/_2\mathbf{x}\Gamma\mathbf{x}], \tag{A.5}$$

and matrix λ satisfies the condition $\lambda\lambda^*=\lambda^+\lambda$. Due to this condition matrix W is symmetrical. Matrix Γ is positively definite and can be analytically continued to the complex plane so that function $\exp(-\frac{1}{2}x\Gamma x)$ turns into zero at infinity.

To prove (A.4) let us notice that due to (A.1) the integral in the left-hand side of (A.4) is the coefficient at

$$\frac{\alpha_1^{m_1}}{m_1!}\cdots\frac{\alpha_n^{m_n}}{m_n!}\frac{\beta_1^{k_1}}{k_1!}\cdots\frac{\beta_n^{k_n}}{k_n!} \tag{A.6}$$

in the expression

$$(2\pi)^{-n/2}(\det\Gamma^{-1})^{-1/2}\int d\mathbf{x}\exp[WA\mathbf{x}\alpha - {}^1/_2\alpha W\alpha + \\ + W^*A^*\mathbf{x}\beta - {}^1/_2\beta W\beta - {}^1/_2\mathbf{x}\Gamma\mathbf{x}]. \tag{A.7}$$

Due to the relation

$$\int d\mathbf{x}\exp\left[-\frac{1}{2}\mathbf{x}\Gamma\mathbf{x} + C\mathbf{x}\right] = \frac{(2\pi)^{n/2}}{(\det\Gamma)^{1/2}}\exp\left[\frac{1}{2}C\Gamma^{-1}C\right]$$

expression (A.7) is equal to

$$\exp[{}^1/_2(\alpha WA + \beta W^*A^*)\Gamma^{-1}(\alpha WA + \beta W^*A^*)]\exp[-{}^1/_2(\alpha W\alpha + \beta W\beta)]. \tag{A.8}$$

As the consequence of (A.5) we have the relations

$$\alpha W A\Gamma^{-1}\tilde{A}W\alpha = -\alpha\lambda^*\lambda^{-1}\alpha = \alpha W\alpha,$$
$$\beta W^* A^*\Gamma^{-1}\tilde{A}^*W^*\beta = -\beta\lambda^*\lambda^{-1}\beta = \beta W\beta,$$
$$\alpha W A\Gamma^{-1}A^{T}W^*\beta = \alpha\beta,$$
$$\beta W^* A^*\Gamma^{-1}\tilde{A}W\alpha = \alpha\beta,$$

due to which (A.8) assumes the form

$$\exp[\alpha\beta] = \sum \frac{(\alpha_1\beta_1)^{m_1}}{m_1!}\cdots\frac{(\alpha_n\beta_n)^{m_n}}{m_n!}. \qquad (A.9)$$

Comparing (A.6) and (A.9) one obtains (A.4). We see that no restrictions were imposed on matrix W; the only requirement is the positive definiteness of matrix Γ.

As an example let us consider two-dimensional Hermite's polynomials. Let $W=\begin{Vmatrix}0&1\\1&0\end{Vmatrix}$. It is not positively definite matrix. Then other matrices are as follows:

$$\lambda = \frac{1}{2}\begin{Vmatrix}\pm ib & b\\ \pm ib^* & -b^*\end{Vmatrix}, \quad |b|^2=1, \quad A=\mp\begin{Vmatrix}-b & \pm ib\\ -b^* & \pm ib^*\end{Vmatrix}, \quad \Gamma = 2E.$$

The orthogonality relation

$$(2\pi)^{-1}\int dx_1\,dx_2 H^{W}_{n_1n_2}(A\mathbf{x})\,H^{*W}_{m_1m_2}(A\mathbf{x})\exp[-\mathbf{x}^2] = \delta_{n_1m_1}\delta_{n_2m_2}n_1!n_2!$$

permits to expand any function $f(\mathbf{x})\in L^2$ in a series over $H^{W}_{n_1n_2}(\mathbf{x})$ with $W=\begin{Vmatrix}0&1\\1&0\end{Vmatrix}$, which cannot be done on the basis of the biorthogonality relation (A.2).

Another example is W=E (i.e., unit matrix). Then λ=iE, Γ=A=E, and since $(H^E(\mathbf{x}))^*=G^E(\mathbf{x})$, the orthogonality relation (A.4) coincides with biorthogonality relation (A.2).

In Reference [11] the following sum rule for multidimensional Hermite's polynomials was proved,

$$\sum_{m_1=0}^{n_1}\cdots\sum_{m_N=0}^{n_N} C^{m_1}_{n_1}\cdots C^{m_N}_{n_N}(-x_1)^{n_1-m_1}\cdots(-x_N)^{n_N-m_N}H^{\sigma}_{m_1\ldots m_N}(\mathbf{x}\sigma^{-1}) =$$
$$= H^{\sigma}_{n_1\ldots n_N}(0)$$

Here we wish to work out the details of this formula in the special case of two dimensions. In this case the defining matrix σ can be reduced to one of three canonical forms [54,55]:

1. $\sigma=\begin{Vmatrix}0&1\\1&0\end{Vmatrix}$,

$$H^{00}_{mn}(x_1,x_2) = (-1)^{\mu_{mn}}\mu_{mn}!\,x_1^{(n-m+|n-m|)/2}x_2^{(m-n+|m-n|)/2}L^{|m-n|}_{\mu_{mn}}(x_1x_2),$$

where $\mu_{mn}=\min(m,n)$, $\nu_{mn}=\max(m,n)$, and $L^q_p(z)$ is the associated Laguerre polynomial [33]. Then the sum rule reads

$$\sum_{m_1=0}^{n_1}\sum_{m_2=0}^{n_2} C^{m_1}_{n_1}C^{m_2}_{n_2}(-x_1)^{n_1-m_1}(-x_2)^{n_2-m_2}H^{00}_{m_1m_2}(x_2,x_1) = \delta_{n_1n_2}n_1!(-1)^{n_1}.$$

2. $\sigma=\begin{Vmatrix}1&1\\1&0\end{Vmatrix}$. (A.10)

In this case from

$$H^{10}_{mn}(x_1, x_2) = \sum_{p+l=m} \frac{m!n!}{\nu_{\gamma n}!} 2^{-1/2} (-1)^{\mu_{pn}} \times$$
$$\times\, x_1^{(n-p+|n-p|)/2} x_2^{(p-n+|p-n|)/2} L^{|p-n|}_{\mu_{pn}}(x_1 x_2)\, H_l\left(\frac{x_1}{\sqrt{2}}\right)$$

we obtain

$$\sum_{m_1=0}^{n_1} \sum_{m_2=0}^{n_2} C^{m_1}_{n_1} C^{m_2}_{n_2} (-x_1)^{n_1-m_1} (-x_2)^{n_2-m_2} H^{10}_{m_1 m_2}(x_2, x_1 - x_2) =$$
$$= \frac{n_1!\,(n_1-n_2)!\,(-1)^{n_1}}{2^{(n_1-n_1)/2}\,((n_1-n_2)/2)!}, \tag{A.11}$$

provided n_1 and n_2 have the same parity (otherwise the right-hand side of (A.11) equals zero).

3. $\sigma = \begin{Vmatrix} 1 & \beta \\ \beta & 1 \end{Vmatrix}$,

$$H^{\beta}_{mn}(0) = \mu_{mn}!\,(\beta^2 - 1)^{(m+n)/4} (-1)^{\nu_{nm}} (-1)^{|n-m|/4} P^{|m-n|/2}_{(m+n)/2}\left(\frac{\beta}{\sqrt{\beta^2 - 1}}\right).$$

m,n ought to have the same parity, otherwise $H^{\beta}_{mn}(0)=0$. Then

$$\sum_{m_1=0}^{n_1} \sum_{m_2=0}^{n_2} C^{m_1}_{n_1} C^{m_2}_{n_2} (-x_1)^{n_1-m_1} (-x_2)^{n_2-m_2} H^{3}_{m_1 m_2}\left(\frac{x_1 - \beta x_2}{1-\beta^2}, \frac{x_2 - \beta x_1}{1-\beta^2}\right) =$$
$$= \mu_{n_1 n_2} (\beta^2 - 1)^{(n_1+n_2)/4} (-1)^{\nu_{n_1 n_2}} (-1)^{|n_1-n_2|/4} P^{|n_1-n_2|/2}_{(n_1+n_2)/2}\left(\frac{\beta}{\sqrt{1-\beta^2}}\right). \tag{A.12}$$

APPENDIX B

On the Completeness of the Set of Quadratic Exponentials

Let us consider the set of coherent states of a one dimensional harmonic oscillator $|\alpha\rangle$. Its completeness is well known [5,39]. From the expansion of CS over the Fock basis $|n\rangle$

$$|\alpha\rangle = \exp\left[-\frac{1}{2}|\alpha|^2\right] \sum_n \frac{\alpha^n}{\sqrt{n!}} |n\rangle$$

and the integral equality

$$\int d^2\alpha\, e^{-|\alpha|^2} (\alpha^*)^n \alpha^m = \int_0^\infty d|\alpha| \cdot |\alpha|^{n+m+1} e^{-|\alpha|^2} \int_0^{2\pi} d\theta\, e^{i(m-n)\theta} = \pi n!\, \delta_{nm},$$

with $\alpha^2\alpha$=dReα•dImα one gets immediately

$$\int d^2\alpha\, |\alpha\rangle\langle\alpha| = \pi \sum_n |n\rangle\langle n|.$$

Since the n-quantum states form the complete set, $\sum_n |n\rangle\langle n| = 1$ and consequently,

$$\pi^{-1} \int d^2\alpha\, |\alpha\rangle\langle\alpha| = 1,$$

which is nothing but the completeness relation for coherent states. Any function $f(x) \in L^2$ can be expanded over coherent states, i.e.,

$$|f\rangle = \pi^{-1} \int d^2\alpha\, |\alpha\rangle\langle\alpha|f\rangle,$$

or

$$\langle x \mid f\rangle = f(x) = \pi^{-1} \int d^2\alpha \, \langle x \mid \alpha\rangle \langle \alpha \mid f\rangle = (\pi)^{-1} \int d^2\alpha \, \langle x \mid \alpha\rangle f(\alpha), \quad \text{(B.1)}$$

where $\langle x \mid \alpha\rangle$ is the x-representation of the coherent state:

$$\langle x \mid \alpha\rangle = \pi^{-1/4} \exp\left[-{}^1/_2 x^2 + \sqrt{2} x\alpha - \alpha^2/2 - |\alpha|^2/2\right]$$

(vacuum fluctuations are supposed to normalize to unity), and

$$\langle \alpha \mid f\rangle = f(\alpha) = \int dx \, \langle \alpha \mid x\rangle \langle x \mid f\rangle = \int dx \, \langle \alpha \mid x\rangle f(x).$$

Remember, CS form overcomplete set due to their nonorthogonality. Designating $\mathrm{Re}\alpha=\alpha_1$, $\mathrm{Im}\alpha=\alpha_2$, let us rewrite (B.1) in the form

$$\begin{aligned} f(x) &= \pi^{-5/4} \int d\alpha_1 \, d\alpha_2 f(\alpha) \exp\left[-{}^1/_2 x^2 + \sqrt{2}\, x(\alpha_1 + i\alpha_2) - \alpha_1^2 - i\alpha_1\alpha_2\right] = \\ &= \pi^{-5/4} \int d\alpha_1 \, d\alpha_2 f(\alpha) \exp\left[-\boldsymbol{\alpha} W \boldsymbol{\alpha} + \mathbf{S}\boldsymbol{\alpha} + C(x)\right], \end{aligned} \quad \text{(B.2)}$$

where

$$\boldsymbol{\alpha} = \begin{pmatrix} \alpha_1 \\ \alpha_2 \end{pmatrix}, \quad W = \begin{Vmatrix} 1 & i/2 \\ i/2 & 0 \end{Vmatrix}, \quad \mathbf{S} = \sqrt{2}\, x \begin{pmatrix} 1 \\ i \end{pmatrix}, \quad C(x) = -\frac{x^2}{2}.$$

Introducing new variables

$$\alpha_1^2 + \alpha_1\alpha_2 i = y_2, \quad \sqrt{2}\,(\alpha_1 + i\alpha_2) = y_1, \quad \text{(B.3)}$$

we get

$$\alpha_1 = \sqrt{2}\, y_2/y_1, \quad \alpha_2 = -i\left(y_1/\sqrt{2} - \sqrt{2}\, y_2/y_1\right).$$

Now let us make this change of variables in integral (B.2):

$$\iint_{-\infty}^{\infty} d\alpha_1 \, d\alpha_2 \to \iint_C dy_1 \, dy_2 J,$$

where the Jacobian of transformation is $J=i/y_1$, and the integration is performed over the complex plane determined by Equations (B.3). Thus we have the decomposition

$$f(x) = \iint_C dy_1 \, dy_2 f(y) \exp\left(-\frac{1}{2}x^2 - y_1 x - y_2\right), \quad \text{(B.4)}$$

with the image function $f(y)=f(y_1, y_2)$ determined from

$$f(y) = \frac{i}{\pi^{3/2} y_1} \int dx f(x) \exp\left[-\frac{x^2}{2} + x\left(\frac{4y_2}{y_1} - y_1\right) - \frac{4y_2^2}{y_1^2} + y_2\right]. \quad \text{(B.5)}$$

Formulas (B.4) and (B.5) prove the completeness of the set of quadratic exponentials and the possibility of the expansion of an arbitrary function into an integral over such exponentials.

The set of quadratic exponentials thus constructed is not orthogonal. Designating the basis matrix elements by $\langle y_1 y_2 \mid x\rangle$ we get

$$\begin{aligned} &\int dx \, \langle y_1 y_2 \mid x\rangle \langle x \mid \tilde{y}_1 \tilde{y}_2\rangle = \\ &= \exp\left\{-\frac{1}{2}\left[\frac{2y_2^2}{y_1^2} + \frac{1}{2}\left(y_1 + \frac{2y_2}{y_1}\right)^2 + \frac{2\tilde{y}_2^2}{y_1^2} + \frac{1}{2}\left(\tilde{y}_1 - \frac{2\tilde{y}_2}{\tilde{y}_1}\right)^2 - y_1\tilde{y}_1\right]\right\}. \end{aligned}$$

REFERENCES

1. Dodonov, V.V., Kurmyshev, E.V., Man'ko, V.I. "Generalized uncertainty relation and correlated coherent states" *Phys. Lett. A*, 1980, vol. 79, no. 2/3, pp. 150-152.
2. Dodonov, V.V., Kurmyshev, E.V., Man'ko, V.I. *"The more exact uncertainty relation and correlated coherent states"* Proc. First Intern. Seminar on Group Theoretical Methods in Physics, Zvenigorod (Russia), 1979, Moscow: Nauka, 1980, vol. 1, pp. 227-232.
3. Schrodinger, E. *"Zum Heisenbergschen Unscharfeprinzip"* Ber. Kgl. Acad. Wiss., Berlin, 1930, S. 296-303.
4. Robertson, H.P. "A general formulation of the uncertainty principle and its classical interpretation" *Phys. Rev. A*, 1930, vol. 35, no. 5, p. 667.
5. Glauber, R.J. "Coherent and incoherent states of the radiation field" *Phys. Rev. A*, 1963, vol. 131, no. 6, pp. 2766-2788.
6. Stoler, D. "Generalized coherent states" *Phys. Rev. D*, 1971, vol. 4, no. 8, pp. 1309-1312.
7. Yuen, H.P. "Two-photon coherent states of the radiation field" *Phys. Rev. A*, 1976, vol. 13, no. 6, pp. 2226-2243.
8. Hollenhorst, J.N. "Quantum limits on resonant-mass gravitational-radiation detection" *Phys. Rev. D*, 1979, vol. 19, no. 6, pp. 1669-1679.
9. Caves, C.M. "Quantum-mechanical noise in an interferometer" *Phys. Rev. D*, 1981, vol. 23, no. 8, pp. 1693-1708.
10. Man'ko, V.I. "Correlated states of quantized fields" Izmeritel'naya Tekhnika [*Measurement Technics*] 1987, no. 7, pp. 20-23.
11. Dodonov, V.V., Man'ko, V.I. *"Invariants and correlated states of nonstationary quantum systems"* Proc. Lebedev Phys. Inst., Moscow: Nauka, 1987, vol. 183, pp. 71-181. [Translation: Nova Science, Commack, 1989, pp. 103-261].
12. Akhiezer, A.I., Berestetskiy, V.B. "Kvantovaya elektrodinamika" [*Quantum electrodynamics*], Moscow: Nauka, 1969, p. 623.
13. Landau, L.D., Lifshits, E.M. "Teoriya polya" [*Field theory*], Moscow: Nauka, 1973, p. 323.
14. Bohr, N., Rosenfeld, L. "Field and charge measurement in quantum electrodynamics" *Phys. Rev.*, 1950, vol. 78, pp. 794-798.
15. Landau, L.D., Lifshits, E.M. "Teoriya uprugosti" [*Theory of elasticity*], Moscow: Nauka, 1987, p. 255.
16. Landau, L.P., Lifshits, E.M. "Gidrodinamika" [*Hydrodynamics*], Moscow: Nauka, 1986, p. 631.
17. Dodonov, V.V., Kurmyshev, E.V., Man'ko, V.I. *"Correlated coherent states"* Proc. Lebedev Phys. Inst., Moscow: Nauka, 1986, vol. 176, pp. 128-150. [Translation: Nova Science, Commack, 1988, pp. 169-199].
18. Miller, M.M., Mishkin, E.A. "Characteristic states of the electromagnetic radiation field" *Phys. Rev.*, 1966, vol. 152, no. 4, pp. 1110-1114.
19. Bertrand, P.P., Moy,K., Mishkin, E.A. "Minimum uncertainty states and the states of the electromagnetic field" *Phys. Rev. D*, 1971, vol. 4, no. 6, pp. 1909-1912.
20. Canivell, V., Seglar, P. "Minimum uncertainty states and pseudo-classical dynamics" *Phys. Rev. D*, 1977, vol. 15, no. 4, pp. 1050-1054.
21. Stoler, D. "Equivalence classes of minimum uncertainty packets" *Phys. Rev. D*, 1970, vol. 1, no. 12, pp. 3217-3219.
22. Lu, E.Y.C., "New coherent states of electromagnetic field" *Lett. Nuovo Cim.*, 1971, vol. 2, no. 24, pp. 1241-1244.
23. Walls, D.F. "Squeezed states of light" *Nature*, 1983, vol. 306, no. 5939, pp. 141-146.

24. Schumaker, B.L. "Quantum mechanical pure states with gaussian wave functions" *Phys. Repts.*, 1986, vol. 135, no. 6, pp. 317-408.
25. Takahasi, H. "Information theory of quantum-mechanical channels" *Adv. Commun. Syst.*, 1965, vol. 1, pp. 227-310.
26. Raiford, M.T., "Statistical dynamics of quantum oscillators and parametric amplification in a single mode" *Phys. Rev. A*, 1970, vol. 2, no. 4, pp. 1541-1558.
27. Raiford, M.T. "Degenerate parametric amplification with time-dependent pump amplitude and phase" *Phys. Rev. A*, 1974, vol. 9, no. 5, pp. 2060-2069.
28. Husimi, K. "Miscellanea in elementary quantum mechanics. II" *Progr. Theor. Phys.*, 1953, vol. 9, no. 4, pp. 381-402.
29. Malkin, I.A., Man'ko, V.I. "Dinamicheskie simmetrii i kogerentnye sostoyaniya kvantovykh sistem" *[Dynamical symmetries and coherent states of quantum systems]* Moscow: Nauka, 1979, p. 320.
30. Smirnov, D.F., Troshin, A.C. *"New phenomena in quantum optics: anti-bunching and sub-Poissonian photon statistics; squeezed states"* UFN, 1987, vol. 153, no. 2, pp. 233-272.
31. Paul, H. "Interference between independent photons" *Rev. Mod. Phys.*, 1986, vol. 58, no. 1, pp. 209-231.
32. "Squeezed states of the electromagnetic field" special issue of *J. Opt. Soc. Amer. B*, 1987, vol. 4, no. 10.
33. Bateman, H., Erde'lyi, A. *"Higher transcendental functions "* New York; Toronto; London: McGraw Hill Book Company, Inc., 1953, vol. 2, 1955, vol. 3.
34. Schubert, M., Vogel, W. "Field fluctuation of two-photon coherent states" *Phys. Lett. A*, 1978, vol. 68, no. 33, pp. 321-322.
35. Schubert, M., Vogel, W. "Quantum statistical properties of the radiation field in the degenerate two-photon emission process" *Opt. Communs.*, 1981, vol. 36, no. 2, pp. 164-168.
36. Schleich, W., Wheeler, A. "Oscillations in photon distribution of squeezed states" *J. Opt. Amer. B*, 1987, vol. 4, no. 10, pp. 1715-1719.
37. Dodonov, V.V., Klimov, A.B., Man'ko, V.I. *"Photon-number oscillations in correlated light"* Preprint Lebedev Phys. Inst., no. 178, Moscow, 1988, p. 24; *Journal of Soviet Laser Research*, Plenum Publ. Co., 1989, vol. 10, no. 1, pp. 35-48.
38. Carruthers, P., Nieto, M. "The phase-angle variables in quantum mechanics" *Rev. Modern. Phys.*, 1968, vol. 40, no. 2, p. 411.
39. Loudon, R. *"The quantum theory of light"*, Oxford: Clarendon Press, 1973.
40. Prudnikov, A.P., Brychkov, Yu.A., Marichev, O.I. "Integraly i ryady. Elementarnye funkzii" *[Integrals and series. Elementary functions]* Moscow, Nauka, 1981, p. 711.
41. Dodonov, V.V., Man'ko, V.I. *"Universal invariants of quantum systems and generalized uncertainty relation"* Proc. Second. Inter. Seminar on Group Theoretical Methods in Physics. Zvenigorod, Moscow: Nauka, 1983, vol. 2, pp. 11-33. [Translated by Harwood Academic Publishers, Chur-London-Paris-New York, 1985, v. 1, pp. 591-612].
42. Baz', A.I., Zel'dovich, Ya.B., Perelmov, A.N. "Rasseyaniye, reaktsii, raspady v nerelyativistskoy kvantovoy mekhanike" *[Scattering, reactions and decays in nonrelativistic quantum mechanics]* Moscow, Nauka, 1966, p. 339.
43. Affleck, I. "Quantum-statistical metastability" *Phys. Rev. Lett.*, 1981, vol. 46, no. 2, pp. 388-391.
44. Likharev, K.K. "Vvedenie v dinamiku dzhozefsonovckikh perekhodov" *[Introduction in dynamics of Josephson junctions]* Moscow: Nauka, 1985, p. 320.
45. Gol'danskiy, V., Trakhtenberg, L., Flerov, V. "Tunnel'nye yavleniya v kvantovoy fizike" *[Tunnel phenomena in quantum physics]*, Moscow: Nauka, 1986, p. 294.

46. Berry, M.V. "Quantal phase factors accompanying adiabatic changes" *Proc. Roy. Soc. London A.* 1984, vol. 392, no. 1802, pp. 45-47.
47. *"Topological Phases in Quantum Theory"*, Eds. B. Markovski and S.I. Vinitsky, Singapore: World Sci. Publ., 1989.
48. Dodonov, V.V., Man'ko, V.I. *"Adiabatic invariants, correlated states and Berry's phase" Topological Phases in Quantum Theory*, Singapore: World Sci. Publ., 1989, pp. 74-83.
49. Malkin, I.A., Man'ko, V.I., Trifonov, D.A. "Linear adiabatic invariants and coherent states" *J. Math. Phys.*, 1973, vol. 14, no. 5, pp. 576-582.
50. Berry, M.V. "Classical adiabatic angles and quantal adiabatic phase" *J. Phys. A.*, 1985, vol. 18, no. 1, pp. 15-27.
51. Engineer, M.H., Ghosh, G. "Berry's phase as the asymptotic limit of an exact evolution: An example" *J. Phys. A*, 1988, vol. 21, no. 2, pp. L95-L98.
52. Kurlsrud, R.M. "Adiabatic invariant of the harmonic oscillator" *Phys. Rev.*, 1957, vol. 106, pp. 205-207.
53. Ghosh, G., Dutta-Roy, B. "The Berry phase and the Hannay angle" *Phys. Rev. D*, 1988, vol. 37, pp. 1709-1711.
54. Dodonov, V.V., Man'ko, V.I. *"Evolution of multidimensional systems. Magnetic properties of ideal gases of charged particles"* Proc. Lebedev Phys. Inst., Moscow: Nauka, 1987, vol. 183, pp. 182-286.[Translation: Commack: Nova Science, 1989, pp. 263-414].
55. Dodonov, V.V., Man'ko, V.I., Semjonov, V.V. *"The density matrix of the canonically transformed multidimensional Hamiltonian in the Fock basis" Nuovo Cimento* B, 1984, vol. 83, pp. 145-161.

SQUEEZED AND CORRELATED LIGHT SOURCES

S.M. Chumakov, M. Kozierowski, A.A. Mamedov, V.I. Man'ko

Abstract: We study the possibilities of generating squeezed and correlated light in different quantum-optical processes. In particular, we consider in detail the harmonics generation, the spontaneous emission by a single two- or three-level atom in an ideal resonator (including multiphoton transitions), as well as the spontaneous radiation from a system of two-level atoms confined in a small volume inside an ideal resonator.

1. INTRODUCTION

Studying the properties of certain non-traditional types of states of the electromagnetic radiation has become one of last years most actual subjects in literature on quantum optics. The special kind of these states are the so-called correlated states of light [1]. Some related objects introduced under different names by different authors were described in [2-5] (see also [6-8] for the detailed list of references). The states concerned demonstrate the wealth of physical properties arising with accounting the quantum nature of the electromagnetic field. In the case of a single field mode all these states are given in the coordinate and momentum representations by wave functions of the Gaussian type. They coincide precisely with wave packets constructed by Schrodinger [9] for a harmonic oscillator as far back as 1926 (see also [10]). However since these wave functions had found their application for describing the oscillators of the electromagnetic field, certain additional aspects were revealed resulting in the detailing and closer definition of the terminology. From the point of view of methodics all the Gaussian wave packets (in particular, coherent, squeezed and correlated states described below) can be considered as examples of continuous representations [11-13]. The corresponding overcomplete basis is marked with continuous indices, which can be considered to a certain extent as analogs of classical variables. To proceed to a more concrete exposition, we introduce the creation and annihilation operators of a single mode of the electromagnetic field α^+, α (bearing in mind, e.g., the principal mode of an ideal resonator). Besides, it is convenient to introduce the so-called field quadrature component operators $Q=(\alpha+\alpha^+)$ and $P=-i(\alpha-\alpha^+)$ corresponding to canonically conjugated coordinate and momentum variables. The variances of quadrature components $\sigma_Q=\langle Q^2\rangle-\langle Q\rangle^2=(\Delta Q)^2$ and $\sigma_P=\langle P^2\rangle-\langle P\rangle^2=(\Delta P)^2$ satisfy the Heisenberg uncertainty relation

$$\sigma_Q\sigma_P \geqslant 1. \tag{1.1}$$

However, a more precise inequality exists relating variances and the correlation coefficient between components:

$$\gamma^2 = \frac{\langle \Delta P \Delta Q + \Delta Q \Delta P\rangle}{2\left[(\Delta Q)^2(\Delta P)^2\right]^{1/2}} = \frac{\langle QP + PQ\rangle - \langle Q\rangle\langle P\rangle}{2\left[(\langle Q^2\rangle - \langle Q\rangle^2)(\langle P^2\rangle - \langle P\rangle^2)\right]^{1/2}}.$$

This is the so-called Schrodinger-Robertson uncertainty relation:

$$\sigma_Q\sigma_P\,(1-\gamma^2) \geqslant 1. \tag{1.2}$$

The usual (Glauber's) coherent states [14] are determined by two requirements: first, they minimize the Heisenberg relation (1.1), second, they possess equal variances of the quadrature components, $\sigma_P=\sigma_Q$. Hence, the correlation coefficient turns into zero for coherent states. Moreover, the coherent state is the eigenstate of the annihilation operator $\alpha=(Q+iP)/2$. Following Reference [1] (see also [7,8]) let us demand the generalized uncertainty relation to minimize the Schrodinger-Robertson uncertainty relation (1.2). Then the properties of coherent states are generalized in the following way: the new state is the eigenstate of the operator

$$a(\mu,\chi) = \frac{Q + i\mu e^{i\chi}P}{2\sqrt{\mu\cos\chi}}, \tag{1.3}$$

μ and x being arbitrary parameters ($\mu>0$, $-\pi/2 < x < \pi/2$). The wave function of this state in the coordinate and momentum representation, as well as in the mixed Wigner representation, is an exponential of a quadratic form (over respective arguments). If phase x=0, then the correlation coefficient γ^2=sinx=0, and variances of the components are given by the expressions

$$\sigma_Q = \mu, \quad \sigma_P = \mu^{-1}. \tag{1.4}$$

We see again the variances are not equal to each other. Quantum fluctuations in one of two quadrature components turn out diminished μ times in comparison with coherent states, whereas the fluctuations in another component are increased to the same extent. Therefore these states [2-5] are often called *squeezed states* [6]. Sometimes they are also called "two-photon states" [4]. This name reflects the fact that the vacuum squeezed state (corresponding to zero eigenvalue of operator (1.3) is certain superposition of the usual Fock states with even numbers of photons. For a nonzero phase x, the states constructed possess the nonzero correlation coefficient γ^2, therefore they were named in Reference [1] *correlated coherent states.*

A clear geometrical interpretation of correlated states can be obtained from the form of the corresponding Wigner functions in the phase space. In the general case they are Gaussians of two variables (with the form of a "hat" in the three-dimensional space), the "equal probability curves" (i.e. the level lines of the Wigner function) being ellipses whose principal axis constitute certain nonzero angle with x-axis dependent on the correlation coefficient. For large or small values of μ the ellipses look like very narrow and long lines corresponding to the wave functions with almost exactly determined momentum and very spread coordinate or vice versa. One may treat the squeezed states with $\mu \to \infty$ or $\mu \to 0$ as physically natural regularizations for the coordinate or momentum eigenfunctions. Accordingly, the squeezed states wave functions, e.g., in the coordinate representation, tend with $\mu \to \infty$ or $\mu \to 0$ to the plane wave exponential or to the δ-function.

The evolution of initially Gaussian states (squeezed or correlated) under the action of an arbitrary quadratic Hamiltonian describing many physical effects

$$H = \omega_f a^+ a + F_1(t) a + F_1^*(t) a^+ + F_2(t) a^2 + F_2^*(t) a^{+2}, \tag{1.5}$$

was studied in paper [4] (see also [1,7]). (Here $F_1(t)$, $F_2(t)$ may be arbitrary c-number functions of time.) It appears that Hamiltonian (1.5) transforms any initial Gaussian state again to a Gaussian state, but with another set of parameters. The correlated state is determined by three parameters: real numbers μ and x entering operator (1.3), and the complex number α designating the corresponding eigenvalue of this operator:

$$a(\mu, \chi) | \alpha; \mu, \chi\rangle = \alpha | \alpha; \mu, \chi\rangle$$

Consequently, the time-dependent state vector corresponding to the correlated state reads

$$| \psi(t)\rangle = e^{-itH} | \alpha_0; \mu_0 \chi_0\rangle = | \alpha(t); \mu(t), \chi(t)\rangle. \tag{1.6}$$

The equivalent statement is that the evolution governed by quadratic Hamiltonian (1.5) preserves the relation between the parameters given by the Schrodinger-Robertson uncertainty relation (1.2).

Real physical processes are described with Hamiltonians like (1.5), c-number functions $F_1(t)$, $F_2(t)$ being replaced by some operators. These may be photon operators describing another field mode (as in the harmonics generation process), or certain atomic operators in studying the interaction of radiation with one or many atoms. Then initially coherent (correlated) states lose their coherence (i.e. formula (1.6, generally speaking, does not hold any more), and the equality in (1.2) cannot be maintained for all time instants. Therefore variances σ_P, σ_Q, and the correlation coefficient γ^2 become independent characteristics. In this case it is useful to speak on squeezing and correlation of quadrature components, when one of the variances becomes less than unity, or, respectively, the correlation coefficient differs from zero. Since uncertainty relations (1.1) and (1.2) are not minimized in general, the corresponding quantum state is no longer coherent.

Our study is devoted to the problems of squeezed and correlated light generation in different physical processes, namely, in the harmonics generation and in the spontaneous emission by a single atom (the generalized Jaynes-Cummings model) or by an ensemble of two-level atoms (Dicke's model). The problem of squeezed light production, as well as obtaining photons antibunching and sub-Poissonian statistics, lately has attracted a great attention both from the theoretical point of view and in the light of possible applications of squeezed light in communication systems with reduced noise [15] and in quantum non-demolition measurements [16]. The photons antibunching was observed for the first time in the experiment on the resonance fluorescence of a single two-level atom [17]. The antibunching of photons in fluorescence from dye molecules was discovered in [18]. In Reference [19] the simulating experiment was performed, the results being interpreted as the observation of sub-Poissonian statistics in the second harmonic generation. The direct measurements of sub-Poissonian statistics were carried out for the resonance fluorescence process and for "down-conversion" by Mandel with co-authors [20,21]. In that measurement both two- and single-photon states were obtained. References to many papers relating to photons antibunching and sub-Poissonian statistics can be found in reviews [22,23].

During the 1980s intensive attempts at experimentally observing squeezed light were undertaken (see Reviews [24,25]). Squeezed states were obtained by means of four-wave mixing in atomic vapors [26] and waveguides [27], optical parametric oscillations [28], and optical bi-stability effects [29].

Because of real possibilities of obtaining squeezed fields, it is interesting, naturally, to investigate the interaction of such fields with matter. Then certain new quantitative and qualitative phenomena may arise. So, the collapses and revivals effects in population oscillations for a single two- or three-level atom interacting with squeezed light were discussed in studies [30] and [31], respectively. The radiation by an excited atom interacting with the squeezed vacuum was considered in [32]. The resonance fluorescence and collective resonance fluorescence in the presence of the squeezed vacuum were investigated in References [33] and [34], respectively. The clear difference from the usual resonance fluorescence spectra in the limits of strong and weak guiding fields were found.

Let us proceed to a brief description of the content of our paper. Section 2 is devoted to studying squeezing and correlation in the principal field mode for the k-th harmonic generation process. We show that in the shot time approximation the principal mode can experience squeezing and correlation of its components, but only in the second harmonic generation case ($k=2$) the principal mode remains in the state minimizing the uncertainty relation, at least for the beginning of the process. With $k\geq 3$ the product of the quadrature component variances exceeds the lower bound established by relation (1.2) already in the beginning of the process. We shall also touch the problem of higher orders squeezing in the harmonics generation processes.

In Section 3 squeezing (including higher orders squeezing) in the multiphoton Jaynes-Cummings model (JC-model) describing the interaction of a two-level atom with a field mode was discussed under the assumption of the multiphoton resonance.

In Section 4 the process of interaction between a three-level atom and one or two field modes is investigated. The possibility of producing squeezed light is demonstrated for different configurations of atomic levels. The next section is devoted to checking the fulfillment of the uncertainty relations in JC-models. It is shown that the field mode remains in the minimal uncertainty product ((1.1) or (1.2)) state only for JC-model with the two-photon transition. The main mathematical tool in Sections 2-5 is solving Heisenberg's equations of motion by means of power series expansions over time t (with the accuracy varying from terms of the order of t^2 to t^8, depending on necessity). Thus, results of Sections 2-5 are valid for the initial stages of the process discussed.

Sections 6 and 7 are devoted to studying squeezing in spontaneous emission by a system of many two-level atoms placed into an ideal resonator. This situation is described with the well known Dicke model Hamiltonian. Here we confine ourselves to the case when only a small part of the whole number of atoms is excited at the initial time instant, the initial state being symmetric with respect to any permutations of atoms. Let N be the whole number of atoms, s of them being excited at the initial instant. Ever since paper [51] containing numerical computer calculations appeared, it has been known that the spectrum of the Dicke Hamiltonian eigenvalues (see (6.1)) is close to equidistant in the case of N>>s. In this limit some approximate expressions for the averages of Heisenberg's operators, e.g., for the mean energy stored in the atomic subsystem, were obtained in terms of the elliptic functions [52-54]. But the equidistant spectrum and, consequently, the harmonic time dependences of Heisenberg's operators indicate the existence of the dynamical symmetry group in this limit. Really, in Section 6 we show that with N>>s the Hamiltonian of the system is a generator of the (s+1)-dimensional representation of SU(2) group. This fact enables us to obtain very simple approximate expressions for the transition amplitudes similar to those found in the exactly solvable JC-model. Assuming as the zeroth order approximation the motion over the SU(2) group representation, we develop in Section 6 the perturbation theory with respect to the small parameter $\varepsilon=[N-(s-1)/2]^{-1}$, the energy eigenvalues being calculated with an accuracy to the second order, and the transition amplitudes - to the first order of ε. In our opinion, the solutions obtained in Section 6 have two advantages in comparison with the elliptic functions solutions of References [52-54]. First, they can be used in the case of significant deviations from the purely equidistant spectrum. Second, utilizing our solutions one can employ the powerful apparatus of calculations in the SU(2) representations. The latter is illustrated in Section 7, where the zeroth approximation of Section 6 is used to calculate squeezing of the field quadrature components in the spontaneous radiation by a system of two-level atoms in an ideal resonator. (Note that squeezing in the Dicke model in the framework of the semiclassical consideration was studied in Reference [55]. The consistent quantum examination in Section 7 shows that the degree of squeezing increases with the increase of the number of initially excited atoms s. For example, with s=20 it reaches 69%. It should be noted that the perturbation theory of Section 6 is valid in fact not only for small times t, but at the sufficiently remote stages of the process as well. The limited applicability of results obtained in Section 7 (they hold for several first periods of quadrature components variances oscillations) is caused not by the method itself, but it is due to some additional assumptions made in order to simplify the calculations. For the sake of convenience in reading the paper, we have transferred many technical details in the appendices.

In conclusion we would like to thank E.A. Akhundova, V.V. Dodonov, V.P. Karassiov, and E.V. Kurmyshev for helpful discussions and their interest in the paper.

2. SQUEEZING IN K-TH HARMONIC GENERATION

In the electric dipole approximation and for perfect phase matching, the generation of the k-th harmonic in the absence of damping is described in terms of the following Hamiltonian:

$$H = \hbar\omega a_f^+ a_f + \hbar k\omega a_k^+ a_k + c\hbar g_k (a_k^+ a_f^k + a_f^{+k} a_k). \qquad (2.1)$$

where the suffixes f and k refer to the fundamental and k-th harmonic mode respectively, and g_k is the mode coupling constant depending on the order of the process and involving appropriate nonlinear characteristics of the medium. The symbol c denotes the light velocity which is equal for both beams due to the assumption on the phase matching, α^+_i and α_i (i=f, k) are obviously the photon creation and annihilation operators: $[\alpha_i, \alpha^+_k]=\delta_{ik}$; $[\alpha_i, \alpha_k]=0$.

In the Heisenberg picture, the time evolution of any operator A is given by the equation

$$i\hbar\dot{A} = [\dot{A}, H]. \qquad (2.2)$$

where H is the Hamiltonian of a system. For A equal to α_i, the commutator (2.2) with the free-field part of the Hamiltonian (2.1) gives the "rapid" oscillations of the operator α_i. The commutator with the interaction part of the Hamiltonian (2.1) leads to the additional "slow" evolution of the operator α_i, resulting from nonlinear coupling between the beams. Generally, we can write

$$a_f(t) = a_{fs}(t)\, e^{-i\omega t}, \qquad a_k(t) = a_{ks}(t)\, e^{-ik\omega t}. \qquad (2.3)$$

The slowly varying parts of the operators $\alpha_{fs}(t)$ and $\alpha_{ks}(t)$ satisfy the same commutation rules as operators $\alpha_i(t)$ and $\alpha^+_i(t)$. The equations of motion for the slowly varying parts of the photon operators have the form:

$$\dot{a}_{fs}(t) = -ikg_k (a_{fs}^+(t))^{k-1} a_{ks}(t), \qquad \dot{a}_{ks}(t) = -ig_k a_{fs}^k(t). \qquad (2.4)$$

The above equations cannot be strictly solved.

We assume that the time t which the beams take to traverse the nonlinear medium is sufficiently short. For times greater than t, interacting field modes evolve as they were free, non-interacting modes. Thus we apply the short-time approximation procedure [35]. Namely, we expand the operators in a Taylor series about t=0. Within a sufficient accuracy we find that

$$a_{fs}^{(k)}(t) = a_{f0} - ikg_k t (a_{f0}^+)^{k-1} a_{f0} + \\ + \frac{1}{2} k g_k^2 t^2 \left[k \sum_{m=0}^{k-2} \sum_{r=0}^{m} r! \binom{m}{r} \binom{k-1}{r} (a_{f0}^+)^{k-2-r} a_{f0}^{k-1-r} n_{k0} - (a_{f0}^+)^{k-1} a_{f0}^k \right]. \qquad (2.5)$$

The superscript (k) following the quantity α_{fs} (t) denotes the order of the process. The symbol 0 at the suffixes f and k denotes that the operators are taken at t=0, i.e. $\alpha_{io}=\alpha_{is}(0)=\alpha_i(0)$. Moreover, $n_{io}=\alpha^+_{io}\alpha_{io}$ is the photon-number operator of the i-th mode

and $\binom{m}{r} = m!/r!(m-r)x!$. To get Equation (2.5) we have used the fact that, as always, operators associated with two modes commute at the same time t, in particular at t=0. In what follows, we restrict our attention to squeezing for the fundamental mode. Hence we present the solution for the photon operator $a^{(k)}_{fs}$ (t) only.

Phase-sensitive detectors respond to the envelope of the optical field only. Hence, we introduce two slowly varying Hermitian canonical quantities

$$Q = a_s + a_s^+, \quad P = -i\,(a_s - a_s^+), \tag{2.6}$$

corresponding to two quadrature components of the field. Their variances may be presented in the form

$$\sigma_Q = 1 + A + B, \quad \sigma_P = 1 + A - B, \tag{2.7}$$

where

$$A = 2\,(\langle n\rangle - \langle a_s^+\rangle\langle a_s\rangle), \quad B = \langle(\Delta a_s^+)^2\rangle + \langle(\Delta a_s)^2\rangle. \tag{2.8}$$

The quantity A is non-negative, while B can take positive as well as negative values. From the Heisenberg uncertainty relation and (2.7) we have

$$A\,(A + 2) - B^2 \geqslant 0. \tag{2.9}$$

In accordance with the definition given in Section 1, one of the quadrature components is squeezed if $A/|B|<1$. Then Q is squeezed for B<0 while P is squeezed for B>0. Squeezing increases as the ratio $A/|B|$ decreases.

In turn, from the Schrodinger-Robertson uncertainty relation (1.2) and (2.7) one has

$$A\,(A + 2) - B^2 \geqslant \gamma^2\,(1 - \gamma^2). \tag{2.10}$$

The above relation requires for $\gamma \neq 0$ enhanced fluctuations, at least in one of the quadratures in comparison with the relation (2.9); the ratio $A/|B|$ is simply implied to be greater than that allowed by (2.9). This means that correlations of the quadrature components degrade squeezing.

Returning to the harmonic generation processes, we assume that the harmonic mode is in the vacuum state at the input to the medium, so that the following initial conditions are satisfied: $\alpha_{k0}|0\rangle = n_{k0}|0\rangle = 0$. The fundamental field mode may be initially in an arbitrary state. Here we take this mode initially in a coherent state $|\alpha\rangle$,

$$a_{f0}\,|\,\alpha\rangle = \alpha\,|\,\alpha\rangle, \quad \alpha = \langle n_{f0}\rangle^{1/2} e^{i\theta}.$$

where $\langle n_{f0}\rangle$ is the mean number of photos incident on the medium and ϑ is the initial phase of α.

At the above initial conditions, within an accuracy to t^2, we have

$$A_f^{(k)} = 0, \quad B_f^{(k)} = -k\,(k-1) g_k^2 t^2\,\langle n_{f0}\rangle^{k-1} \cos 2\theta. \tag{2.11}$$

Hence, from (2.7) we get

$$\sigma_{Q_f}^{(k)} = 1 - k\,(k-1)\,g_k^2 t^2\,\langle n_{f0}\rangle^{k-1} \cos 2\theta,$$

$$\sigma_{P_f}^{(k)} = 1 + k\,(k-1)\,g_k^2\,t^2\,\langle n_{f0}\rangle^{k-1} \cos 2\theta. \tag{2.12}$$

It is clear that, depending on the initial phase ϑ, one or the other quadrature component may be squeezed. Maximum squeezing, i.e. the smallest values of σ_{Qf} and α_{Pf} occur for $\vartheta=0$, π and $\vartheta=\pi/2$, $3\pi/2$, respectively. Then the maximal squeezing in the fundamental mode is equal to the scaled response in the Hanbury-Brown and Twiss correlation experiment [36].

It is obvious from the forms of the variances (2.12) that their product is less than unity. This does not mean a violation of the uncertainty principles at all. Namely, the deviation from unity is proportional to t^4. In the solutions (2.5) and (2.12) the terms proportional to t^4 are omitted. They ought to be taken into account if one wishes to obtain the correct form of the uncertainty relation within a higher-order accuracy. Within an accuracy of t^2 we get $\sigma_{Qf}\,\sigma_{Pf}=1$. Then the fundamental mode might seem to be in a minimum uncertainty squeezed state. However, the correlation coefficient is in general non-zero (what will be shown further on) already in the second-order approximation in t. This suggests that the field of the fundamental mode is transformed into a correlated coherent state rather than in a minimum Heisenberg's uncertainty state. Since the correlation coefficient appears in the Schrodinger-Robertson uncertainty relation in the second power, the fourth-order approximation in t is sufficient to examine the uncertainty relations.

2.1 Second-harmonic generation as a source of correlated coherent states

In what follows, the uncertainty relations will be examined in detail for the fundamental mode in the course of second-harmonic generation. Then k=2, and from (2.12) Mandel's results are recovered [36]. Within an accuracy up to t^4, for $\alpha^{(2)}{}_{fs}(t)$ we have [37]

$$\begin{aligned} a_{fs}^{(2)}(t) = {} & a_{f0} - 2ig_2 t a_{f0}^{+} a_{20} + g_2^2 t^2 (2n_{20} - n_{f0})\, a_{f0} - \\ & - {}^2\!/_3\, i g_2^3 t^3 (2n_{20} a_{20} a_{f0}^{+} - 3a_{20} n_{20} a_{f0}^{+} + 2a_{20}^{+} a_{f0}^{3} - a_{20} a_{f0}^{+}) + \\ & + {}^1\!/_6\, g_2^4 t^4 (4a_{20}^{+2} a_{20}^{2} a_{f0} - 28 n_{20} n_{f0} a_{f0} - 20 n_{20} a_{f0} + 8a_{20}^{2} a_{f0}^{+3} + \\ & + 5a_{f0}^{+2} a_{f0}^{3} + n_{f0} a_{f0}) + \ldots \end{aligned} \tag{2.13}$$

At the same initial conditions as previously assumed we now get that $A^{(2)}{}_f$ is different from zero and amounts to

$$A_f^{(2)} = 2g_2^4 t^4 \langle n_{f0}\rangle^2, \tag{2.14}$$

while $B^{(2)}{}_f$ takes the form

$$B_f^{(2)} = -2g_2^2 t^2 \langle n_{f0}\rangle \cos 2\theta\, [1 - {}^1\!/_6\, g_2^2 t^2 (6\langle n_{f0}\rangle + 1)]. \tag{2.15}$$

Hence from (2.7) we arrive at

$$\begin{aligned} \sigma_{Q_f}^{(2)} &= 1 - 2g_2^2 t^2 \langle n_{f0}\rangle \cos 2\theta + 2g_2^4 t^4 \langle n_{f0}\rangle^2 (1 + \cos 2\theta) + {}^1\!/_3\, g_2^4 t^4 \langle n_{f0}\rangle \cos 2\theta, \\ \sigma_{P_f}^{(2)} &= 1 + 2g_2^2 t^2 \langle n_{f0}\rangle \cos 2\theta + 2g_2^4 t^4 \langle n_{f0}\rangle^2 (1 - \cos 2\theta) - {}^1\!/_3\, g_2^4 t^4 \langle n_{f0}\rangle \cos 2\theta. \end{aligned} \tag{2.16}$$

After simple algebra, retaining the terms only of order t^4, we find that the product of the above variances is for the majority of phases ϑ greater than unity and amounts to

$$\sigma_{Q_f}^{(2)} \sigma_{P_f}^{(2)} = 1 + 4g_2^4 t^4 \langle n_{f0}\rangle^2 \sin^2 2\theta. \tag{2.17}$$

In turn, for the correlation coefficient we have

$$\gamma_f^{(2)} = -2g_2^2 t^2 \langle n_{f0}\rangle \sin 2\theta\, (1 + 4g_2^4 t^4 \langle n_{f0}\rangle^2 \sin^2 2\theta)^{-1/2}. \tag{2.18}$$

Thus, within an accuracy at t^4, from (2.17) and (2.18) we find

$$\sigma_{Q_f}^{(2)} \sigma_{P_f}^{(2)} (1 - (\gamma_f^{(2)})^2) = 1. \tag{2.19}$$

This means that the fundamental mode is in general in a correlated coherent state (which minimizes the Schrodinger-Robertson uncertainty relation) with the correlation coefficient given by (2.18). The optimum choice of ϑ for correlation is $\vartheta=\pi/4 + m\pi/2$ (m=0,1,2,3). Then the field is unsqueezed, namely

$$\sigma_{Q_f}^{(2)} = \sigma_{P_f}^{(2)} = 1 + 2g_2^4 t^4 \langle n_{f0}\rangle^2.$$

and we deal with equal variances of the quadrature components. In turn, in the case of the earlier discussed maximum squeezing $\gamma^{(2)}{}_f=0$. Only then the field of the fundamental mode, even within an accuracy up to t^4, is in a squeezed state corresponding to the minimum of the Heisenberg uncertainty relation. One of the quadratures is still squeezed if additionally $\vartheta \neq \pi/4 + m\pi/2$, but squeezing is not then maximal. This shows once more that correlations of the quadrature components degrade squeezing.

To conclude, by control over the phase ϑ we can get the field of the fundamental mode in a squeezed or unsqueezed correlated coherent state or in a minimum Heisenberg's uncertainty squeezed state.

In general, for k-th harmonic generation the correlation coefficient $\gamma^{(k)}{}_f$ is also proportional to t^2 in the first nonvanishing approximation in t. Hence, the Schrodinger-Robertson uncertainty relation for the fundamental mode in the course of k-th harmonic generation, within an accuracy of t^4, reads

$$\sigma_{Q_f}^{(k)}\sigma_{P_f}^{(k)}\left(1-(\gamma_f^{(k)})^2\right) = 1 + k^2 g_k^4 t^4 \langle n_{f0}\rangle^{2(k-1)} \times \times \left[\sum_{s=1}^{k-1} s!\binom{k-1}{s}\binom{k-1}{s}\langle n_{f0}\rangle^{1-s} - (k-1)^2\right]. \tag{2.20}$$

It is easily seen that only for k=2 this product equals unity. In other words, depending on the initial phase the fundamental mode, with an accuracy of t^4, does not remain for $k \geq 3$ in a minimum uncertainty state, at least at the beginning of the process. For short times, the field is in squeezed state if $\gamma^{(k)}{}_f=0$ and in a correlated squeezed or unsqueezed state if $\gamma^{(k)}{}_f \neq 0$, but none of these states is a minimum uncertainty state.

So, for short times, within an accuracy of t^4 we can finally write

$$\sigma_{Q_f}^{(k)}\sigma_{P_f}^{(k)}\left(1-\gamma_f^{(k)2}\right) = \begin{cases} 1 \quad \text{при} \quad k=2, \\ 1 + k^2 g_k^2 t^4 \langle n_{f0}\rangle^{2k-1} \sum_{s=2}^{k-1} s!\binom{k-1}{s}^2 \langle n_{f0}\rangle^{-s} \quad \text{при} \quad k \geqslant 3. \end{cases} \tag{2.21}$$

2.2 Squeezing by multiple higher-harmonic generation

The time evolution of the photon operators $\alpha_f(t)$ and $\alpha_k(t)$ has been solved analytically in the short-time approximation. Hence sub-Poissonian photon statistics [35] and squeezing [36] have been predicted solely at the beginning of the process. On the other hand, numerical calculations [38,39] show that sub-Poissonian photon statistics of the generated second-harmonic is absent in the long-time approximation due to the spontaneous re-emission of photons between the harmonic and fundamental mode. The same can be cautionally suspected for squeezing for every generated harmonic as well as for the fundamental mode. Hence, we cannot expect a decrease of quantum fluctuations by using longer and longer media as in the case of multiphoton absorption, which is a one-mode process. However, when higher-harmonic generation takes place in a cascade of thin plates and the harmonic is filtered out behind every plate, the process runs in a way similar to multiphoton absorption [40]. Obviously, the short-time approximation provides an almost realistic description of generation in each thin plate. It was then shown [41] that sub-Poissonian photon statistics of the initially coherent beam increases considerably.

We assume that the thickness L of all the plates is identical. Thus the time of nonlinear interaction in each plate is also the same and equal to t=L/c. The boundary conditions for the m-th plate are

$$a^{(k)}_{f,m}(0) = a^{(k)}_{f,m-1}(t), \qquad n_{k,m}(0) = 0, \tag{2.22}$$

where $\alpha_{fm}{}^{(k)}$ is the photon annihilation operator of the fundamental mode in the m-th plate. In the one-step higher-harmonic generation process, the short-time solution for $\alpha^{(k)}{}_{fs}(t)$, within an accuracy of t^2, is given by Equation (2.5). Using this solution and the boundary conditions (2.22) for calculating the variances of the quadrature components after the m-th step of k-th harmonic generation at coherent input radiation we get [42]:

$$\sigma^{(k)}_{Qf,m} = 1 - mk(k-1)\, g^2_k t^2 \langle n_{f0}\rangle^{k-1} \cos 2\theta,$$

$$\sigma^{(k)}_{Pf,m} = 1 + mk(k-1)\, g^2_k t^2 \langle n_{f0}\rangle^{k-1} \cos 2\theta.$$

The deviation from unity is now m times greater than that in Equation (2.12). In other words, squeezing grows with increasing number m of k-th harmonic passages. As for the experimental demonstration of the squeezing phenomena, they can be considerably enhanced further in destructive interference when measured by means of increased sub-Poissonian photon statistics.

2.3 Higher-order squeezing

Hong and Mandel [43] have introduced the concept of higher-order squeezing. According to the definitions of these authors the relations

$$\sigma_Q<1 \text{ or } \sigma_P<1$$

represent second-order squeezing. Equivalently, the above relations may be presented with the help of the normally ordered variances in the form

$$\langle:(\Delta Q)^2:\rangle<0 \text{ or } \langle:(\Delta P)^2:\rangle<0, \tag{2.23}$$

where the symbol :...: denotes normal ordering of the photon creation and annihilation operators within their products.

According to Hong and Mandel [43] the field is squeezed to the N-th order, if for the quadratures Q or P

$${}_N\sigma_Q = \langle(\Delta Q)^N\rangle < (N-1)!! \quad \text{или} \quad {}_N\sigma^P = \langle(\Delta P)^N\rangle < (N-1)!!. \tag{2.24}$$

The squeezing effect is uniquely nonclassical effect for even N. The moments (2.24) may be expressed by their normally ordered even moments as follows [44] (R=Q or P):

$${}_N\sigma_R = (N-1)!! + \sum_{r=0}^{N/2-1} \binom{N}{2r} \frac{(2r)!}{r!\,2^r} \langle:(\Delta R)^{N-2r}:\rangle \tag{2.25}$$

The N-th order *intrinsic squeezing* is connected with the negative value of the normally ordered moment $\langle:(\Delta R)^N:\rangle$. As was shown [43], one of the inequalities (2.24) may be fulfilled at $\langle:(\Delta R)^2:\rangle<0$ even in the case of nonnegative terms $\langle:(\Delta R)^{N-2r}:\rangle$ ($r\in[0,N/2-2]$). In this case, the field is said to be squeezed to the N-th order but is not *intrinsically squeezed* to this order. The N-th order squeezing is then related to the intrinsic second-order squeezing, as in the case of resonance fluorescence of a single two-level atom or the fundamental mode in second-harmonic generation [45]. On the other hand, in particular cases, one quadrature component may reveal higher-order squeezing even in the absence of second-order squeezing in it [45].

2.4 Higher-order squeezing in k-th harmonic generation

Second-order squeezing has been obtained in the quadratic approximation in t. We shall calculate here all expectations within the same accuracy. With respect to (2.5) for the normally ordered moments $Q^{(k)}_f$ of the fundamental mode when it is initially in a coherent state we get (M even):

$$\langle :(\Delta Q_f^{(k)})^M :\rangle = \sum_{r=0}^{M}\binom{M}{r}\Big\langle\Big\{\overset{+}{a}_{f0}-\alpha^* - \frac{k}{2}g_k^2t^2\,[\overset{+}{a}_{f0}\,(:n_{f0}^{k-1}:) - |\alpha|^{2(k-1)}\alpha^*]\Big\}^r \times$$
$$\times\Big\{a_{f0}-\alpha-\Big(\frac{k}{2}\Big)g_k^2t^2\,[(:n_{f0}^{k-1}:)\,a_{f0} - |\alpha|^{2(k-1)}\alpha\Big\}^{M-r}\Big\rangle.$$

We have neglected here all terms containing the operators of the generated k-th harmonic mode, since they contribute nothing to the second order in t. Inspection of the above equation shows that only terms r=0 and r=M can lead to non-vanishing expectations, and hence

$$\langle :(\Delta Q_f^{(k)})^M :\rangle = -\frac{k^2g_k^2t^2}{2}\sum_{s=0}^{M-1}\{\langle(a_{f0}-\alpha)^{M-1-s}\,[(:n_{f0}^{k-1}:)\,a_{f0} - |\alpha|^{2(k-1)}\alpha]\times$$
$$\times\,(a_{f0}-\alpha)^s\rangle + \langle(\overset{+}{a}_{f0}-\alpha^*)^s\,[\overset{+}{a}_{f0}\,(:n_{f0}^{k-1}:) - |\alpha|^{2(k-1)}\alpha^*]\,(\overset{+}{a}_{f0}-\alpha)^{M-1-s}\rangle\}.$$

From this formula it is easily seen that nonzero expectations may be obtained only at s=0. To evaluate these expectations we use the commutation relation

$$(a_{f0}-\alpha)^{M-1}(:n_{f0}^{k-1}:) = \sum_{m=0} m!\binom{k-1}{m}\binom{M-1}{m}(:n_{f0}^{k-1-m}:)\,(a_{f0}-\alpha)^{M-1-m},$$

where, in general the upper limit of summation should be equal to the smaller at the numbers k-1 or M-1. However, the averaging of the above equation over a coherent state reduces to zero all terms excepting those with m=M-1. Simultaneously the condition M≤k becomes apparent. In other words, the order of the highest nonzero normally ordered moment may be at the most equal to the order of the process if the latter is caused by coherent input radiation. Finally, we have

$$\langle :(sQ_f^{(k)})^M :\rangle = -\,kg_k^2t^2\,\frac{(k-1)!}{(k-M)!}\,|\alpha|^{2k-M}\cos M\theta. \tag{2.26}$$

By the same procedure

$$\langle :(\Delta P_f^{(k)})^M :\rangle = -\,kg_k^2t^2\,\frac{(k-1)!}{(k-M)!}\,|\alpha|^{2k-M}\cos M\Big(\theta-\frac{\pi}{2}\Big). \tag{2.27}$$

From the last two equations it is obvious that, depending on the phase, either the component $Q^{(k)}_f$ (optimum choice of $\vartheta=n\pi$, n=0,1) or $P^{(k)}_f$ (optimum choice of $\vartheta=(n+1/2)\pi$) is squeezed intrinsically for all allowed M(M-even). In accordance with the inequality M≤k, in the case of an even-order process the field of the fundamental mode may be intrinsically squeezed up to the k-th order (2,4,...,k) while in the case of an odd process intrinsic squeezing is to the (k-1)-th order (2,4,...,k-1). If $Q^{(k)}_f$ is intrinsically squeezed for all allowed M, then $P^{(k)}_f$ is intrinsically squeezed in the orders M=4m (m any integer), and vice versa (obviously M≤k for even-order processes or M≤k-1 for odd-order processes).

On substitution of Equation (2.26) into Equation (2.25) we have

$${}_N\sigma_{Q_f}^{(k)} = (N-1)!! - k\,(k-1)!\,g_k^2t^2\sum_{r}^{N/2-1}\binom{N}{2r}\frac{(2r)!\,|\alpha|^{2k-N+2r}}{r!\,2^r\,(k-N+2r)!}\cos(N-2r)\,\theta, \tag{2.28}$$

The summation in (2.28) starts from (N-k)/2 for even k and from (N+1-k)/2 for odd-order processes [44]. The N-th order squeezing (now, in general, N may exceed k) is a manifestation of intrinsic lower-order squeezing. Obviously, higher-order intrinsic squeezing (M≥4) takes place for k≥4, as it arises from the above considerations.

In the considerations of higher-order squeezing, it is useful to introduce the normalized quantity r_N, which represents the so called *degree of squeezing* [43] (r=q or p)

$$r_N = \frac{\langle(\Delta R)^N\rangle}{(N-1)!!} - 1 \tag{2.29}$$

Obviously, the field is squeezed to the N-th order if either q_N or p_N is negative. For N=2 the quantities q_N and p_N are simply equal to the normally ordered variances (2.23) of the quadrature components.

In particular, let us discuss in detail the quantities $q^{(k)}{}_2$ and $q^{(k)}{}_4$ for the fundamental mode in the course of k-th harmonic generation. The first of them, from (2.12) or from (2.28) and (2.29) is

$$q_{f2}^{(k)} = -k\,(k-1)\,g_k^2 t^2 \langle n_{f0}\rangle^{k-1} \cos 2\theta. \tag{2.30}$$

In turn

$$q_{f4}^{(k)} = 2q_{f2}^{(k)} - {}^1\!/_3 k\,(k-1)(k-2)(k-3)\,g_k^2 t^2 \langle n_{f0}\rangle^{k-2} \cos 4\theta. \tag{2.31}$$

The last term in Equation (2.31) is responsible for the fourth-order intrinsic squeezing. It is once more readily seen that the fourth-order intrinsic squeezing can take place only for k≥4. For $\vartheta=\pi/2$ or $3\pi/2$, as it was discussed earlier, there is no second-order squeezing in the quadrature component Q_f. Then

$$q_{f2}^{(k)} = k\,(k-1)\,g_k^2 t^2 \langle n_{f0}\rangle^{k-1} \tag{2.32}$$

and it has a positive value. In turn,

$$q_{f4}^{(k)} = 2q_{f2}^{(k)} - {}^1\!/_3 k\,(k-1)(k-2)(k-3)\,g_k^2 t^2 \langle n_{f0}\rangle^{k-2}, \tag{2.33}$$

i.e. this quadrature reveals the fourth-order intrinsic squeezing. On insertion of (2.32) into (2.33) we have

$$q_{4f}^{(k)} = {}^1\!/_3 k\,(k-1)\langle n_{f0}\rangle^{k-2} g_k^2 t^2 \left[6\langle n_{f0}\rangle - (k-2)(k-3)\right].$$

In particular, for

$$\langle n_{f0}\rangle < (k-2)(k-3)/6, \tag{2.34}$$

due to the fourth-order intrinsic squeezing the quadrature components $Q^{(k)}{}_f$ will also reveal for k>4 the fourth-order squeezing ($Q^{(k)}{}_{4f}<0$) even in the absence of second-order squeezing. The reality of the condition (2.34) grows as k increases. This case shows unambiguously that the concept of higher-order squeezing is more than purely mathematical.

3. SQUEEZING IN THE MULTIPHOTON JAYNES-CUMMINGS MODEL

3.1. The second order squeezing

It has already been shown by Meystre and Zubairy [46] that light squeezing is possible in the simple Jaynes-Cummings model of a two-level atom coupled to an initially coherent single-mode radiation field in a high-Q cavity. Squeezing is also possible in the case of spontaneous emission in a cavity by a single, specially prepared n a superposition state, two- or three-level atom [47]. Shumovsky, et al [48] have eneralized the results of Meystre and Zubairy [46] to the multiphoton Jaynes-Cummings model. The Jaynes-Cummings model with multiphoton transitions describes he interaction of a single-mode cavity field with a two-level atom via a k-photon rocess. Higher-order squeezing in this model has been considered by Fam Le Kien, et al 45].

The Hamiltonian for the model in question in the rotating-wave approximation eads

(3.1)

where g_k denotes the multiphoton atom-field coupling coefficient and k is the ransition multiplicity, ω and ω_α are the frequencies of the cavity mode and the atom, respectively. In the case of exact resonance $\omega_\alpha = k\omega$. S^z, S^+ and S^- are atomic pseudospin operators. Let us denote by $|+\rangle$ and $|-\rangle$ the excited and ground state of the atom.

If, initially, the atom starts in its lower state and the field is in coherent state $|\alpha\rangle$ with mean photon number $\langle n\rangle = |\alpha|^2$, the resulting interaction-picture wave function of the atom-field system is [45]

$$|\psi(t)\rangle = \sum [\,|-, n\rangle C_-^{(n)}(t) + |+, n-k\rangle C_+^{(n)}(t)]\, e^{-\langle n\rangle/2} \alpha^n (n!)^{-1/2}, \tag{3.2}$$

where the probability amplitudes $C^{(n)}_-$ and $C^{(n)}_+$ have the form

$$C_-^{(n)} = \cos\,[g_k t\,(n!/(n-k)!)^{1/2}], \qquad C_+^{(n)} = -i \sin\,[g_k t\,(n!/(n-k)!)^{1/2}]. \tag{3.3}$$

According to (2.6)-(2.8) the variances of the quadratures may be presented in the following forms:

$$\sigma_Q = 1 + 2\langle a_s^+ a_s\rangle + 2\,\mathrm{Re}\,\langle a_s^2\rangle - 4(\mathrm{Re}\,\langle a_s\rangle)^2,$$

$$\sigma_P = 1 + 2\langle a_s^+ a_s\rangle - 2\,\mathrm{Re}\,\langle a_s^2\rangle - 4(\mathrm{Im}\,\langle a_s\rangle)^2. \tag{3.4}$$

Using the wave function (3.2) we arrive at the following exact solutions,

$$\sigma_Q^{(k)} = 1 + 2\,[\Phi(t) - \langle n\rangle \Psi^2(t)] + 2\langle n\rangle\,[\chi(t) - \Psi^2(t)]\cos 2\theta,$$

$$\sigma_P^{(k)} = 1 + 2\,[\Phi(t) - \langle n\rangle \Psi^2(t)] - 2\langle n\rangle\,[\chi(t) - \Psi^2(t)]\cos 2\theta \tag{3.5}$$

in full accordance with the results of Shumovsky, et al [48]. Here we have introduced the auxiliary functions

$$\Phi(t) = \langle a_s^+ a_s\rangle = \langle n\rangle - k \sum_n P_n |C_-^{(n)}(t)|^2,$$

$$\mathrm{X}(t) = \frac{\langle a_s\rangle}{\alpha} = \sum_n P_n \Big[C_-^{(n)}(t)\, C_-^{(n+2)}(t) + {} + C_+^{(n)*}(t)\, C_+^{(n+2)}(t) \Big(\frac{(n-k+1)(n-k+2)}{(n+1)(n+2)} \Big)^{1/2} \Big],$$

$$\Psi(t) = \frac{\langle a_s^2\rangle}{\alpha^2} = \sum_n P_n \Big[C_-^{(n)}(t)\, C_-^{(n+1)}(t) + {} + C_+^{(n)*}(t)\, C_+^{(n+1)}(t) \Big(\frac{n-k+1}{n+1} \Big)^{1/2} \Big]. \tag{3.6}$$

$P_n = |\alpha|^{2n} e^{-\langle n\rangle}/n!$ denotes Poissonian distribution. In general, the time evolution of the above functions has to be studied numerically. However, one can obtain simplified analytical expressions in the short-time approximation. For instance, for $\vartheta = 0$ and π within an accuracy of t^2 we get

$$\sigma_Q^{(k)} = 1 - k\,(k-1)\,\langle n\rangle^{k-1}\, g_k^2 t^2. \tag{3.7}$$

This result for k≥2 coincides in form with expression (2.12) for the variances of the fundamental mode in the course of k-th harmonic generation. Here, however, k may be equal to unity as well. For k=1, within this accuracy of t, the quantum fluctuations in the quadrature component Q would seem unchanged at the beginning of the interaction. However, within an accuracy of t^4 for k=1 we find

$$\sigma_Q^{(1)} = 1 - {}^1/_3\langle n\rangle\, g_1^4 t^4. \tag{3.8}$$

So, in fact, the quantum fluctuations in this quadrature are also suppressed even for k=1. The negative sign in (3.7) indicates the immediate appearance of squeezing (in the quadrature Q) for any k after switching on the atom-field interaction. Certainly, the quantum fluctuations in the quadrature P will then be enhanced comparing to those for coherent radiation.

When the atom is initially in its upper state, none of the quadratures is squeezed in short times [46]. As time goes on the quantities (3.5) start to oscillate and, periodically, one or the other quadrature will be squeezed for the phases ϑ different from $\pi/4+m\pi/2$ (m=0,1,2,3). It was shown, that due to the appropriate choice of the initial parameters the large magnitudes of squeezing, amounting to 57% can be obtained within the model under consideration [48].

3.2. Fourth-order squeezing

Here, we are going to study this effect for the quadrature Q only. Since all the central moments of Q may be expressed by real functions (compare with (3.4)), for the sake of brevity let us introduce the following auxiliary functions:

$$\Phi_{\gamma\delta} = \mathrm{Re}\,\langle a_s^{\dagger\gamma} a_s^{\delta}\rangle, \qquad \Phi_{\gamma 0} = \Phi_{0\gamma} = \Phi_\gamma. \tag{3.9}$$

Using the wave function (3.2) one finds

$$\Phi_{\gamma\delta} = \langle n\rangle^{(\gamma+\delta)/2}\cos[(\gamma-\delta)\,\theta]\sum_n\Big[P_n C_-^{(n+\gamma)}(t)\,C_-^{(n+\delta)}(t) + \\ + C_+^{(n+\gamma)*}(t)\,C_+^{(n+\delta)}(t)\,\frac{n!}{(n-k)!}\left(\frac{(n-k+\gamma)!\,(n-k+\delta)!}{(n+\gamma)!\,(n+\delta)!}\right)^{1/2}\Big], \tag{3.10}$$

The function Φ_{11} is identical with the previously introduced function Φ (3.6). It is only necessary to note that due to the properties of the Poissonian distribution

$$\sum_n P_n |C_+^{(n)}|^2 = \langle n\rangle \sum_n [P_n |C_+^{(n+1)}|^2/(n+1)].$$

In turn, the function $\Phi_2=\langle n\rangle \times \cos 2\vartheta$ while $\Phi_1=\langle n\rangle^{1/2}\Psi\cos\vartheta$. The normalized factors $q^{(k)}{}_2$ and $q^{(k)}{}_4$ (2.29), describing the second-and fourth order squeezing, for the model under consideration read [45]

$$\begin{aligned} q_2^{(k)} &= 2\,(\Phi_2 - \Phi_{11} - 2\Phi_1^2), \\ q_4^{(k)} &= 2q_2^{(k)} + {}^2/_3\,(\Phi_4 + 4\Phi_{13} + 3\Phi_{22}) - {}^{16}/_3\Phi_1\,(\Phi_3 + 3\Phi_{12}) + \\ &+ 16\Phi_1^2\,(\Phi_2 + \Phi_1 - \Phi_1^2). \end{aligned} \tag{3.11}$$

It is certain that $\sigma^{(k)}{}_Q=1+q^{(k)}{}_2$.

In short-time approximation, within an accuracy of t^2, from (3.10) and (3.3) we fin[d] that

$$\Phi_{\gamma\delta} = \langle n\rangle^{(\gamma+\delta)/2} \cos[(\gamma - \delta)\theta]\Big\{1 - \frac{1}{2} g_k^2 t^2 \sum_n P_n \Big[\frac{(n+\gamma)!}{(n-k+\gamma)!} + \frac{(n+\delta)!}{(n-k+\delta)!} - 2\frac{n!}{(n-k)!}\Big]\Big\}. \quad (3.12)$$

With respect to the properties of Poissonian distribution, we have

$$\sum_n P_n \frac{(n+l)!}{(n-k+l)!} = \langle n\rangle^{k-l} \sum_{s=0}^{l} \langle n\rangle^s \frac{k!\, l!}{s!\,(l-s)!\,(k-l+s)!}. \quad (3.13)$$

From Equations (3.12) and (3.13) we get

$$q_2^{(k)} = -k(k-1)\, g_k^2 t^2 \langle n\rangle^{k-1} \cos 2\theta. \quad (3.14)$$

$$q_4^{(k)} = 2q_2^{(k)} - {}^1/_3 k(k-1)(k-2)(k-3)\, g_k^2 t^2 \langle n\rangle^{k-2} \cos 4\theta. \quad (3.15)$$

In addition to the already mentioned similarity of the formulas for $q^{(k)}{}_2$ in the model under consideration, and in the course of k-th harmonic generation of the fundamental mode, we have the similarity of the forms of $q^{(k)}{}_4$ for both mentioned phenomena (3.15) and (2.31). In the case of one-photon transitions, k=1, both factors (3.14) and (3.15) are equal to zero. However, within an accuracy up to t^4, for k=1, we find that

$$\Phi_{\gamma\delta} = \langle n\rangle^{(\gamma+\delta)/2} \cos[(\gamma-\delta)\theta]\{1 - {}^1/_2(\gamma+\delta) g_1^2 t^2 + {}^1/_6[(\gamma+\delta)\langle n\rangle + {}^1/_4(\gamma^2 + 6\gamma\delta + \delta^2)]\, g_1^4 t^4\}. \quad (3.16)$$

Hence, from (3.11) we arrive at

$$q_2^{(1)} = -\frac{1}{3}\langle n\rangle g_1^4 t^4 \cos 2\theta, \quad (3.17)$$

what for ϑ=0 and π coincides with formula (3.8). Besides,

$$q_4^{(1)} = 2q_2^{(1)}. \quad (3.18)$$

From Equations (3.14), (3.15) and (3.17), (3.18) it is easily seen that, for arbitrary k, the quadrature component Q may be squeezed not only to the second order ($q^{(k)}{}_2<0$), but to the fourth order ($q^{(k)}{}_4<0$) as well, at least at the beginning of the interaction. Moreover, for k≥4, the field is then also intrinsically squeezed to the fourth order, which is related with the last negative term in Equation (3.15). For short times, one may cautiously suspect, that the order to which the field may be intrinsically squeezed depends on k similarly as intrinsic squeezing for the fundamental mode in k-th harmonic generation. This conjecture, however, has been proven here for fourth-order intrinsic squeezing only.

Like in the case of the fundamental mode in k-th harmonic generation, in the present model due to fourth-order intrinsic squeezing we can get fourth-order squeezing, even in the absence of second-order squeezing in the chosen quadrature component. Namely, for instance for $\vartheta=\pi/2$ and $3\pi/2$

$$q_2^{(k)} = k(k-1)\langle n\rangle^{k-1} g_k^2 t^2 \geqslant 0,$$

i.e., in fact, there is no second-order squeezing in this quadrature. In turn, from (3.15) one then finds that

$$q_4^{(k)} = {}^1/_3 k(k-1)\langle n\rangle^{k-2} g_k^2 t^2\, [6\langle n\rangle - (k-2)(k-3)].$$

For k≥4, we still deal with the fourth-order intrinsic squeezing. In particular, for

$$\langle n\rangle < (k-2)(k-3)/6 \tag{3.19}$$

the quadrature component Q manifests the fourth-order squeezing in the absence of the second-order squeezing in this component. The condition (3.19) is analogous to that in the case of k-th harmonic generation (2.34).

To conclude briefly, the simple Jaynes-Cummings model appears capable of producing higher-order squeezing, whereas the multiphoton Jaynes-Cummings model can yield for k≥4 higher-order intrinsic squeezing.

4. SQUEEZING BY INTERACTION OF COHERENT LIGHT WITH THREE-LEVEL ATOMS

Here we will consider three-level atoms in "ladder" and "lambda" configurations of the atomic levels interacting with one or two field modes. To start with, let us study a three-level atom in lambda configuration interacting with a single-mode one-photon resonant field (degenerate case). Two lower levels are denoted by 1 and 3, while the upper, common level is denoted by 2. The atom-field coupling coefficients g_1 and g_2 describe the transitions between levels 1↔2 and 2↔3, respectively.

The Hamiltonian for the model in question reads

$$H = \hbar\omega_f a^+ a + \hbar \sum_{j=1}^{3} \omega_{0j} R_{jj} + \hbar g_1 (aR_{21} + a^+ R_{12}) + \hbar g_2 (aR_{23} + a^+ R_{32}). \tag{4.1}$$

The operator $R_{jj}=|j\rangle\langle j|$ represents the population of the level j, and $\hbar\omega_{\alpha j}$ is the corresponding energy. In turn, the operator $R_{ij}=|i\rangle\langle j|\,(i\neq j)$ describes transition of the atom from level j to level i. The transition frequencies are equal for a degenerate single-mode model

$$\hbar\,(\omega_{02}-\omega_{01}) = \hbar\,(\omega_{02}-\omega_{03}) = \hbar\omega_f.$$

If, initially, the atom starts in its level 1 and the field is in a coherent state $|\alpha\rangle$, the resulting interaction-picture wave function of the system is

$$|\psi(t)\rangle = \sum_n [C_1^{(n)}(t)\,|1,n\rangle + C_2^{(n)}(t)\,|2,n\rangle + C_3^{(n)}(t)\,|3,n\rangle]\, e^{-\langle n\rangle/2}\alpha^n (n!)^{-1/2}, \tag{4.2}$$

with the probability amplitudes

$$C_1^{(n)}(t) = \frac{1}{g_1^2+g_2^2}(g_2^2 + g_1^2\cos\Omega t), \qquad C_2^{(n)}(t) = -\frac{ig_1}{\sqrt{g_1^2+g_2^2}}\sin\Omega t,$$

$$C_3^{(n)}(t) = -\frac{g_1^2 g_2}{g_1(g_1^2+g_2^2)}(1-\cos\Omega t) = \frac{g_1 g_2}{g_1^2+g_2^2}(1-\cos\Omega t), \tag{4.3}$$

where

$$\Omega = [n\,(g_1^2+g_2^2)]^{1/2} \tag{4.4}$$

is the Rabi frequency of the oscillations of the system.

In this case, the auxiliary functions Φ, Ψ and X (3.6) have the form,

$$\Phi(t) = \langle n \rangle - \sum_n P_n |C_2^{(n)}(t)|^2,$$

$$\Psi(t) = \sum_n P_n \left[C_1^{(n)}(t) C_1^{(n+1)}(t) + C_2^{(n)*}(t) C_2^{(n+1)}(t) \left(\frac{n}{n+1} \right)^{1/2} + \right.$$

$$\left. + C_3^{(n)}(t) C_3^{(n+1)}(t) \right],$$

$$\mathrm{X}(t) = \sum_n P_n \left[C_1^{(n)}(t) C_1^{(n+2)}(t) + C_2^{(n)*}(t) C_2^{(n+2)}(t) \left(\frac{n}{n+2} \right)^{1/2} + \right.$$

$$\left. + C_3^{(n)}(t) C_3^{(n+2)}(t) \right], \tag{4.5}$$

where P_n, as previously, is the Poissonian distribution. It is certain that for $g_2=0$ the results for the Jaynes-Cummings model with one-photon transitions are recovered. The general forms of the variances of the quadrature components Q and P are identical with those formerly presented (3.5).

In the short-time approximation, within an accuracy of t^4, for $\vartheta=0$ or π we find that the quadrature Q is squeezed, namely

(4.6)

Squeezing is here diminished compared to the Jaynes-Cummings model of a two-level atom, if $g^2_1 > 2g^2_2$. In principle, for $g^2_1 < 2g^2_2$, there could be squeezing in the quadrature P for these initial phases.

Let us consider now a three-level *ladder* atom, with levels in the sequence 1-2-3, interacting with a *single* resonant initially coherent field-mode. Such a degenerate three-level ladder atom can be realized by the presence of an external magnetic field which modifies the atomic transition frequencies, making them equal ($\omega_{1\to 2} = \omega_{2\to 3} = \omega$). For such a system the Hamiltonian reads

$$H = \hbar\omega_f a^+ a + \sum_{j=1}^{3} \hbar\omega_{0j} R_{jj} + \sum_{\alpha=1}^{2} \hbar g_\alpha (a R_{\alpha+1,\,\alpha} + a^+ R_{\alpha,\,\alpha+1}). \tag{4.7}$$

Again, we assume that the atom is initially in its lower level 1. The interaction-picture wave function has the form

$$|\psi(t)\rangle = \sum_n [C_1^{(n)}(t)\,|1, n\rangle + C_2^{(n)}(t)\,|2, n-1\rangle +$$

$$+ C_3^{(n)}(t)\,|3, n-2\rangle]\, e^{-\langle n \rangle/2} \alpha^n (n!)^{-1/2}, \tag{4.8}$$

and the probability amplitudes are found to be

$$C_1^{(n)}(t) = \Omega^{-2}\,[g_2^2\,(n-1) + g_1^2 n \cos \Omega t],$$
$$C_2^{(n)}(t) = -i g_1 \Omega^{-1} n^{1/2} \sin \Omega t,$$
$$C_3^{(n)}(t) = -g_1 g_2 \Omega^{-2}\,(n\,(n-1))^{1/2}\,(1 - \cos \Omega t), \tag{4.9}$$

where the Rabi frequency is now

$$\Omega = [g_1^2 n + g_2^2\,(n-1)]^{1/2}. \tag{4.10}$$

The variances of the quadratures are expressed through functions Φ, Ψ and X, which with the help of the wave-function (4.8), are found to be

$$\Phi(t) = \langle n\rangle - \sum_n P_n[|C_2^{(n)}(t)|^2 + 2C_3^{(n)*}(t)],$$

$$\Psi(t) = \sum_n P_n\Big\{C_1^{(n)}(t)C_1^{(n+1)}(t) + \Big[\Big(\frac{n}{n+1}\Big)^{1/2} C_2^{(n)*}(t)C_2^{(n+1)}(t) +$$

$$+ \Big(\frac{n-1}{n+1}\Big)^{1/2} C_3^{(n)}(t)C_3^{(n+1)}(t)\Big]\Big\},$$

$$X(t) = \sum_n P_n\Big\{C_1^{(n)}(t)C_1^{(n+2)}(t) + \Big(\frac{n}{n+2}\Big)^{1/2}\Big[C_2^{(n)*}(t)C_2^{(n+2)}(t) +$$

$$+ \Big(\frac{n-1}{n+1}\Big)^{1/2} C_3^{(n)}(t)C_3^{(n+2)}(t)\Big]\Big\}. \tag{4.11}$$

For short times and ϑ=0 and π, within an accuracy of t^4 we arrive at

$$\sigma_Q = 1 - {}^1/_3\langle n\rangle\, g_1^2 t^4\, (g_1^2 - {}^1/_2 g_2^2). \tag{4.12}$$

In this model the initial squeezing is also degraded compared to the Jaynes-Cummings model and vanishes in Q if $g^2_1<g^2_2/2$. If the atom starts initially in the upper level 3 there is no squeezing at the beginning of the interaction [49].

Let us now consider a three-level atom in *lambda* configuration of levels interacting with *two resonant* cavity-field modes with frequencies ω_1 and ω_2. The first mode interacts with levels 1 and 2 through the coupling coefficient g_1. In turn, the second mode couples the atomic levels 2 and 3; the corresponding coupling parameter is g_2. The Hamiltonian describing the system in the rotating wave approximation reads

$$H = \sum_{\beta=1}^{2} \hbar\omega_\beta a_\beta^+ a_\beta + \sum_{j=1}^{3} \hbar\omega_{0j} R_{jj} + \hbar g_1 (a_1 R_{21} + a_1^+ R_{12}) + \hbar g_2 (a_2 R_{23} + a_2^+ R_{32}). \tag{4.13}$$

The operators α_β and $\alpha^+{}_\beta$ describe two one-photon resonant modes (β=1,2) of the radiation field:

$$\omega_{02} - \omega_{01} = \omega_1, \qquad \omega_{02} - \omega_{03} = \omega_2.$$

If the atom is initially in its lower state and the field modes are in coherent states $|\alpha_1>$ and $|\alpha_2>$ with the amplitudes

$$\alpha_k = <n_k>^{1/2} \exp(i\vartheta_k),\ k = 1,2,$$

the resulting interaction-picture wave function of the atom-field system is:

$$|\psi(t)\rangle = \sum_{n_1, n_2} [C_1^{(n_1, n_2)}(t)|1, n_1, n_2\rangle + C_2^{(n_1, n_2)}(t)|2, n_1 - 1, n_2\rangle +$$

$$+ C_3^{(n_1, n_2)}(t)|3, n_1 - 1, n_2 + 1\rangle]\, e^{-(\langle n_1\rangle + \langle n_2\rangle)/2} \frac{\langle n_1\rangle^{n_1}}{n_1!} \frac{\langle n_2\rangle^{n_2}}{n_2!}, \tag{4.14}$$

and the probability amplitudes read

$$C_1^{(n_1, n_2)}(t) = \Omega^{-2}[g_2^2(n_2 + 1) + g_1^2 n_1 \cos\Omega t],$$

$$C_2^{(n_1, n_2)}(t) = -ig_1\Omega^{-1} n_1^{1/2} \sin\Omega t,$$

$$C_3^{(n_1, n_2)}(t) = -g_1 g_2 \Omega^{-2}(n_1(n_2 + 1))^{1/2}(1 - \cos\Omega t), \tag{4.15}$$

where the Rabi frequency Ω is as follows

$$\Omega = [g_1^2 n_1 + g_2^2(n_2 + 1)]^{1/2}. \tag{4.16}$$

With the help of the wave function (4.14) the auxiliary functions $\Phi^{(\beta)}$, $\Psi^{(\beta)}$ and $X^{(\beta)}$ for each mode are found to be

$$\Phi^{(1)}(t) = \langle n_1\rangle - 1 + \sum_{n_1, n_2} P(n_1) P(n_2) C_1^{(n_1, n_2)^2}(t),$$
$$\Phi^{(2)}(t) = \langle n_2\rangle + \sum_{n_1, n_2} P(n_1) P(n_2) C_0^{(n_1, n_2)^2}(t),$$
$$\Psi^{(1)}(t) = \sum_{n_1, n_2} P(n_1) P(n_2) \Big\{ C_1^{(n_1, n_2)}(t) C_1^{(n_1+1, n_2)}(t) + \left(\frac{n_1}{n_1+1}\right)^{1/2} \times$$
$$\times [C_2^{*(n_1, n_2)}(t) C_2^{(n_1+1, n_2)}(t) + C_3^{(n_1, n_2)}(t) C_3^{(n_1+1, n_2)}(t)]\Big\},$$
$$\Psi^{(2)}(t) = \sum_{n_1, n_2} P(n_1) P(n_2) \Big[C_1^{(n_1, n_2)}(t) C_1^{(n_1, n_2+1)}(t) + C_2^{*(n_1, n_2)}(t) C_2^{(n_1, n_2+1)}(t) +$$
$$+ \left(\frac{n_2+2}{n_2+1}\right)^{1/2} C_3^{(n_1, n_2)}(t) C_3^{(n_1, n_2+1)}(t)\Big],$$
$$X^{(1)}(t) = \sum_{n_1, n_2} P(n_1) P(n_2) \Big\{ C_1^{(n_1, n_2)}(t) C_1^{(n_1+2, n_2)}(t) + \left(\frac{n_1}{n_1+2}\right)^{1/2} \times$$
$$\times [C_2^{*(n_1, n_2)}(t) C_2^{(n_1+2, n_2)}(t) + C_3^{(n_1, n_2)}(t) C_3^{(n_1+2, n_2)}(t)]\Big\},$$
$$X^{(2)}(t) = \sum_{n_1, n_2} P(n_1) P(n_2) \Big[C_1^{(n_1, n_2)}(t) C_1^{(n_1, n_2+2)}(t) + C_2^{*(n_1, n_2)}(t) C_2^{(n_1, n_2+2)}(t) +$$
$$+ \left(\frac{n_2+3}{n_2+1}\right)^{1/2} C_3^{(n_1, n_2)}(t) C_3^{(n_1, n_2+2)}(t)\Big], \tag{4.17}$$

$P(n_1)$ and $P(n_2)$ are the Poissonian distributions of the photon numbers of modes 1 and 2, respectively. To obtain the above presented form of the function $\Phi^{(1)}(t)$ we have used the normalization condition for the probability amplitudes.

The variances of the quadrature components Q_β and P_β (β=1,2) are given by (3.5) on insertion of the functions (4.17). In particular, when the mode 2 is initially in the vacuum state, i.e. $P(n_2)=\delta_{n_2, 0}$, then

$$\langle n_2\rangle = 0,$$
$$\Phi^{(2)}(t) = \sum_{n_1} P(n_1) C_3^{(n_1)^2}(t) > 0. \tag{4.18}$$

Hence

$$\sigma_{Q_2} = \sigma_{P_2} = 1 + 2\Phi^{(2)}(t) > 1 \tag{4.19}$$

and there is no squeezing in mode 2 because of lack of any initial phase information for this mode.

For short times, within an accuracy of t^4 for the mode 1 we find

$$\sigma_{Q_1} = 1 - {}^1/_3\langle n_1\rangle g_1^4 t^4 \cos 2\theta_1,$$
$$\sigma_{P_1} = 1 + {}^1/_3\langle n_1\rangle g_1^4 t^4 \cos 2\theta_1, \tag{4.20}$$

irrespective of the magnitude of α_2. In (4.20) ϑ_1 is the initial phase of the amplitude α_1. Depending on ϑ_1, one or the other quadrature may be squeezed, and the squeezing does not depend on $\langle n_2\rangle$ and g_2. In turn, for mode 2 in the same order of approximation in t we get the enhanced quantum fluctuations in both quadratures, namely

$$\sigma_{Q_2} = \sigma_{P_2} = 1 + {}^1/_2\langle n_1\rangle g_1^2 g_2^2 t^4 \tag{4.21}$$

irrespective of $<n_2>$ and the initial phase ϑ_2 of the amplitude α_2. For $g_2=0$ mode 2 would obviously remain in a coherent state since it did not interact with the atom.

To complete our discussion of squeezing by interaction of coherent light with a three-level atom, let us discuss a three-level *ladder* atom interacting with *two one-photon resonant* cavity-field modes. Previously, the mode 1 couples the atomic levels 1 and 2 while the mode 2 couples the intermediate level 2 and the upper level 3. The corresponding coupling coefficients are g_1 and g_2, respectively. The Hamiltonian for the system in question, in the rotating wave approximation reads

$$H = \sum_{\beta=1}^{2} \hbar\omega_\beta a_\beta^+ a_\beta + \sum_{j=1}^{3} \hbar\omega_{0j} R_{jj} + \sum_{\beta=1}^{2} \hbar g_\beta (a_\beta R_{\beta+1,\,\beta} + a_\beta^+ R_{\beta,\,\beta+1}). \tag{4.22}$$

If, initially, the atom starts in the lower state 1 and the field modes are in coherent states $|\alpha_1>$ and $|\alpha_2>$, the interaction-picture wave function of the system is

$$|\psi(t)\rangle = \sum_{n_1,\,n_2} [C_1^{(n_1,\,n_2)}(t)\,|1, n_1, n_2\rangle + C_2^{(n_1,\,n_2)}\,|2, n_1 - 1, n_2\rangle +$$
$$+ C_3^{(n_1,\,n_2)}(t)\,|3, n_1 - 1, n_2 - 1\rangle]. \tag{4.23}$$

The probability amplitudes have the form

$$C_1^{(n_1,\,n_2)}(t) = \Omega^{-2}(g_2^2 n_2 + g_1^2 n_1 \cos\Omega t),$$
$$C_2^{(n_1,\,n_2)}(t) = -\,i g_1 \Omega^{-1} (n_1)^{1/2} \sin\Omega t,$$
$$C_3^{(n_1,\,n_2)}(t) = -\,g_1 g_2 \Omega^{-2} (n_1 n_2)^{1/2} (1 - \cos\Omega t), \tag{4.24}$$

where the Rabi frequency reads

$$\Omega = (g_1^2 n_1 + g_2^2 n_2)^{1/2}. \tag{4.25}$$

Finally, we find that the general form of the auxiliary functions $\Phi^{(1)}$, $\Psi^{(1)}$ and $X^{(1)}$ for the mode 1 is given by Equations (4.17 but, obviously, the probability amplitudes are different and they are presented in the formulas (4.24). For mode 2 we arrive at

$$\Phi^{(2)}(t) = \langle n_2\rangle - \sum_{n_1,\,n_2} P(n_1)\,P(n_2)\,C_3^{(n_1,\,n_2)^2}(t),$$

$$\Psi^{(2)}(t) = \sum_{n_1,\,n_2} P(n_1)\,P(n_2)\Big[C_1^{(n_1,\,n_2)}(t)\,C_1^{(n_1,\,n_2+1)}(t) + C_2^{*(n_1,\,n_2)}(t)\,C_2^{(n_1,\,n_2+1)}(t) +$$
$$+ \left(\frac{n_2}{n_2+1}\right)^{1/2} C_3^{(n_1,\,n_2)}(t)\,C_3^{(n_1,\,n_2+1)}(t)\Big],$$

$$X^{(2)}(t) = \sum_{n_1,\,n_2} P(n_1)\,P(n_2)\Big[C_1^{(n_1,\,n_2)}(t)\,C_1^{(n_1,\,n_2+2)}(t) + C_2^{*(n_1,\,n_2)}(t)\,C_2^{(n_1,\,n_2+2)}(t) +$$
$$+ \left(\frac{n_2}{n_2+2}\right)^{1/2} C_3^{(n_1,\,n_2)}(t)\,C_3^{(n_1,\,n_2+2)}(t)\Big]. \tag{4.26}$$

Using the above functions, within an accuracy of t^4 we get

$$\sigma_{Q_1} = 1 - {}^1/_3 \langle n_1\rangle g_1^4 t^4 \cos 2\theta_1,$$
$$\sigma_{P_1} = 1 + {}^1/_3 \langle n_1\rangle g_1^4 t^4 \cos 2\theta_1. \tag{4.27}$$

In this model the short-time squeezing is also independent of the mean photon number $<n_2>$ and of the coupling coefficient g_2. In turn, the quantum fluctuations in mode 2 appear unchanged in short times within an accuracy up to t^4, i.e. $\sigma_{Q_2}=\sigma_{P_2}=1$.

5. UNCERTAINTY RELATIONS FOR THE JAYNES-CUMMINGS - TYPES MODELS

From our considerations carried out in former sections, it is easily seen that for any initially coherent resonant cavity field mode interacting with a two- or multi-level atom (including degenerate atoms and multiphoton transitions), the time evolution of the variances of the quadratures Q and P is to be [50]

$$\sigma_Q(t) = 1 + 2\,[\Phi(t) - \langle n\rangle \Psi^2(t)] + 2\langle n\rangle [X(t) - \Psi^2(t)] \cos 2\theta,$$

$$\sigma_P(t) = 1 + 2\,[\Phi(t) - \langle n\rangle \Psi^2(t)] - 2\langle n\rangle [X(t) - \Psi^2(t)] \cos 2\theta, \tag{5.1}$$

where <n> denotes the initial mean photon-number and ϑ-the initial phase of coherent light. The forms of the functions Φ, Ψ and X are obviously dependent on the structure of the atomic levels and on the photon multiplicity of the atomic transitions.

The product of the variances (5.1) reads

$$\sigma_Q(t)\,\sigma_P(t) = [1 + 2(\Phi - \langle n\rangle \Psi^2)]^2 - 4\langle n\rangle^2 (X - \Psi^2)^2 \cos^2 2\theta, \tag{5.2}$$

where, for simplicity, the argument t of the functions has been omitted. The correlation coefficient may be presented in the following equivalent form

$$\gamma(t) = 2\,(\mathrm{Im}\,\langle a_s^2\rangle - 2\,\mathrm{Re}\,\langle a_s\rangle\,\mathrm{Im}\,\langle a_s\rangle)\,(\sigma_Q\sigma_P)^{-1/2} \tag{5.3}$$

and, hence, for an initially coherent field-mode it can be expressed by means of the auxiliary functions Ψ and X as follows:

$$\gamma(t) = 2\langle n\rangle (X - \Psi^2) \sin 2\theta\,(\sigma_Q\sigma_P)^{-1/2}. \tag{5.4}$$

Finally, we have

$$\sigma_Q(t)\,\sigma_P(t)\,[1 - \gamma^2(t)] = [1 + 2(\Phi - \langle n\rangle \Psi^2)]^2 - 4\langle n\rangle^2 (X - \Psi^2)^2. \tag{5.5}$$

i.e. the above quantity for the models under consideration is independent of the initial phase ϑ of coherent light for any time t.

In general, the quantity (5.5) has to be studied numerically. For a great <n> it can be calculated with the help of saddle point technique. We are going to present short-time solutions for the multiphoton Jaynes-Cummings model.

For the photon multiciplicity k=1, from (3.6) we find that the first nonvanishing approximation in t for the function $X^{(1)}$-$\Psi^{(1)2}$ is t^4 (compare to (3.8), namely

$$X^{(1)} - \Psi^{(1)2} = -{}^1/_6\,\langle n\rangle\, g_1^4 t^4. \tag{5.6}$$

Hence, the correlation coefficient $\gamma^{(1)}(t)$ amounts to

$$\gamma^{(1)} = -{}^1/_3\langle n\rangle\, g_1^4 t^4 \sin 2\theta\,(\sigma_Q^{(1)}\sigma_P^{(1)})^{-1/2} \tag{5.7}$$

and the formula (5.5) must be studied within an accuracy up to t^8. In fact, however, the function $\Phi^{(1)}$-<n>$\Psi^{(1)2}$ is already different from zero in the sixth-order approximation in t:

$$\Phi^{(1)} - \langle n\rangle \Psi^{(1)2} = {}^1/_9\langle n\rangle^2\, g_1^6 t^6. \tag{5.8}$$

Finally, we have

$$\sigma_Q^{(1)}\sigma_P^{(1)}\,(1 - \gamma^{(1)})^2 = 1 + {}^4/_9\langle n\rangle^2 g_1^6 t^6 + O(t^8), \tag{5.9}$$

and the field does not then remain in a minimum uncertainty state. For $\vartheta=\pi/4+m\pi/2$ (m=0,1,2,3) we have maximal correlation of the quadratures, and their variances are equal to

$$\sigma_Q^{(1)} = \sigma_P^{(1)} = 1 + {}^2/_9\langle n\rangle^2\, g_1^6 t^6.$$

For the phases $\vartheta\neq\pi/4+m\pi/2$ and $\vartheta\neq m\pi/2$ the quadratures are correlated as well but one of them is squeezed. For $\vartheta=m\pi/2$ the quadratures become uncorrelated and squeezing in one of them then reaches its maximal value. So, in the simple Jaynes-Cummings model of a single two-level atom coupled to a single one-photon resonant mode, an initially coherent field is transformed into a squeezed state or in a squeezed and correlated state, but none of them appears a minimum uncertainty state.

In turn, for $k\geq 2$ we arrive at

$$X^{(k)} - \Psi^{(k)2} = -{}^1/_2 k\,(k-1)\,\langle n\rangle^{k-2} g_k^2 t^2 \tag{5.10}$$

(compare to (3.7)). The correlation coefficient has the following form,

$$\gamma^{(k)}(t) = -k\,(k-1)\,\langle n\rangle^{k-1} g_k^2 t^2 \sin 2\theta. \tag{5.11}$$

Thus, in order to check the fulfillment of the uncertainty relations, the function $\Phi^{(k)}-\langle n\rangle\Psi^{(k)2}$ has to be calculated up to the fourth-order in t. After some algebra we find that

$$\Phi^{(k)} - \langle n\rangle\,\Psi^{(k)2} = 1 + \frac{1}{4}\langle n\rangle^k g_k^4 t^4 \left\{\sum_{s=0}^{k}\langle n\rangle^s \times \right.$$
$$\left. \times \frac{(k!)^2}{s!\,(s+1)!\,(k-s)!}\left[k\,(s+1) - \langle n\rangle\,(k-s)\right] - k^2\langle n\rangle^{k-1}\right\}. \tag{5.12}$$

On insertion of (5.12) and (5.10) into (5.5), within an accuracy of t^4, we get

$$\sigma_Q^{(k)}\sigma_P^{(k)}\,(1-\gamma^{(k)2}) = 1 + k^2\langle n\rangle^k g_k^4 t^4 \times$$
$$\times\left\{\sum_{s=0}^{k}\langle h\rangle^s \frac{[(k-1)!]^2}{s!\,(s+1)!\,(k-s)!}\left[k\,(s+1) - \langle h\rangle\,(k-s)\right] - \right.$$
$$\left. - \langle n\rangle^{k-1} - (k-1)^2\langle n\rangle^{k-2}\right\}. \tag{5.13}$$

Careful inspection of the above formula shows that within the calculated accuracy

$$\sigma_Q^{(k)}\sigma_P^{(k)}[1-(\gamma^{(k)})^2] = \begin{cases}1 \text{ for } k=2\\ >1 \text{ for } k\geq 3\end{cases}, \tag{5.14}$$

In other words, the two-photon Jaynes-Cummings model distinguishes itself from the others. For short times, irrespective of <n>, an initially coherent field remains for this model in a minimum Heisenberg's uncertainty squeezed state for the phases ϑ for which $\gamma^{(2)}=0$, i.e. for $\vartheta=m\pi/2$ (m-as previously), or remains in a minimum Schrodinger's-Robertson's uncertainty correlated state if $\gamma^{(2)}\neq 0$; then, the quantum fluctuations in both quadratures are enhanced for $\vartheta=\pi/4+m\pi/2$ and squeezed in one of them for $\vartheta\neq\pi/4+m\pi/2$. But squeezing is not maximal.

6. DICKE MODEL. CONSTRUCTING SU(2) - INVARIANT PERTURBATION THEORY

Consider the generation of squeezed states by spontaneous radiation of a system of two-level atoms in an ideal resonator. This process can be described with the aid of the Dicke model Hamiltonian [56]. It is well known that such a system exhibits cooperative effects which are absent in the exactly solvable Jaynes-Cummings model. In the first place, this is the "superradiance" effect in the case of initial conditions symmetrical with respect to permutations of atoms [56], and "radiation capture" for asymmetrical initial conditions (see, e.g., [57]). On the other hand, one is unable to write down an exact solution even for the simplest forms of the Dicke model based on the "small volume approximation" (see below) unless the replacement of the quantized electromagnetic field by some classical field is made. The distinction of the Dicke model from the Jaynes-Cummings one causes troubles investigating the physical effects for which the quantized nature of field is significant.

In the present section we construct the mathematical apparatus suitable for the current situation and obtain approximate expressions for the energy levels of the system and for average values of the atomic subsystem energy at arbitrary instants of time. As will be shown further, our approximation is rather exact in the case when a small number $s \ll N$ of atoms is excited at the initial moment (N is the whole number of atoms), provided the excitation is symmetrical with respect to arbitrary permutations of atoms.

In the next section the scheme constructed will be applied for calculating the degree of squeezing of the spontaneous radiation from the system of two-level atoms.

The Hamiltonian of the standard Dicke model describing a system of N identical two-level atoms (molecules) placed in an ideal resonator and interacting with a single quantized mode of the electromagnetic field reads [56-59]

$$H = \omega_f a^+ a + \varepsilon_0 \sum_{j=1}^{N} s_3^{(j)} + g\left(\sum_{j=1}^{N} a^+ s_-^{(j)} + \sum_{j=1}^{N} a s_+^{(j)}\right) \qquad (\hbar = c = 1). \tag{6.1}$$

Here α^+, α means the creation and annihilation operators of quanta corresponding to the field mode with frequency ω_f, $s^{(j)}{}_3 = \frac{1}{2}\begin{Vmatrix} 1 & 0 \\ 0 & -1 \end{Vmatrix}$ is the Hamiltonian of the isolated j-th atom; $s^{(j)}{}_+ = \begin{Vmatrix} 0 & 1 \\ 0 & 0 \end{Vmatrix}$ and $s^{(j)}{}_- = \begin{Vmatrix} 0 & 0 \\ 1 & 0 \end{Vmatrix}$ are the Pauli matrices describing the transitions between the resonance levels of the j-th atom; ε_0 is the distance between levels which is assumed the same for all atoms; g is the coupling constant determined by the atomic dipole matrix element. Note that operators $s^{(j)}{}_\pm$ and $s^{(j)}{}_3$ satisfy the standard commutation relations of the SU(2) group. Hamiltonian (6.1) is based on the following widely used approximations (for the details see [58-61]):

1) the resonator losses have not been taken into account, i.e. it is supposed that the characteristic times at which the cooperative effects manifest themselves are at least an order of magnitude less than the time of photons absorption by the resonator walls;

2) the dimension of the domain inside the resonator occupied by atoms is assumed much less than the field wavelength, so that all atoms feel the same field, and coupling constant g turns out the same for all atoms (this assumption is usually called the "small volume approximation");

3) the interatomic interactions are neglected; it should be mentioned that although this assumption apparently contradicts the previous one, nonetheless their widespread joint usage, as a rule, yields reasonable results (see, e.g., a discussion in [59]);

4) the transitions to the rest (nonresonance) atomic levels are not considered, as well as the influence of the resonator higher modes;

5) the "rotating wave approximation" is used, i.e. the terms like $\alpha^+ s^{(j)}{}_+$ and $\alpha s^{(j)}{}_-$ appearing in the higher orders of the perturbation theory and describing the virtual processes not conserving energy are excluded from the Hamiltonian;

6) further we assume the exact resonance condition: the common for all atoms transition frequency ε_0 equals exactly the principal resonator eigenfrequency ω_f.

It is convenient to rewrite Hamiltonian (6.1) introducing the collective operators

$$S_3 = \sum_{j=1}^{N} s_3^{(j)}, \qquad S_\pm = \sum_{j=1}^{N} s_\pm^{(j)},$$

also satisfying the SU(2) group commutation relations

$$[S_3,\ S_\pm] = \pm S_\pm,\ [S_+,\ S_-] = 2S_3.$$

Then

$$H = \varepsilon_0\,(a^+a + S_3) + g\,(a^+S_- + aS_+). \tag{6.2}$$

First of all, note that this Hamiltonian, as well as the Jaynes-Cummings model describing the case of N=1, preserves the "number of excitations." This means that operator

$$\hat{N} = a^+a + S_3$$

commutes with operator (6.2). It is easy to see that the "excitation number" conservation is the consequence of the rotating wave approximation. Under the strict resonance condition the Hamiltonian can be written as follows:

$$H = \varepsilon_0\hat{N} + g\hat{V}, \qquad \hat{V} \equiv (a^+S_- + aS_+), \qquad [\hat{N},\ \hat{V}] = 0. \tag{6.3}$$

Consequently, only transitions inside subspaces corresponding to the fixed eigenvalues of operator $\hat{N}$ are possible in the process of evolution.

Further, we shall consider only those initial states which are symmetrical with respect to all permutations of atoms. Then this symmetry will be conserved during the evolution due to the invariance of Hamiltonian (6.1) with respect to permutations of atoms. Therefore, we may introduce the following orthonormal basis in the space of states of the system:

$$|\,s,\ m\rangle \equiv |\,s - m\rangle_a \otimes |\,m\rangle_f \qquad (0 \leqslant m \leqslant s). \tag{6.4}$$

Here $|m\rangle_f$ is the m-quantum field state, $|s\text{-}m\rangle_\alpha$ is the state of the atomic system with (s-m) excited atoms symmetrized over all permutations:

$$|\,k\rangle_a = \left[\frac{N!}{k!\,(N-k)!}\right]^{-1/2} \sum |+\rangle_{j_1} \otimes |+\rangle_{j_2} \otimes \cdots \otimes |+\rangle_{j_k} \otimes \prod_{j \neq j_1,\, j_2, \ldots, j_k} |-\rangle_j,$$

the summation being performed over all ways of choosing k numbers $j_1, j_2, \ldots, j_k$ from N. Symbols $|+\rangle_j$ and $|-\rangle_j$ designate the excited and ground states of the j-th atom, respectively.

The term $\varepsilon_0\hat{N}$ in Hamiltonian (6.3) describing noninteracting atoms and field leads only to a phase factor in transition amplitudes. Consider the action of the interaction operator $\hat{V}$ on the basis states $|s,m\rangle$:

$$\begin{aligned}\hat{V}\,|\,s,\ m\rangle &= ((m+1)\,(s-m)\,(N-s+m+1))^{1/2}\,|\,s,\ m+1\rangle + \\ &+ (m\,(s-m+1)\,(N-s+m))^{1/2}\,|\,s,\ m-1\rangle \qquad (0 \leqslant m \leqslant s).\end{aligned} \tag{6.5}$$

We see that the evolution is set inside the (s+1)- dimensional subspace. Let us choose as the initial state the vector

$$|\,s,\ 0\rangle \equiv |\,s\rangle_a \otimes |\,0\rangle_f,$$

describing s symmetrically excited atoms in the absence of photons. Let s<<N. It is convenient to choose the quantity

$$\varepsilon = (N - (s-1)/2)^{-1} \ll 1 \tag{6.6}$$

as the small parameter of the problem. Then the square roots in (6.5) can be expanded in Taylor's series as follows,

$$((m+1)(s-m)(N-s+m+1))^{1/2} = \left(N - \frac{s-1}{2}\right)^{1/2} ((s-m)(m+1))^{1/2} \times$$
$$\times \left\{1 + \frac{\varepsilon}{2}\left(m - \frac{s-1}{2}\right) - \frac{\varepsilon^2}{2}\left(m - \frac{s-1}{2}\right)^2 + \dots\right\},$$
$$(m(s-m+1)(N-s+m))^{1/2} = \left(N - \frac{s-1}{2}\right)^{1/2} (m(s-m+1))^{1/2} \times$$
$$\times \left\{1 + \frac{\varepsilon}{2}\left(m - \frac{s+1}{2}\right) - \frac{\varepsilon^2}{8}\left(m - \frac{s+1}{2}\right)^2 + \dots\right\} \quad (0 \leqslant m \leqslant s).$$

Consequently, the interaction Hamiltonian can be represented in a series form over the small parameter ε:

$$V = \varepsilon^{-1/2} (H^{(0)} + \varepsilon H^{(1)} + \varepsilon^2 H^{(2)} + \dots) \equiv \varepsilon^{-1/2} H'. \tag{6.7}$$

Let us consider in detail the structure of matrices $H^{(0)}$, $H^{(1)}$, $H^{(2)}$ in basis (6.4).

$$H^{(0)} | s, m\rangle = \sqrt{(m+1)(s-m)}\, | s, m+1\rangle +$$
$$+ \sqrt{m(s-m+1)}\, | s, m-1\rangle \quad (0 \leqslant m \leqslant s). \tag{6.8}$$

One can easily see that operator $H^{(0)}$ acts as the generator of the (s+1)-dimensional representation of SU(2) group. This fact makes it possible to construct the perturbation theory based on the evolution in the frame of the representation of SU(2) as the zeroth approximation. The first- and second-order corrections correspond to a small deviation of the exact evolution governed by Hamiltonian (6.3) from the motion through the representation of SU(2). The utilization of the properties of SU(2) representations matrix elements enables to calculate all first- and second-order corrections in an explicit form. For the sake of convenience, we give some well-known properties of SU(2) group representations in Appendix A. (For details see N.Ya. Vilenkin's book [62].)

For $H^{(1)}$ we have

$$H^{(1)} | s, m\rangle = {}^1/_2 (m - (s-1)/2) \sqrt{(m+1)(s-m)}\, | s, m+1\rangle +$$
$$+ {}^1/_2 (m - (s+1)/2) \sqrt{m(s-m+1)}\, | s, m-1\rangle \equiv$$
$$\equiv H^{(1)}_{m,\, m+1} | s, m+1\rangle + H^{(1)}_{m,\, m-1} | s, m-1\rangle, \quad H^{(1)}_{m,\, m+1} = H^{(1)}_{m+1,\, m}. \tag{6.9}$$

Finally, matrix elements of $H^{(2)}$ are determined by the relations

$$H^{(2)} | s, m\rangle = -{}^1/_8 (m - (s-1)/2)^2 \sqrt{(m+1)(s-m)}\, | s, m+1\rangle -$$
$$- {}^1/_8 (m - (s+1)/2)^2 \sqrt{m(s-m+1)}\, | s, m-1\rangle \equiv$$
$$\equiv H^{(2)}_{m,\, m+1} | s, m+1\rangle + H^{(2)}_{m,\, m-1} | s, m-1\rangle, \qquad H^{(2)}_{m,\, m+1} = H^{(2)}_{m+1,\, m}. \tag{6.10}$$

Let us emphasize once more that both $H^{(1)}$ and $H^{(2)}$ do not send the evolution out the space of the (s+1)-dimensional representation of SU(2).

Let us first start calculating energy eigenvalues and eigenvectors of Hamiltonian (6.1)-(6.5) up to the second-order terms (with respect to ε) for energy and the first-order terms for eigenvectors. Since operator $H^{(0)}$ is a generator of the SU(2) representation, it possesses the integer spectrum

$$H^{(0)}\,|m\rangle^{(0)} = \lambda_m^{(0)}\,|m\rangle^{(0)},$$
$$\lambda_0^{(0)} = s, \quad \lambda_1^{(0)} = s-2, \ldots, \quad \lambda_m^{(0)} = s-2m, \ldots, \lambda_s^{(0)} = -s. \tag{6.11}$$

The eigenvectors are as follows:

$$|m\rangle^{(0)} = \begin{pmatrix} \alpha_{0m}^{(s)} \\ \alpha_{1m}^{(s)} \\ \vdots \\ \alpha_{sm}^{(s)} \end{pmatrix}, \qquad {}^{(0)}\langle k\,|\,m\rangle^{(0)} = \delta_{km} \qquad (0 \leqslant m \leqslant s), \tag{6.12}$$

$$\alpha_{nm}^{(s)} = \left(\frac{n!\,m!}{(s-n)!\,(s-m)!\,2^s}\right)^{1/2} \sum_{j=0}^{\min(n,\,m)} \frac{(s-j)!\,(-1)^j\,2^j}{j!\,(n-j)!\,(m-j)!}\,. \tag{6.13}$$

The unitary transition matrix $\alpha^{(s)}$ relating basis $|s,k\rangle$ to basis $|m\rangle^{(0)}$ consisting of eigenvectors of $H^{(0)}$ has the dimension (s+1)x(s+1), and vectors $|m\rangle^{(0)}$ are just its columns

$$(\hat{\alpha}^{(s)})_{k,\,m} = \alpha_{km}^{(s)} = \langle s,k\,|\,m\rangle^{(0)}, \qquad |m\rangle^{(0)} = \hat{\alpha}^{(s)}\,|\,s,m\rangle. \tag{6.14}$$

Matrix $\hat{\alpha}^{(s)}$ is real and symmetrical due to the special choice of the phases of eigenvectors. The proof of formulas (6.11)-(6.13) is given in Appendix A, as well as the detailed properties of matrix $\hat{\alpha}^{(s)}$. Further we shall sometimes write the matrix elements $\alpha^{(s)}{}_{km}$ without the superscript s. The stationary Schrodinger equation reads

$$H'\,|\psi\rangle \equiv (H^{(0)} + \varepsilon H^{(1)} + \varepsilon^2 H^{(2)} + \ldots)\,|\psi\rangle = \lambda\,|\psi\rangle,$$
$$|\psi\rangle = \sum_{m=0}^{s} c_m\,|m\rangle^{(0)}, \qquad c_m^{(0)} = \delta_{mp} \qquad (0 \leqslant m, p \leqslant s),$$
$$\lambda = \lambda^{(0)} + \varepsilon\lambda^{(1)} + \varepsilon^2\lambda^{(2)} + \ldots, \qquad c_m = c_m^{(0)} + \varepsilon c_m^{(1)} + \varepsilon^2 c_m^{(2)} + \ldots \tag{6.15}$$

The corrections to the p-th eigenvalue and the corresponding eigenfunction are given by the well-known formulas

$$\lambda_p^{(1)} = {}^{(0)}\langle p\,|\,H^{(1)}\,|\,p\rangle^{(0)}, \tag{6.16}$$

$$\lambda_p^{(2)} = \sum_{m=0}^{s} \frac{|\,{}^{(0)}\langle p\,|\,H^{(1)}\,|\,m\rangle^{(0)}\,|^2}{\lambda_p^{(0)} - \lambda_m^{(0)}} + {}^{(0)}\langle p\,|\,H^{(2)}\,|\,p\rangle^{(0)}, \tag{6.17}$$

$$c_k^{(1)} = \frac{{}^{(0)}\langle k\,|\,H^{(1)}\,|\,p\rangle^{(0)}}{\lambda_p^{(0)} - \lambda_k^{(0)}}\,, \tag{6.18}$$

$$|p\rangle = |p\rangle^{(0)} + \varepsilon \sum_{k=0}^{s} \frac{{}^{(0)}\langle k\,|\,H^{(1)}\,|\,p\rangle^{(0)}}{\lambda_p^{(0)} - \lambda_k^{(0)}}\,|k\rangle^{(0)}. \tag{6.19}$$

The matrix elements of operator $H^{(1)}$ in the basis $|m\rangle^{(0)}$ of eigenstates of operator $H^{(0)}$ are as follows:

$${}^{(0)}\langle p\,|\,H^{(1)}\,|\,m\rangle^{(0)} = \langle s,p\,|\,H^{(1)}\,|\,s,m\rangle = \langle s,p\,|\,\hat{\alpha}^{-1}H^{(1)}\hat{\alpha}\,|\,s,m\rangle,$$

where the transition matrix between bases (6.14) and (6.4) was used. As proved in Appendix B, matrices $\hat{\alpha}$ and $H^{(1)}$ commute

$$[\hat{\alpha}, H^{(1)}] = 0, \qquad \hat{\alpha}^{-1}H^{(1)}\hat{\alpha} = H^{(1)}.$$

We see that the transformation to the new basis of eigenstates of $H^{(0)}$ preserves the form (6.9) of matrix $H^{(1)}$. This fact demonstrates the advantage of the expansion (6.6)

(6.7), (6.15). Specifically, for the first-order corrections to the energy (6.16), which are equal to the diagonal matrix elements of $H^{(1)}$, we have in the old basis (6.9)

$$\lambda_p^{(1)} = 0. \tag{6.20}$$

The diagonal matrix elements of operator $H^{(2)}$ in the new basis ${}^{(0)}\langle p | H^{(2)} | p \rangle^{(0)}$ are given in Appendix C. According to (6.17) calculating the second order corrections to the energy we get

$$\lambda_p^{(2)} = -\frac{(s-2p)}{16}\left\{5p(s-p) - \frac{(s-1)(s-2)}{2}\right\}. \tag{6.21}$$

Taking into account the results of Appendix B, one can derive from (6.18) and (6.19) the following expressions for eigenstates in the first order with respect to ε:

$$A_{mp}^{(1)} = \alpha_{mp} - {}^1/_2\varepsilon\,\{H^{(1)}_{p-1,\,p}\alpha_{m,\,p-1} - H^{(1)}_{p+1,\,p}\alpha_{m,\,p+1}\}, \tag{6.22}$$

the vector $|p\rangle^{(0)}$ being given by (6.12) and (6.13).

Now we are proceeding to calculating the average energy stored in the atomic subsystem to time instant t. In the Schrodinger picture this quantity equals the matrix element of operator $S_3 = \sum_{j=1}^{N} s_3^{(j)}$ between the corresponding time dependent state vectors

$$E_{\text{ат}} = \langle \psi(t) | S_3 | \psi(t) \rangle = \langle s, 0 | e^{i\hat{H}t} S_3 e^{-i\hat{H}t} | s, 0 \rangle. \tag{6.23}$$

The average amount of energy in the photon subsystem equals under the same conditions

$$E_{\text{фот}} = \langle \psi(t) | a^+ a | \psi(t) \rangle = \langle \psi(t) | (\hat{N} - S_3) | \psi(t) \rangle = \left(s - \frac{N}{2}\right) - E_{\text{ат}}.$$

Let $|p\rangle$ be an exact eigenvector of the whole Hamiltonian $H' = \varepsilon^{1/2}V$, i.e. $H'|p\rangle = \lambda_p |p\rangle$, and A_{mn} be the matrix element of the transition from basis $|s,m\rangle$ to $|p\rangle$-basis: $|p\rangle = A|\hat{s},p\rangle$, $A_{mp} = \langle s,m|p\rangle$. Since the interaction Hamiltonian matrix (6.5) is real, the phases of eigenvectors $|p\rangle$ can be chosen in such a way that matrix $\|A_{mp}\|$ is real and orthogonal: $(A^{-1})_{mp} = A^*_{pm} = A_{pm}$. Then introducing the new time scale $\tau = gt\varepsilon^{-1/2}$ we get

$$\begin{aligned}
E_{\text{ат}} &= \sum_{p=0}^{s}\sum_{q=0}^{s} A_{0q}A^*_{0p} e^{i(\lambda_q-\lambda_p)\tau} \langle q | S_3 | p \rangle = \\
&= \sum_{p=0}^{s}\sum_{q=0}^{s}\sum_{m=0}^{s}\sum_{n=0}^{s} A^*_{op}A_{0q}A^*_{nq}A_{mp} \langle s,n | S_3 | s,m \rangle e^{i(\lambda_q-\lambda_p)\tau} = \\
&= \sum_{p=0}^{s}\sum_{q=0}^{s}\sum_{m=0}^{s} (s/2-m)\, A_{0p}A_{0q}A_{mp}A_{mq} e^{i(\lambda_q-\lambda_p)\tau} = \\
&= \sum_{m=0}^{s}\sum_{p=0}^{s} (s/2-m)\, |A_{0p}A_{mp}|^2 + \frac{s}{2}\sum_{p=0}^{s}\sum_{\substack{q=0\\(q\neq p)}}^{s} A_{0p}A_{0q} \times \\
&\times \left(\sum_{m=0}^{s} A_{mq}A_{mp}\right) e^{i(\lambda_q-\lambda_p)\tau} - 2\sum_{p=0}^{s}\sum_{q=p+1}^{s}\sum_{m=0}^{s} m A_{0p}A_{0q}A_{mp}A_{mq}\cos((\lambda_q-\lambda_p)\tau).
\end{aligned} \tag{6.24}$$

The second term in the right-hand side vanishes due to the orthogonality of matrix $\|A_{mp}\|$, and we arrive at the following exact expression for the energy of atomic subsystem at time t:

$$E_{\text{at}} = \frac{s}{2} - \sum_{p=0}^{s} A_{0p}^2 \sum_{m=0}^{s} m A_{mp}^2 - $$
$$-2 \sum_{p=0}^{s} \sum_{q=p+1}^{s} A_{0p} A_{0q} \Big(\sum_{m=0}^{s} m A_{mp} A_{mq} \Big) \cos\left((\lambda_q - \lambda_p)\tau\right). \tag{6.25}$$

The atomic subsystem energy variation rate is given by

$$\frac{dE_{\text{at}}}{dt} = -\frac{dE_f}{dt} = 2g\left(N - \frac{s-1}{2}\right)^{1/2} \sum_{p=0}^{s} \sum_{q=p+1}^{s} A_{0p} A_{0q} \times$$
$$\times \sum_{m=0}^{s} m(\lambda_q - \lambda_p) A_{mp} A_{mq} \sin\left((\lambda_q - \lambda_p)\tau\right). \tag{6.26}$$

We shall give the explicit expressions for these quantities in the zeroth and the first approximations. In the zeroth approximation matrix $||A_{ij}||$ coincides with matrix $||\alpha_{ij}||$ defined by (6.13). Then (6.25) can be converged to the following form (for the details see Appendix D):

$$E_{\text{at}}^{(0)} = \frac{1}{2^s} \sum_{p=0}^{s} \frac{s!}{p!\,(s-p-1)!} \cos\left((\lambda_{p+1} - \lambda_p)\tau\right). \tag{6.27}$$

Note that only the terms oscillating with frequencies close to the main frequency $\omega^{(0)}=2$ survive in sum (6.25) in the zeroth approximation. Coefficients at the terms oscillating with frequencies close to $2\omega^{(0)}$, $3\omega^{(0)}$, etc., turn into zero. Neglecting the second order corrections to the frequencies, we obtain the final expression for the atomic subsystem average energy in the zeroth approximation:

$$E_{\text{at}}^{(0)} = \frac{s}{2} \cos 2\tau = \frac{s}{2} \cos\left(2g\left(N - \frac{s-1}{2}\right)^{1/2} t\right). \tag{6.28}$$

The corresponding intensity of radiation equals

$$-\frac{dE_{\text{at}}}{dt} = \frac{sg}{\sqrt{\varepsilon}} \sin 2\tau = sg\left(N - \frac{s-1}{2}\right)^{1/2} \sin\left(2g\left(N - \frac{s-1}{2}\right)^{1/2} t\right). \tag{6.29}$$

Formulas (6.28) and (6.29) have the same simple form as in the Jaynes-Cummings model. The presence of N atoms tells on the change of the time scale [Equation 1, page 132] and the amplitude of vibrations [Equation 2, page 132].

The oscillating character of the process is quite natural, since the energy of the system does not decrease, provided the lossless resonator is considered. At very large times, individual terms in sum (6.27) are dephased due to the second order corrections to eigenfrequencies. As a consequence, the behavior of the system becomes chaotic.

At the account of the first order terms in the coefficients at the trigonometrical functions in (6.24) we get for the atomic subsystem average energy the expression

$$E_{\text{at}}^{(1)} = \frac{s}{2} \cos 2\tau + \frac{\varepsilon s(s-1)}{16} (1 - \cos 4\tau). \tag{6.30}$$

Here the oscillations at the twice frequency $2\omega^{(0)}=4$ appear. The subsequent orders of the perturbation theory contain the terms with higher frequencies. The intensity of radiation in the first order approximation reads

$$-\frac{dE_{\text{at}}^{(1)}}{dt} = \frac{gs}{\sqrt{\varepsilon}} \sin 2\tau - \frac{\sqrt{\varepsilon}\, s(s-1)}{4} g \sin 4\tau. \tag{6.31}$$

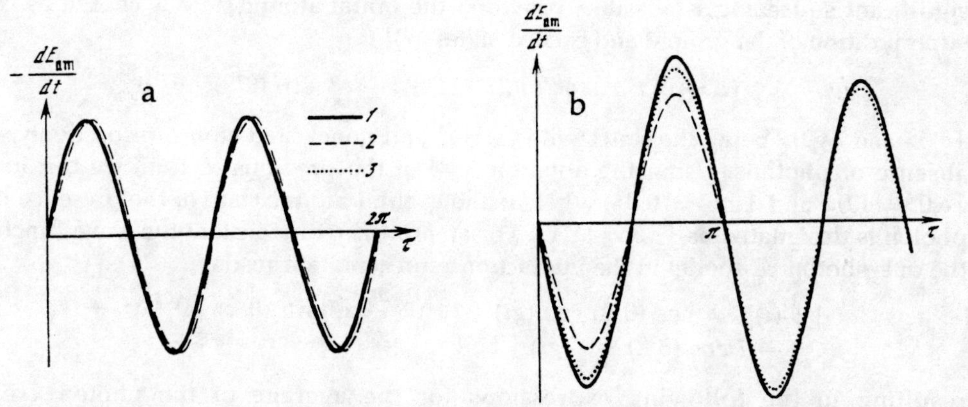

Figure 1. Rate of atomic subsystem energy variation: a) s=5, N=10; b) s=5, N=5. 1-exact result; 2-zeroth approximation; 3-first approximation. The time scale $\tau = gt/\sqrt{\varepsilon}$.

To check the quality of the approximation, let us compare the results obtained above with exact ones resulting from computer numerical diagonalizing Hamiltonian matrix (6.1) and subsequently calculating the transition amplitudes. Figure 1a represents the graph of quantity dE_{at}/dt as the function of time in the scale [Equation 1, page 133] for the case of N=10 atoms, half of which (s=5, i.e. the smallness parameter $\varepsilon=1/8$) being excited at the initial time instant. We see that the curves corresponding to the zeroth approximation (6.29) and to the numerical calculations coincide with a good accuracy, so that there is no need in the first approximation. The worse of our approximation case is N=s=5, when we have only five atoms, all of them being excited at the initial moment. Then the expansion parameter is $\varepsilon=1/3$, so that, generally speaking, one cannot expect a good agreement between approximate and exact results. Nonetheless, Figure 1b shows that the zeroth and the first approximations give rise to quite a good accuracy even in this unfavorable case. However, the question on the convergence of the series over ε in the worst case s=N>>1 corresponding to the initial condition most frequently considered in the literature, when all atoms are excited initially, requires special investigation.

7. SQUEEZING IN THE DICKE MODEL

In this section we study in the Hamiltonian frame (6.1) the possibility of generating squeezing light in the process of spontaneous emission by an ensemble of N two-level atoms in an ideal resonator. The calculations will be made with the aid of the approximate method described in the previous section. The wave functions will be taken in the zeroth order of the perturbation theory expansion (6.6) and (6.7). Remember that the scheme of the previous section is applicable at least in the case of N-s>>1, s being the number of atoms excited at the initial moment of time. Consequently, if only a small part of the whole number of atoms was excited at t=0, then one may expect a good approximation in a wide interval of time (not only in the beginning of the process) already in the zeroth approximation.

Squeezing in the spontaneous emission by a single and by two two-level atoms was investigated, respectively, in studies [63,47] and [64]. We start with discussing the results of these papers. In the single atom case (i.e., for the one-photon JC-model), a significant squeezing is possible, provided the initial atomic state is chosen as certain superposition of the ground and excited states [63]

$$|\psi(t=0)\rangle = \cos(\theta/2)\,|e,0\rangle + e^{i\varphi}\sin(\theta/2)\,|g,0\rangle, \tag{7.1}$$

$|e,0\rangle$ and $|g,0\rangle$ being the states with excited and nonexcited atoms, respectively, in the absence of photons. Using the notation (6.4) of the previous section, we can identify $|e,0\rangle=|1,0\rangle$ and $|g,0\rangle=|0,0\rangle$, whereas the ground atomic state in the presence of one photon is designated as $|g,1\rangle=|1,1\rangle$. For an arbitrary time instant the wave function of the one-photon JC-model in the interaction representation reads

$$|\psi(t)\rangle = \cos(\theta/2)\cos(gt)\,|1,0\rangle + e^{i\varphi}\sin(\theta/2)\,|0,0\rangle - \\ - i\cos(\theta/2)\sin(gt)\,|1,1\rangle, \tag{7.2}$$

resulting in the following expressions for the average of the photon creation, annihilation, and number operators:

$$\langle a\rangle = \langle\psi(t)\,|\,a\,|\,\psi(t)\rangle = -e^{-i\varphi}\sin(\theta/2)\sin(gt)\cos(\theta/2), \\ \langle a^{+}a\rangle = \cos^2(\theta/2)\sin^2(gt), \\ \langle a^2\rangle = 0. \tag{7.3}$$

(Here α and α^{+}, as before, correspond to the *slowly varying operators* (2.3).) For the field quadrature components designated in accordance with [63,64] as $\alpha_1=2Q=1/2(\alpha+\alpha^{+}$, $\alpha_2=2P=(1/2i)(\alpha-\alpha^{+})$ we get

$$(\Delta a_1)^2 = \langle a_1^2\rangle - \langle a_1\rangle^2 = {}^1/_4 + \cos^2(\theta/2)\sin^2(gt)\,[{}^1/_2 - \\ - \sin^2\varphi\cdot\sin^2(\theta/2)], \\ (\Delta a_2)^2 = \langle a_2^2\rangle - \langle a_2\rangle^2 = {}^1/_4 + \cos^2(\theta/2)\sin^2(gt)\,[{}^1/_2 - \\ - \cos^2\varphi\cdot\sin^2(\theta/2)]. \tag{7.4}$$

The maximal squeezing for the quadrature component α_1 is achieved for the initial state parameters $\theta=2\pi/3$, $4\pi/3$ and $\psi=\pi/2$. It reaches 25%,

$$(\Delta a_1)^2 = {}^1/_4 - {}^1/_{16}\sin^2(gt). \tag{7.5}$$

The system of two atoms in a resonator was considered in Reference [64], where the following initial superposition of symmetric (with respect to permutations) atomic states was taken:

$$|\psi(t=0)\rangle = \{\cos(\theta/2)\,|-,-\rangle - \sin(\theta/2)\,e^{i\varphi}\,|+,+\rangle\}\otimes|0\rangle_f, \tag{7.6}$$

symbols $|+,+\rangle$ and $|-,-\rangle$ designating the states when both atoms are excited or not, respectively. In our notation (6.4) $|+,+\rangle=|2,0\rangle$ and $|-,-\rangle\otimes|0\rangle_f=|0,0\rangle$. Note that the state with one excited atom $|1,0\rangle=2^{-1/2}\{|+,-\rangle+|-,+\rangle\}\otimes|0\rangle_f$ is absent in the superposition (7.6). For $t>0$ state (7.6) turns into (we use the interaction picture)

$$|\psi(t)\rangle = \cos(\theta/2)\,|0,0\rangle - \sin(\theta/2)\,e^{i\varphi}\,\{{}^1/_3\,[2+\cos(\sqrt{6}\,gt)]\,|2,0\rangle - \\ - (i/\sqrt{3})\sin(\sqrt{6}\,gt)\,|2,1\rangle - (\sqrt{2/3})\,[1-\cos(\sqrt{6}\,gt)]\,|2,2\rangle. \tag{7.7}$$

In contradistinction to the single atom case (7.3), the averages $\langle\alpha^2\rangle$ and $\langle\alpha^{+2}\rangle$ turn out to be nonzero. The variance of the quadrature component α_1 equals

$$(\Delta a_1)^2 = {}^1/_4 + {}^2/_9\sin^2(\theta/2)\,[1-\cos(\sqrt{6}gt)]^2 + {}^1/_6\sin^2(\theta/2)\sin^2(\sqrt{6}gt) + \\ + {}^1/_6\sin\theta\cos\varphi\,[1-\cos(\sqrt{6}gt)]. \tag{7.8}$$

The maximal squeezing now is 45%. It is achieved for $\cos\varphi=-1$, $\cos\theta=4/5$;

$$(\Delta a_1)^2_{\min} = {}^1/_4\,(1 - {}^4/_9). \tag{7.9}$$

However, the question on the possibility of the experimental realization of states like (7.6) is still open.

Bearing in mind these examples, one may expect a significant degree of squeezing in spontaneous emission by a large system of atoms, provided its initial state is a certain superposition of states with different numbers of excited atoms. Therefore let us choose the initial state as follows:

$$|\psi(t=0)\rangle = \sum_{k=0}^{s} b_k\,|k,0\rangle, \tag{7.10}$$

where vector $|k,0\rangle$ defined by (6.4) describes the symmetric state of the atomic subsystem with k excited atoms from the whole number N, photons being absent. Suppose that amplitude b_k equals

$$b_k \equiv e^{i\varphi_k}\,|b_k| = e^{i\varphi(k+1)} \cos^k\theta \sin^{s-k}\theta \left(\frac{s!}{k!\,(s-k)!}\right)^{1/2}. \tag{7.11}$$

The wave function (7.10) and (7.11) is derived from the vector

$$\prod_{j=1}^{s} (\cos\theta\,|+\rangle_j + e^{i\varphi}\sin\theta\,|-\rangle_j) \otimes \prod_{k=s+1}^{N} |-\rangle_k \otimes |0\rangle_f$$

(with the first s atoms in state (7.2), the rest N-s atoms in the ground state, and with no photons) by means of the symmetrization over all permutations of atoms. If one takes the density matrix corresponding to the pure state (7.10) and (7.11) and calculates its trace with respect to variables of all atoms excepting the first one, then the following probabilities of finding the first atom in the excited $|\rho_+\rangle$ or ground $|\rho_-\rangle$ states can be obtained,

$$\rho_+^{(1)} = (s/N)\cos^2\theta, \quad \rho_-^{(1)} = 1 - (s/N)\cos^2\theta.$$

Remember once again that the maximal number of excited atoms is believed small in comparison with N.

According to the results of the previous section, the wave function obeying the initial condition (7.10) has the following form in the zeroth approximation:

$$|\psi(t)\rangle^{(0)} = \sum_{l=0}^{s} b_l \sum_{p=0}^{l} \alpha_{op}^{(l)} e^{-igt\lambda_{p(l)}^{(0)}/\sqrt{\varepsilon_l}}\,|p\rangle_{(l)}^{(0)},$$

where the eigenvector $|p\rangle^{(0)}_{(1)}$ has the components $\alpha^{(1)}_{0p}$ defined by (6.13), and $\lambda^{(0)}_{p(1)}=\iota-2p$ means the frequency in the zeroth approximation. Besides, we have designated

$$\varepsilon_l \equiv (N - (l-1)/2)^{-1}.$$

Index ι labels subspaces corresponding to different eigenvalues of the "excitation number" operation (see Section 6). The dimension of every subspace is $\iota+1$.

The wave function in basis $|s,k\rangle$ (6.4) reads

$$|\psi(t)\rangle^{(0)} = \sum_{l=0}^{s} b_l \sum_{p=0}^{l} \sum_{n=0}^{l} \alpha_{op}^{(l)} \alpha_{np}^{(l)} e^{-igt\lambda_{p(l)}^{(0)}/\sqrt{\varepsilon_l}}\,|l,n\rangle. \tag{7.12}$$

Note that expression (7.12) is suitable for the approximate description of the process both at small and large times. But to simplify the further calculations of the quadrature components variances, we shall ignore the ι-dependences of the factors $\varepsilon^{-1/2}{}_1=[N-(\iota-1)/2]^{1/2}$ entering the arguments of the exponential functions in (7.12). As pointed out in Appendix E, this way of calculating quadrature components averages gives a good approximation for the first several periods of oscillations (with frequency $gN^{1/2}$), provided the inequality N>>s/2. Calculations performed in Appendix E give rise to the following formulas for the average values of the photon operators:

$$\langle a\rangle = {}^{(0)}\langle\psi(t)\,|\,a\,|\,\psi(t)\rangle^{(0)} = -\,i\sin\frac{gt}{\sqrt{\varepsilon}}\sum_{l=1}^{s}\sqrt{l}\,e^{i\varphi}\,|\,b_{l-1}^{*}b_l\,|,$$

$$\langle a^2\rangle = -\sin^2\frac{gt}{\sqrt{\varepsilon}}\sum_{l=2}^{s}\sqrt{l(l-1)}\,e^{2i\varphi}\,|\,b_{l-2}^{*}b_l\,|,$$

$$\langle a^{+}a\rangle = \sin^2\frac{gt}{\sqrt{\varepsilon}}\sum_{l=1}^{s} l\,|\,b_l\,|^2. \tag{7.13}$$

The averages of the quadrature components are as follows:

$$\langle a_1\rangle = \left\langle\frac{a+a^{+}}{2}\right\rangle = \sin\frac{gt}{\sqrt{\varepsilon}}\sin\varphi\sum_{l=1}^{s}\sqrt{l}\,|\,b_{l-1}^{*}b_l\,|,$$

$$\langle a_2\rangle = \left\langle\frac{a-a^{+}}{2i}\right\rangle = -\sin\frac{gt}{\sqrt{\varepsilon}}\cos\varphi\sum_{l=1}^{s}\sqrt{l}\,|\,b_{l-1}^{*}b_l\,|.$$

It is convenient to introduce the functions dependent only on the initial "mixing angle" θ and the maximal number s of excited atoms

$$\alpha_1 = 2\sum_{l=2}^{s}\sqrt{l(l-1)}\,|\,b_{l-2}^{*}b_l\,| = 2\sum_{l=2}^{s}\frac{s!\,\sqrt{(s-l+1)(s-l+2)}}{(l-2)!\,(s-l+2)!}(\cos\theta)^{2l-2}\times$$
$$\times(\sin\theta)^{2s-2l+2},$$

$$\alpha_2 = 2\Big[\sum_{l=1}^{s}\sqrt{l}\,|\,b_{l-1}^{*}b_l\,|\Big]^2 = 2\Big[\sum_{l=1}^{s}\frac{s!\,\sqrt{s-l+1}}{(l-1)!\,(s-l+1)!}(\cos\theta)^{2l-1}(\sin\theta)^{2s-2l+1}\Big]^2,$$

$$\alpha_3 = 2\sum_{l=1}^{s} l\,|\,b_l\,|^2 = 2\sum_{l=1}^{s}\frac{s!}{(l-1)!\,(s-l)!}(\cos\theta)^{2l}(\sin\theta)^{2(s-l)}. \tag{7.14}$$

Then the variances of the quadrature components can be expressed as follows:

$$(\Delta a_1)^2 = {}^1/_4 + {}^1/_4\sin^2(gt/\sqrt{\varepsilon})\,\{(\alpha_3-\alpha_2)-(\alpha_1-\alpha_2)\cos 2\varphi\},$$
$$(\Delta a_2)^2 = {}^1/_4 + {}^1/_4\sin^2(gt/\sqrt{\varepsilon})\,\{(\alpha_3-\alpha_2)+(\alpha_1-\alpha_2)\cos 2\varphi\}. \tag{7.15}$$

We see that the variances have very simple dependences on time and the initial phase φ in the zeroth approximation.

In the case of a single atom in the cavity we have N=s=ε=1, and the zeroth approximation (7.14) and (7.15) coincides with the exact solution obtained in [63] (see (7.5)]. To test the quality of the approximate expressions for the variances (7.15) and (7.14) we have compared them with the exact results (obtained with the aid of the numerical computer calculations) for the case of s=3, N=10. The initial state was chosen in accordance with the conditions (7.10) and (7.11). The maximal squeezing of the first quadrature component is achieved for the initial parameters φ=π/2, θ=0.864. It is equal to 40.4% (the expression in the figure brackets for the first formula in (7.15) equals 0.404). Graphs illustrating the exact time dependences of the variances together with the approximate results (7.15) are given in Figure 2 (for the values of parameters φ and

θ corresponding to the maximal squeezing). As expected (see Appendix E), approximation (7.15) yields a satisfactory result for the first several periods of oscillations. Let us emphasize once more that the discrepancy between the approximate and exact solutions at large times is connected not with the structure of the zeroth order wave functions (7.12), but with additional simplifying assumptions made in calculating variances (7.15) (see Appendix E)[1]. Note that for the case considered in Figure 2 the numerical result represented demonstrates the maximal squeezing just in the first period of oscillations.

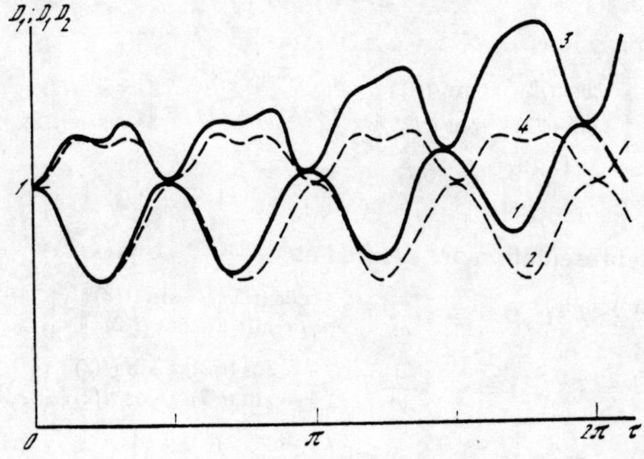

Figure 2. The variance of the first quadrature component $D_1=4(\Delta\alpha_1)^2$ (1,2) and the product of variances $D_1D_2=(4\Delta\alpha_1\Delta\alpha_2)^2$ (3,4) as the functions of time. 1,3-exact results; 2,4-zeroth approximation. N=10, s=3. The time scale τ=3.1gt.

Formulas (7.14) and (7.15) yield in the case of s=10 the maximal degree of squeezing 65% for the choice of parameters $\varphi=\pi/2$, $\theta=0.65$; in the case of s=20 squeezing reaches 69.6% for $\varphi=\pi/2$, $\theta=0.53$. As we see, one may expect the increase of squeezing with the increase of the number of initially excited atoms (at least under the condition $s \ll N$).

APPENDICES

Appendix A

For reading convenience we give several well known definitions from the theory of SU(2) group representations. The details can be found in book [62]. Besides, we shall discuss the properties of matrix $||\alpha_{ij}||$ consisting of eigenvectors of Hamiltonian $H^{(0)}$ from (6.8).

SU(2) group contains the matrices of the form

$$g = \begin{Vmatrix} \alpha & \beta \\ \gamma & \delta \end{Vmatrix} \qquad \gamma = -\beta^*, \quad \delta = \alpha^*, \quad \alpha\delta - \gamma\beta = 1.$$

[1] With the given accuracy $N \gg s/2$ one can take $\varepsilon=N^{1/2}$ in (7.15). Real values of ε_1 vary in the limits $N^{-1/2} \le \varepsilon_1 \le [[N-(s-1)/2]^{-1/2}$. Since in the example considered (s=3, N=10) ε is not very small in comparison with N, we have use the approximate graph in Figure 2 for constructing formula (7.15) with $\varepsilon^{-1}=3.1$.

Let L be an integer or half-integer. Designate by $|n\rangle$ (n=-L, -L+1, -L+2,...,L) the basis of the unitary irreducible (2L+1)-dimensional representation of SU(2) The matrix elements of the representation in this basis have the form (provided $\delta\neq 0$)

$$t^L_{mn}(g) = \left[\frac{(L-m)!\,(L-n)!}{(L+m)!\,(L+n)!}\right]^{1/2} \frac{\delta^{m+n}}{\beta^m \gamma^n} \sum_{j=\max(m,\,n)}^{L} \frac{(L+j)!\,(\beta\gamma)^j}{(L-j)!\,(j-m)!\,(j-n)!}\,. \tag{A.1}$$

One-parametric subgroups corresponding to spin operators

$$\hat{s}_x = \frac{1}{2}\begin{Vmatrix} 0 & 1 \\ 1 & 0 \end{Vmatrix}, \quad \hat{s}_y = \frac{1}{2}\begin{Vmatrix} 0 & -i \\ i & 0 \end{Vmatrix}, \quad \hat{s}_z = \frac{1}{2}\begin{Vmatrix} 1 & 0 \\ 0 & -1 \end{Vmatrix},$$

are given by the expressions

$$\exp(it\hat{s}_x) = \begin{Vmatrix} \cos(t/2) & i\sin(t/2) \\ i\sin(t/2) & \cos(t/2) \end{Vmatrix}, \quad \exp(it\hat{s}_y) = \begin{Vmatrix} \cos(t/2) & \sin(t/2) \\ -\sin(t/2) & \cos(t/2) \end{Vmatrix},$$

$$\exp(it\hat{s}_z) = \begin{Vmatrix} e^{it/2} & 0 \\ 0 & e^{-it/2} \end{Vmatrix}$$

Generators of the representation are defined as

$$\hat{s}^{(L)}_x \equiv T^L(s_x) = -i\frac{\partial}{\partial t} T^L\left(\begin{Vmatrix} \cos(t/2) & i\sin(t/2) \\ i\sin(t/2) & \cos(t/2) \end{Vmatrix}\right)\Bigg|_{t=0},$$

$$\hat{s}^{(L)}_y \equiv T^L(s_y) = -i\frac{\partial}{\partial t} T^L\left(\begin{Vmatrix} \cos(t/2) & \sin(t/2) \\ -\sin(t/2) & \cos(t/2) \end{Vmatrix}\right)\Bigg|_{t=0},$$

$$\hat{s}^{(L)}_z \equiv T^L(s_z) = -i\frac{\partial}{\partial t} T^L\left(\begin{Vmatrix} e^{it/2} & 0 \\ 0 & e^{-it/2} \end{Vmatrix}\right)\Bigg|_{t=0}.$$

Consequently, in the $|n\rangle$-basis

$$\hat{s}^{(L)}_+|n\rangle = \sqrt{(L-n)(L+n+1)}\,|n+1\rangle,$$

$$\hat{s}^{(L)}_-|n\rangle = \sqrt{(L+n)(L-n+1)}\,|n-1\rangle, \qquad \hat{s}^{(L)}_z|n\rangle = -n|n\rangle, \tag{A.2}$$

where $\hat{s}^{(L)}_{\pm} = \hat{s}^{(L)}_x \pm i\hat{s}^{(L)}_y$. If s=2L, m=S/2-n, then $|s,m\rangle$-basis (6.4) coincides with $|n\rangle$-basis:

$$|s,\,m\rangle = |s/2-m\rangle = |n\rangle, \qquad 0 \leqslant m \leqslant s.$$

It is clear that the zeroth Hamiltonian approximation (6.8) coincides with the generator $2\hat{s}^{(L)}_x = \hat{s}^{(L)}_+ + \hat{s}^{(L)}_-$ of the (2L+1)-dimensional representation of the SU(2) group: $H^{(0)} = 2\hat{s}^{(L)}_x = T^L(2s_x)$. Respectively, the evolution operator of the system in the zeroth approximation belongs to the same representation; $\exp(-itH^{(0)}) = T^L(\exp(-itL\hat{s}_x))$. To diagonalize $H^{(0)}$ we note that

$$g_0^{-1} s_x g_0 = -s_z, \quad g_0 = \frac{1}{\sqrt{2}}\begin{Vmatrix} 1 & 1 \\ -1 & 1 \end{Vmatrix}, \quad g_0^{-1} = \frac{1}{\sqrt{2}}\begin{Vmatrix} 1 & -1 \\ 1 & 1 \end{Vmatrix}.$$

Writing down this equality in the (2L+1)-dimensional representation and designating $\hat{\underset{\sim}{\alpha}}^L = T^L(g_0)$, we get

$$\hat{\underset{\sim}{\alpha}}^{L-1} H^{(0)} \hat{\underset{\sim}{\alpha}}^L = T^L(-2s_z) = \begin{Vmatrix} L & 0\ldots 0 \\ 0 & L-1\ldots 0 \\ \vdots & \ddots\ \vdots \\ 0 & -L \end{Vmatrix}.$$

Therefore matrix $\hat{\underset{\sim}{\alpha}}$ diagonalizes Hamiltonian $H^{(0)}$, and its columns are eigenvectors of $H^{(0)}$. Using (A.1) one can obtain matrix elements $\hat{\alpha}^L{}_{mn}=t^L{}_{mn}$. It is convenient to perform an additional unitary transformation $\hat{\alpha}^L=A^{-1}\hat{\underset{\sim}{\alpha}}^L A$ with $A_{mn}=(-1)^{L-m}\delta_{mn}$, which changes phases of eigenvectors and makes matrix $\hat{\alpha}^L$ symmetrical. Then we get

$$\alpha^L_{mn}=\left[\frac{(L-m)!\,(L-n)!}{(L+m)!\,(L+n)!}\right]^{1/2}\sum_{j=\max(m,\,n)}^{L}\frac{(L+j)!\,(-1)^{L-j}}{(L-j)!\,(j-m)!\,(j-n)!\,2^j}\,.$$

Making the replacement s=2L, k=L-m, p=L-n, we have

$$\alpha^s_{kp}=\left[\frac{k!\,p!}{(s-k)!\,(s-p)\,2^s}\right]^{1/2}\sum_{j=0}^{\min(p,\,k)}\frac{(s-j)!\,(-1)^j\,2^j}{j!\,(k-j)!\,(p-j)!}\,. \tag{A.3}$$

Matrix $||a^s{}_{kp}||$ is chosen real, which is possible due to the reality of the Hamiltonian matrix $H^{(0)}$. Consequently, the unitarity of $||a^s{}_{kp}||$ is reduced to its orthogonality;

$$\sum_{i=0}^{s}\alpha^s_{ki}\alpha^s_{pi}=\delta_{kp}. \tag{A.4}$$

Moreover, matrix $||\alpha^s{}_{kp}||$ possesses the following symmetry properties:

$$\text{a) } \alpha^s_{pk}=\alpha^s_{kp},\qquad \text{b) } \alpha^s_{pk}=(-1)^p\,\alpha^s_{p,\,s-k},$$

$$\text{c) } \alpha^s_{pk}=(-1)^k\,\alpha^s_{s-p,\,k},\qquad \text{d) } \alpha^s_{pk}=(-1)^{p+k-s}\,\alpha^s_{s-k,\,s-p}. \tag{A.5}$$

Property a) follows from our definition of $||\alpha^s{}_{kp}||$. Property b) is valid also for matrix $||A_{ij}||$ constructed from the eigenvectors of the whole Hamiltonian (6.5). It reflects the fact that for any eigenvector (ε_1, $0\le i\le s$) of a symmetric "three-diagonal" (s+1)-dimensional matrix with an eigenvalue λ, there also exists the eigenvector ($(-1)^1\varepsilon_i$, $0\le i\le s$) with eigenvalue $-\lambda$. Properties c) and d) are the consequences of the two first ones. A lot of different sum rules exists for matrix elements $\alpha^s{}_{kp}$. They are based, in particular, on the following formula reflecting the fact that the column of matrix $||\alpha^s{}_{kp}||$ is the eigenvector of operator $H^{(0)}=2S^{(L)}{}_x$ with eigenvalue $\lambda_p=s-2p$;

$$(s-2p)\,\alpha^s_{kp}=\sqrt{k\,(s-k+1)}\,\alpha^s_{k-1,\,p}+\sqrt{(k+1)\,(s-k)}\,\alpha^s_{k+1,\,p}. \tag{A.6}$$

Similarly,

$$(s-2k)\,\alpha^s_{kp}=\sqrt{p\,(s-p+1)}\,\alpha^s_{k,\,p-1}+\sqrt{(p+1)\,(s-p)}\,\alpha^s_{k,\,p+1}. \tag{A.7}$$

As illustrations we give the explicit forms of matrices $||\alpha^s{}_{kp}||$ for s=4 and s=5;

$$\hat{\alpha}^4=\left\|\begin{matrix}\frac{1}{4} & \frac{2}{4} & \frac{\sqrt{6}}{4} & \frac{2}{4} & \frac{1}{4}\\[4pt] \frac{2}{4} & \frac{2}{4} & 0 & -\frac{2}{4} & -\frac{2}{4}\\[4pt] \frac{\sqrt{6}}{4} & 0 & -\frac{2}{4} & 0 & \frac{\sqrt{6}}{4}\\[4pt] \frac{2}{4} & -\frac{2}{4} & 0 & \frac{2}{4} & -\frac{2}{4}\\[4pt] \frac{1}{4} & -\frac{2}{4} & \frac{\sqrt{6}}{4} & -\frac{2}{4} & \frac{1}{4}\end{matrix}\right\|,$$

$$\hat{\alpha}^5 = \left\| \begin{array}{cccccc} \frac{1}{4\sqrt{2}} & \frac{\sqrt{5}}{4\sqrt{2}} & \frac{\sqrt{10}}{4\sqrt{2}} & \frac{\sqrt{10}}{4\sqrt{2}} & \frac{\sqrt{5}}{4\sqrt{2}} & \frac{1}{4\sqrt{2}} \\ \frac{\sqrt{5}}{4\sqrt{2}} & \frac{3}{4\sqrt{2}} & \frac{\sqrt{2}}{4\sqrt{2}} & -\frac{\sqrt{2}}{4\sqrt{2}} & -\frac{3}{4\sqrt{2}} & -\frac{\sqrt{5}}{4\sqrt{2}} \\ \frac{\sqrt{10}}{4\sqrt{2}} & \frac{\sqrt{2}}{4\sqrt{2}} & -\frac{2}{4\sqrt{2}} & -\frac{2}{4\sqrt{2}} & \frac{\sqrt{2}}{4\sqrt{2}} & \frac{\sqrt{10}}{4\sqrt{2}} \\ \frac{\sqrt{10}}{4\sqrt{2}} & -\frac{\sqrt{2}}{4\sqrt{2}} & -\frac{2}{4\sqrt{2}} & \frac{2}{4\sqrt{2}} & \frac{\sqrt{2}}{4\sqrt{2}} & -\frac{\sqrt{10}}{4\sqrt{2}} \\ \frac{\sqrt{5}}{4\sqrt{2}} & -\frac{3}{4\sqrt{2}} & \frac{\sqrt{2}}{4\sqrt{2}} & \frac{\sqrt{2}}{4\sqrt{2}} & -\frac{3}{4\sqrt{2}} & \frac{\sqrt{5}}{4\sqrt{2}} \\ \frac{1}{4\sqrt{2}} & -\frac{\sqrt{5}}{4\sqrt{2}} & \frac{\sqrt{10}}{4\sqrt{2}} & -\frac{\sqrt{10}}{4\sqrt{2}} & \frac{\sqrt{5}}{4\sqrt{2}} & -\frac{1}{4\sqrt{2}} \end{array} \right\| .$$

Appendix B

Here we shall prove that the form of operator $H^{(1)}$ defined according to (6.9) in the basis consisting of eigenvectors of $H^{(0)}$ coincides with the form of $H^{(1)}$ in the initial basis (6.4). Accounting for (6.14) this means that $\hat{\alpha}^{-1}H^{(1)}\hat{\alpha}=H^{(1)}$ or $H^{(1)}\hat{\alpha}=\hat{\alpha}H^{(1)}$. For the matrix element of $H^{(1)}$ in the turned basis (6.12) we have

$$^{(0)}\langle k \,|\, H^{(1)} \,|\, n\rangle^{(0)} = \sum_{l=0}^{s}\sum_{p=0}^{s} \alpha_{kl} H^{(1)}_{lp} \alpha_{pn} =$$

$$= \frac{1}{2}\sum_{l=0}^{s} \alpha_{kl}\left\{\alpha_{l-1,\,n}\left(l - \frac{s+1}{2}\right)\sqrt{l(s-l+1)} + \right.$$

$$\left. + \alpha_{l+1,\,n}\left(l - \frac{s-1}{2}\right)\sqrt{(l+1)(s-l)}\right\} = \frac{s-2n}{2}\sum_{l=0}^{s} l\alpha_{kl}\alpha_{ln} -$$

$$- \frac{(s-1)(s-2n)}{4}\sum_{l=0}^{s}\alpha_{kl}\alpha_{kn} - \frac{1}{2}\sum_{l=0}^{s}\sqrt{l(s-l+1)}\,\alpha_{kl}\alpha_{n,\,l-1}.$$

Here we have used formula (A.6). Taking into account (A.4) and property a) from (A.5) we get

$$^{(0)}\langle k | H^{(1)} | n\rangle^{(0)} = -{}^1/_4(s-2n)(s-1)\,\delta_{kn} + \frac{s-2n}{2}\sum_{l=0}^{s} l\alpha_{kl}\alpha_{nl} -$$

$$- \frac{1}{2}\sum_{l=0}^{l}\sqrt{(l(s-l+1)}\,\alpha_{kl}\alpha_{n,\,l-1}. \tag{B.1}$$

Further actions will be divided into several steps.

1. First we show that

$$\sum_{l=0}^{s} l\alpha_{kl}\alpha_{nl} = -{}^1/_2\sqrt{k(s-k+1)}\,\delta_{n,\,k-1} +$$

$$+ \frac{s}{2}\delta_{nk} - {}^1/_2\sqrt{(k+1)(s-k)}\,\delta_{n,\,k+1}. \tag{B.2}$$

It is sufficient to consider the case of k≥n. Suppose s to be even. (The proof for odd s is similar.) At first let n=k-2t, with t=0,1,2,... Accounting property b) of (A.5) we have

$$\sum_{l=0}^{s} l\alpha_{kl}\alpha_{k-2t,\,l} = \alpha_{k1}\alpha_{k-2t,\,1} + 2\alpha_{k2}\alpha_{k-2t,\,2} + \ldots + (s-1)\,\alpha_{k,\,s-1}\alpha_{k-2t,\,s-1} +$$

$$+ s\alpha_{ks}\alpha_{k-2t,\,s} = s\alpha_{k0}\alpha_{k-2t,\,0} + s\alpha_{k-2t,\,1}\alpha_{k1} + \ldots + \frac{s}{2}\,\alpha_{k,\,s/2}\alpha_{k-2t,\,s/2} =$$

$$= \frac{s}{2}\sum_{l=0}^{s} \alpha_{kl}\alpha_{k-2t,\,l} = \delta_{t,\,0}.$$

Further let n=-2t-1+k with t=1,2,... Using (A.6) we get

$$\sum_{l=0}^{s} l\alpha_{kl}\alpha_{k-2t-1,\,l} = \sqrt{(k-2t-1)(s-k+2t+2)}\,\sum_{l=0}^{s}{}' \frac{l}{s-2l}\,\alpha_{kl}\alpha_{k-2t-2,\,l} +$$

$$+ \sqrt{(k-2t)(s-k+2t+1)}\,\sum_{l=0}^{s}{}' \frac{l}{s-2l}\,\alpha_{kl}\alpha_{k-2t,\,l}.$$

Here the primed summation mark means the absence of the term with ι=s/2, which disappears because $\alpha_{k,s/2}=0$ for odd k (and even s) due to property c) of (A.5). Grouping terms with ι and s-ι (0≤ι<s/2) in both sums and using property b) of (A.5) we get

$$\sum_{l=0}^{s} l\alpha_{kl}\alpha_{k-2t-1,\,l} = -\sqrt{(k-2t-1)(s-k+2t+2)}\,\sum_{l=0}^{s} \alpha_{kl}\alpha_{k-2t-2,\,l} -$$

$$- \sqrt{(k-2t)(s-k+2t+1)}\,\sum_{l=0}^{s} \alpha_{kl}\alpha_{k-2t,\,l} +$$

$$+ \frac{1}{2}\,\alpha_{ks/2}\,\{\sqrt{(k-2t-1)(s-k+2t+2)}\,\alpha_{k-2t-2,\,s/2} +$$

$$+ \sqrt{(k-2t)(s-k+2t+1)}\,\alpha_{k-2t,\,s/2}\} = 0.$$

Here we have used again (A.6).

Now let n=k-1. Taking into account (A.6) we get

$$\sum_{l=0}^{s} l\alpha_{kl}\alpha_{k-1,\,l} = \sum_{l=0}^{s}{}' l\alpha_{kl}\alpha_{k-1,\,l} + \frac{s}{2}\,\alpha_{ks/2}\alpha_{k-1,\,s/2} =$$

$$= \sqrt{(k-1)(s-k+2)}\,\sum_{l=0}^{s'} \frac{l}{s-2l}\,\alpha_{kl}\alpha_{k-2,\,l} +$$

$$+ \sqrt{k(s-k+1)}\,\sum_{l=0}^{s} \frac{l}{s-2l}\,\alpha_{kl}^{2}.$$

Grouping again in both sums the ι-th and the (s-ι)-th terms and using property b) of (A.5) we obtain

$$\sum_{l=0}^{s} l\alpha_{kl}\alpha_{k-1,\,l} = -\frac{1}{2}\sqrt{(k-1)(s-k+2)}\,\sum_{l=0}^{s} \alpha_{kl}\alpha_{k-2,\,l} -$$

$$- \frac{1}{2}\sqrt{k(s-k+1)}\,\sum_{l=0}^{s} \alpha_{kl}^{2} = -{}^{1}/_{2}k\,(s-k+1).$$

2. Let us show that

$$\sum_{l=0}^{s} \sqrt{l(s-l+1)}\,\alpha_{kl}\alpha_{n,\,l-1} = \frac{s-2k}{2}\,\delta_{kn} - {}^{1}/_{2}\sqrt{k(s-k+1)}\,\delta_{k-1,\,n} +$$

$$+ {}^{1}/_{2}\sqrt{(k+1)(s-k)}\,\delta_{k+1,\,n}. \tag{B.3}$$

At first let n=k-2t with t=1,2,... Using (A.7) we get

$$\sum_{l=0}^{s} \sqrt{l(s-l+1)}\,\alpha_{kl}\alpha_{k-2t,\,l-1} = (s-2k+4t)\sum_{l=0}^{s}\alpha_{kl}\alpha_{k-2t,\,l} -$$
$$-\sum_{l=0}^{s}\sqrt{(l+1)(s-l)}\,\alpha_{kl}\alpha_{k-2t,\,l+1} =$$
$$= -\sum_{l=0}^{s}\sqrt{(l+1)(s-l)}\,\alpha_{k,\,s-l}\alpha_{k-2t,\,s-l-1} =$$
$$= -\sum_{l=0}^{s}\sqrt{l(s-l+1)}\,\alpha_{kl}\alpha_{k-2t,\,l-1}.$$

Lastly, we have used property b) of (A.5) in one equality. Equating the first and the last links in this chain of equalities one can be convinced of the validity of (B.3) in this case.

Further, let n=k-2t-1, where t=1,2,... Using identities (A.6) and (B.3) with n=k-2t we have

$$\sqrt{(k-2t-1)(s-k+2t+2)}\sum_{l=0}^{s}\sqrt{l(s-l+1)}\,\alpha_{kl}\alpha_{k-2t-2,\,l-1} +$$
$$+\sqrt{(k-2t)(s-k+2t+1)}\sum_{l=0}^{s}\sqrt{l(s-l+1)}\,\alpha_{kl}\alpha_{k-2t,\,l-1} =$$
$$= \sum_{l=0}^{s}(s-2l+2)\sqrt{l(s-l+1)}\,\alpha_{kl}\alpha_{k-2t-1,\,l-1}. \tag{B.4}$$

Due to the same identities

$$\sum_{l=0}^{s}(s-2l)\sqrt{l(s-l+1)}\,\alpha_{kl}\alpha_{k-2t-1,\,l-1} =$$
$$= \sqrt{k(s-k+1)}\sum_{l=0}^{s}\sqrt{l(s-l+1)}\,\alpha_{k-1,\,l}\alpha_{k-2t-1,\,l-1} +$$
$$+\sqrt{(k+1)(s-k)}\sum_{l=0}^{s}\sqrt{l(s-l+1)}\,\alpha_{k+1,\,l}\alpha_{k-2t-1,\,l-1}. \tag{B.5}$$

Subtracting (B.5) from (B.4) we arrive at (B.3).

Now let n=k. Using (A.7) we get

$$\sum_{l=0}^{s}\sqrt{l(s-l+1)}\,\alpha_{kl}\alpha_{k,\,l-1} = \sum_{l=0}^{s}(s-2k)\,\alpha_{kl}^2 -$$
$$-\sum_{l=0}^{s}\sqrt{(l+1)(s-l)}\,\alpha_{kl}\alpha_{k,\,l+1} = s-2k-\sum_{l=0}^{s}\sqrt{l(s-l+1)}\,\alpha_{k,\,l-1}\alpha_{kl}.$$

Hence

$$\sum_{l=0}^{s}\sqrt{l(s-l+1)}\,\alpha_{kl}\alpha_{k,\,l-1} = \frac{s-2k}{2}$$

in accordance with (B.3).

Finally let n=k-1. Then the consequence of (A.6) and (B.3) with n=k and n=k-2 is the relation

$$\sqrt{k(s-k+1)}\sum_{l=0}^{s}\sqrt{l(s-l+1)}\,\alpha_{kl}\alpha_{k,\,l-1} +$$
$$+\sqrt{(k-1)(s-k+2)}\sum_{l=0}^{s}\sqrt{l(s-l+1)}\,\alpha_{kl}\alpha_{k-2,\,l-1} =$$
$$= \sum_{l=0}^{s}(s-2l+2)\sqrt{l(s-l+1)}\,\alpha_{kl}\alpha_{k-1,\,l-1} = \frac{s-2k}{2}\sqrt{k(s-k+1)}. \tag{B.6}$$

On the other hand, from the same identities we get

$$\sum_{l=0}^{s}\sqrt{l(s-l+1)}\,\alpha_{k-1,\,l-1}\sqrt{k(s-k+1)}\,\alpha_{k-1,\,l}+$$

$$+\sum_{l=0}^{s}\sqrt{l(s-l+1)}\,\alpha_{k-1,\,l-1}\sqrt{(k+1)(s-k)}\,\alpha_{k+1,\,l}=$$

$$=\sum_{l=0}^{s}(s-2l)\sqrt{l(s-l+1)}\,\alpha_{k-1,\,l-1}\alpha_{kl}=\frac{s-2(k-1)}{2}\sqrt{k(s-k+1)}. \tag{B.7}$$

Subtracting (B.7) from (B.6) we have

$$\sum_{l=0}^{s}\sqrt{l(s-l+1)}\,\alpha_{kl}\alpha_{k-1,\,l-1}=-\frac{1}{2}\sqrt{k(s-k+1)}.$$

Now identity (B.3) is proved completely.

3. As follows from (B.1)-(B.3) with $n \geq k$, matrix elements ${}^{(0)}\langle k \mid H^{(1)} \mid n\rangle^{(0)}$ in the $\mid n\rangle^{(0)}$-basis coincide with matrix elements $H^{(1)}{}_{kn}$ in the $\mid s,n\rangle$-basis defined by (6.9). Moreover, it is clear from the derivation of (B.1) that ${}^{(0)}\langle k \mid H^{(1)} \mid n\rangle^{(0)} = {}^{(0)}\langle n \mid H^{(1)} \mid k\rangle^{(0)}$. Thus the statement of the present section is proved completely.

Appendix C

Here we shall find the diagonal element of operator $H^{(2)}$ on the basis of the eigenvectors of operator $H^{(0)}$. According to (6.10) and (6.12) we have

$$^{(0)}\langle k \mid H^{(2)} \mid k\rangle^{(0)}=\sum_{l=0}^{s}\sum_{p=0}^{s}\alpha_{kl}H^{(2)}_{lp}\alpha_{pk}=$$

$$=-\frac{1}{32}\sum_{l=0}^{s}\sqrt{l(s-l+1)}\,(s-2l+1)^2\,\alpha_{kl}\alpha_{l-1,\,k}-$$

$$-\frac{1}{32}\sum_{l=0}^{s}\sqrt{(l+1)(s-l)}\,(s-2l-1)^2\,\alpha_{kl}\alpha_{l+1,\,k}=$$

$$=-\frac{1}{16}\sum_{l=0}^{s}\sqrt{l(s-l+1)}\,(s-2l+1)^2\,\alpha_{kl}\alpha_{k,\,l-1}.$$

Note that

$$(s-2l+1)^2=(s-2l)(s-2l+2)+1,$$

$$^{(0)}\langle k \mid H^{(2)} \mid k\rangle^{(0)}=-\frac{1}{16}\sum_{l=0}^{s}(s-2l)(s-2l+2)\sqrt{l(s-l+1)}\,\alpha_{kl}\alpha_{kl-1}-$$

$$-\frac{s-2k}{32}, \tag{C.1}$$

where (6.3) has been used. To find the rest of the sum we use (B.6):

$$\sum_{l=0}^{s}(s-2l+2)\sqrt{l(s-l+1)}\,\alpha_{k+1,\,l}\alpha_{k,\,l-1}=\frac{s-2(k+1)}{2}\sqrt{(k+1)(s-k)}. \tag{C.2}$$

Multiplying it by $\sqrt{(k+1)(s-k)}$ and using (A.6) we have

$$\sum_{l=0}^{s}(s-2l)(s-2l+2)\sqrt{l(s-l+1)}\,\alpha_{k,\,l-1}\alpha_{kl}=\frac{s-2(k+1)}{2}(k+1)(s-k)+$$
$$+\sqrt{k(s-k+1)}\sum_{l=0}^{s}(s-2l+2)\sqrt{l(s-l+1)}\,\alpha_{k,\,l-1}\alpha_{k-1,\,l}.$$

Once more using (C.2), we get

$$\sum_{l=0}^{s}(s-2l)(s-2l+2)\sqrt{l(s-l+1)}\,\alpha_{k,\,l-1}\alpha_{kl}=$$
$$=\frac{s-2(k+1)}{2}(k+1)(s-k)+\frac{s-2(k-1)}{2}k(s-k+1), \tag{C.3}$$
$${}^{(0)}\langle k\mid H^{(2)}\mid k\rangle^{(0)}=-{}^{1}/_{32}\,(s-2k)\,[2k\,(s-k)+s-1]. \tag{C.4}$$

Appendix D

Here we calculate the average energy of the atomic subsystem at an arbitrary time instant in the zeroth and the first approximations. According to (6.23) and (6.24)

$$E_{\mathrm{aT}}=\langle s,0\,|\,e^{i\hat{H}t}S_3e^{-i\hat{H}t}\,|\,s,0\rangle=\frac{s}{2}-\sum_{p=0}^{s}A_{0p}^2\sum_{m=0}^{s}mA_{mp}^2-$$
$$-2\sum_{p=0}^{s}\sum_{l=p+1}^{s}A_{0p}A_{0l}\cos\left((\lambda_l-\lambda_p)\tau\right)\left(\sum_{m=0}^{s}mA_{mp}A_{ml}\right), \tag{D.1}$$

where $\tau=gt\sqrt{\varepsilon}$. In the zeroth approximation the orthogonal matrix $||A_{mp}||$ constructed from eigenvectors of the whole Hamiltonian (6.7) is replaced by matrix $||\alpha_{mp}||$ constructed from eigenvectors of Hamiltonian $H^{(0)}$. Using (B.2) and (A.3) we get

$$E_{\mathrm{aT}}^{(0)}=-2\sum_{p=0}^{s}\sum_{l=p+1}^{s}\Big\{-{}^{1}/_{2}\sqrt{p(s-p+1)}\,\delta_{l,\,p-1}+$$
$$+\frac{s}{2}\delta_{lp}-{}^{1}/_{2}\sqrt{(p+1)(s-p)}\,\delta_{l,\,p+1}\Big\}\alpha_{0p}\alpha_{0l}\cos\left((\lambda_l^{(0)}-\lambda_p^{(0)})\tau\right)=$$
$$=\sum_{p=0}^{s}\sqrt{(p+1)(s-p)}\,\alpha_{0p}\alpha_{0,\,p+1}\cos\left((\lambda_{p+1}^{(0)}-\lambda_p^{(0)})\tau\right)=$$
$$=2^{-s/2}\sum_{p=0}^{s}\left(\frac{s!}{p!\,(s-p)!}\right)^{3/2}\cos\left((\lambda_{p+1}^{(0)}-\lambda_p^{(0)})\tau\right)=\frac{s}{2}\cos(2\tau), \tag{D.2}$$

which is in agreement with (6.27) and (6.28). Note that the constant term entering expression (D.1) turns into zero due to property b) from (A.5) for the eigenvectors of the zeroth approximation. Now let us proceed to calculating the first approximation. In accordance with (6.19) and (6.22) matrix $||A_{mp}||$ should be taken in the form

$$A_{mp}=\alpha_{mp}-{}^{1}/_{2}\varepsilon\,\{H_{p-1,\,p}^{(1)}\alpha_{m,\,p-1}-H_{p+1,\,p}^{(1)}\alpha_{m,p+1}\}=$$
$$=\alpha_{mp}+{}^{1}/_{8}\varepsilon\,\{(s-2p+1)\sqrt{p(s-p+1)}\,\alpha_{m,\,p-1}-$$
$$-(s-2p-1)\sqrt{(s-p)(p+1)}\,\alpha_{m,\,p+1}\}. \tag{D.3}$$

In the first place, let us consider the terms containing cosine functions in (D.1). With an accuracy to ε we have

$$\sum_{m=0}^{s} mA_{mp}A_{ml} = \sum_{m=0}^{s} m\left\{\alpha_{mp} + \frac{\varepsilon}{2}[-H^{(1)}_{p-1,\,p}\alpha_{m,\,p-1} + H^{(1)}_{p+1,\,p}\alpha_{m,\,p+1}]\right\} \times$$
$$\times\left\{\alpha_{ml} + \frac{\varepsilon}{2}[-H^{(1)}_{l-1,\,l}\alpha_{m,\,l-1} + H^{(1)}_{l+1,\,l}\alpha_{m,\,l+1}]\right\} = \sum_{m=0}^{s} m\alpha_{mp}\alpha_{ml} -$$
$$-\frac{\varepsilon}{2}H^{(1)}_{p-1,\,p}\sum_{m=0}^{s} m\alpha_{m,\,p-1}\alpha_{ml} + \frac{\varepsilon}{2}H^{(1)}_{p+1,\,p}\sum_{m=0}^{s} m\alpha_{m,\,p+1}\alpha_{ml} -$$
$$-\frac{\varepsilon}{2}H^{(1)}_{l-1,\,l}\sum_{m=0}^{s} m\alpha_{m,\,l-1}\alpha_{mp} + \frac{\varepsilon}{2}H^{(1)}_{l+1,\,l}\sum_{m=0}^{s} m\alpha_{m,\,l+1}\alpha_{mp}. \tag{D.4}$$

Using (B.2) we get rid of sums over m and substitute the result obtained into (D.1). The substitution of the first term from (D.4) into (D.1) gives rise to

$$\sum_{p=0}^{s} A_{0p}A_{0,\,p+1}\sqrt{(p+1)(s-p)}\cos\left((\lambda^{(0)}_{p+1} - \lambda^{(0)}_{p})\tau\right) =$$
$$= \cos(2\tau)\sum_{p=0}^{s}\sqrt{p(s-p+1)}\,A_{0p}A_{0,\,p-1}. \tag{D.5}$$

The second and the fifth terms make no contribution. The third and the fourth terms make equal contributions, and their sum is

$$-\varepsilon s\cos(2\tau)\sum_{p=0}^{s} H^{(1)}_{p-1,\,p}\alpha_{0p}\alpha_{0,\,p-1} +$$
$$+\varepsilon\cos(4\tau)\sum_{p=0}^{s} H^{(1)}_{p-1,\,p}\sqrt{(p+1)(s-p)}\,\alpha_{0,\,p+1}\alpha_{0,\,p-1}. \tag{D.6}$$

Substituting into (D.5) the expression for A_{0p} from (D.3) we get

$$\cos(2\tau)\sum_{p=0}^{s}\sqrt{p(s-p+1)}\,A_{0p}A_{0,\,p-1} =$$
$$= \cos(2\tau)\left\{\sum_{p=0}^{s}\sqrt{p(s-p+1)}\,\alpha_{0p}\alpha_{0,\,p-1} -\right.$$
$$-\frac{\varepsilon}{2}\sum_{p=0}^{s} H^{(1)}_{p-1,\,p-2}\sqrt{p(s-p+1)}\,\alpha_{0p}\alpha_{0,\,p-2} +$$
$$+\frac{\varepsilon}{2}\sum_{p=0}^{s} H^{(1)}_{p,\,p-1}\sqrt{p(s-p+1)}\,\alpha^2_{0p} - \frac{\varepsilon}{2}\sum_{p=0}^{s} H^{(1)}_{p,\,p-1}\sqrt{p(s-p+1)}\,\alpha^2_{0,\,p-1} +$$
$$\left.+\frac{\varepsilon}{2}\sum_{p=0}^{s} H^{(1)}_{p,\,p+1}\sqrt{p(s-p+1)}\,\alpha_{0,\,p+1}\alpha_{0,\,p-1}\right\}. \tag{D.7}$$

The first term here equals

$$\tfrac{1}{2}s\cos(2\tau). \tag{D.8}$$

Consider the sum in the second term

$$\sum_{p=0}^{s} H^{(1)}_{p,\,p-1}\sqrt{(p+1)(s-p)}\,\alpha_{0,\,p+1}\alpha_{0p} =$$

$$= \sum_{p=0}^{s}(s-2p+1)\sqrt{p(s-p+1)(p+1)(s-p)}\,\alpha_{0,\,p-1}\alpha_{0,\,p+1} =$$

$$= s\sum_{p=0}^{s}(s-2p+1)\sqrt{p(s-p+1)}\,\alpha_{0,\,p-1}\alpha_{0p} -$$

$$-\sum_{p=0}^{s}(s-2p+1)\,p\,(s-p+1)\,\alpha^2_{0,\,p-1} =$$

$$= -\sum_{p=0}^{s}(s-2p+1)\,p\,(s-p+1)\,\alpha^2_{0,\,p-1}.$$

In other words, we have the equality

$$\sum_{p=0}^{s}(s-2p+1)\sqrt{p(s-p+1)(p+1)(s-p)}\,\alpha_{0,\,p-1}\alpha_{0,\,p+1} + \sum_{p=0}^{s}p\,(s-p+1)(s-2p+1)\,\alpha^2_{0,\,p-1} = 0. \tag{D.9}$$

Further,

$$-\frac{\sqrt{s}}{2} = \sum_{p=0}^{s}\sqrt{p(s-p+1)}\,\alpha_{1p}\alpha_{0,\,p-1} =$$

$$= \frac{1}{\sqrt{s}}\sum_{p=0}^{s}\sqrt{p(s-p+1)}\,(s-2p)\,\alpha_{0p}\alpha_{0,\,p-1} =$$

$$= \frac{1}{\sqrt{s}}\sum_{p=0}^{s}(s-2p)\sqrt{p(s-p+1)}\left\{\frac{1}{s}\sqrt{p(s-p+1)}\,\alpha_{0,\,p-1} + \frac{1}{s}\sqrt{(p+1)(s-p)}\,\alpha_{0,\,p+1}\right\}\alpha_{0,\,p-1} =$$

$$= \frac{1}{s\sqrt{s}}\sum_{p=0}^{s}(s-2p)\,p\,(s-p+1)\,\alpha^2_{0,\,p-1} + \frac{1}{s\sqrt{s}}\sum_{p=0}^{s}(s-2p)\sqrt{p(s-p+1)(p+1)(s-p)}\,\alpha_{0,\,p+1}\alpha_{0,\,p-1},$$

$$\sum_{p=0}^{s}(s-2p)\,p\,(s-p+1)\,\alpha^2_{0,\,p-1} + \sum_{p=0}^{s}(s-2p)\sqrt{p(s-p+1)(p+1)(s-p)}\,\alpha_{0,\,p+1}\alpha_{0,\,p-1} = -\frac{s^2}{2}. \tag{D.10}$$

Subtracting (D.10) from (D.9) we have

$$\sum_{p=0}^{s}p\,(s-p+1)\,\alpha^2_{0,\,p-1} + \sum_{p=0}^{s}\sqrt{p(s-p+1)(p+1)(s-p)}\,\alpha_{0,\,p+1}\alpha_{0p-1} = \frac{s^2}{2}. \tag{D.11}$$

Further,

$$-\frac{\sqrt{s}}{2}=\sum_{p=0}^{s}\sqrt{p\,(s-p+1)}\,\alpha_{1p}\alpha_{0,\,p-1}=$$

$$=\sum_{p=0}^{s}\sqrt{p\,(s-p+1)}\left\{\frac{1}{s-2}\sqrt{p\,(s-p+1)}\,\alpha_{1,\,p-1}+\right.$$

$$\left.+\frac{1}{s-2}\sqrt{(p+1)\,(s-p)}\,\alpha_{1,\,p+1}\right\}\alpha_{0,\,p-1}=$$

$$=\frac{1}{s-2}\sum_{p=0}^{s}p\,(s-p+1)\,\alpha_{1,\,p-1}\alpha_{0,\,p-1}+$$

$$+\frac{1}{s-2}\sum_{p=0}^{s}\sqrt{p\,(s-p+1)\,(p+1)\,(s-p)}\,\alpha_{1,\,p+1}\alpha_{0,\,p-1}=$$

$$=\frac{1}{s-2}\sum_{p=0}^{s}\frac{s-2p+2}{\sqrt{s}}\,p\,(s-p+1)\,\alpha^{2}_{0,\,p-1}+$$

$$+\frac{1}{s-2}\sum_{p=0}^{s}\frac{s-2p-3}{\sqrt{s}}\sqrt{p\,(s-p+1)\,(p+1)\,(s-p)}\,\alpha_{0,\,p+1}\alpha_{0,\,p-1}=$$

$$=\frac{1}{\sqrt{s}\,(s-2)}\sum_{p=0}^{s}(s-2p+2)\,p\,(s-p+1)\,\alpha^{2}_{0,\,p-1}+$$

$$+\frac{1}{\sqrt{s}\,(s-2)}\sum_{p=0}^{s}(s-2p-2)\sqrt{p\,(s-p+1)\,(p+1)\,(s-p)}\,\alpha_{0,\,p+1}\alpha_{0,\,p-1}.$$

We get

$$\sum_{p=0}^{s}p\,(s-p+1)\,\alpha^{2}_{0,\,p-1}-\sum_{p=0}^{s}\sqrt{p\,(s-p+1)\,(p+1)\,(s-p)}\,\alpha_{0,\,p+1}\alpha_{0,\,p-1}=\frac{s}{2}\,. \tag{D.12}$$

Formulas (D.11) and (D.12) lead to

$$\sum_{p=0}^{s}\sqrt{p\,(s-p+1)\,(p+1)\,(s-p)}\,\alpha_{0,\,p+1}\alpha_{0,\,p-1}=\frac{s\,(s-1)}{4}\,,$$

$$\sum_{p=0}^{s}p\,(s-p+1)\,\alpha^{2}_{0,\,p-1}=\frac{s\,(s+1)}{4}\,. \tag{D.13}$$

Consider the expression

$$\sum_{p=0}^{s}(s-2p)\sqrt{p\,(s-p+1)\,(p+1)\,(s-p)}\,\alpha_{0,\,p+1}\alpha_{0,\,p-1}=$$

$$=\sum_{p=0}^{s}(s-2p-2)\sqrt{p\,(s-p+1)\,(p+1)\,(s-p)}\,\alpha_{0,\,p+1}\alpha_{0,\,p-1}+$$

$$+2\sum_{p=0}^{s}\sqrt{p\,(s-p+1)\,(p+1)\,(s-p)}\,\alpha_{0,\,p+1}\alpha_{0,\,p-1}=$$

$$=\sqrt{s}\sum_{p=0}^{s}\sqrt{p\,(s-p+1)\,(p+1)\,(s-p)}\,\alpha_{1,\,p+1}\alpha_{0,\,p-1}+2\,\frac{s\,(s-1)}{4}=$$

$$= \sqrt{s} \sum_{p=0}^{s} \sqrt{p(s-p+1)}\,\alpha_{0,\,p-1}\{(s-2)\,\alpha_{1p} - \sqrt{p(s-p+1)}\,\alpha_{1,\,p-1}\} +$$

$$+ \frac{s(s-1)}{2} = \sqrt{s}\,(s-2) \sum_{p=0}^{s} \sqrt{p(s-p+1)}\,\alpha_{1p}\alpha_{0,\,p-1} -$$

$$- \sqrt{s} \sum_{p=0}^{s} \sqrt{p(s-p+1)}\,\alpha_{1,\,p-1}\{s\alpha_{0p} - \sqrt{(p+1)(s-p)}\,\alpha_{0,\,p+1}\} +$$

$$+ \frac{s(s-1)}{2} = -\frac{s(s-2)}{2} + \frac{s(s-1)}{2} - \sqrt{s}\,s \sum_{p=0}^{s} \sqrt{p(s-p+1)} \times$$

$$\times \alpha_{1,\,p-1}\alpha_{0p} + \sqrt{s} \sum_{p=0}^{s} \sqrt{p(s-p+1)(p+1)(s-p)}\,\alpha_{1,\,p-1}\alpha_{0,\,p+1} =$$

$$= \frac{s}{2} - s\sqrt{s}\,\frac{\sqrt{s}}{2} + \sqrt{s} \sum_{p=0}^{s} \sqrt{p(s-p+1)(p+1)(s-p)}\,\alpha_{1,\,p-1}\alpha_{0,\,p+1} =$$

$$= -\sqrt{s} \sum_{p=0}^{s} \sqrt{p(s-p+1)(p+1)(s-p)}\,\alpha_{1,\,p+1}\alpha_{0,\,p-1} - \frac{s(s-1)}{2}\,.$$

Thus we have

$$\sqrt{s} \sum_{p=0}^{s} \sqrt{p(s-p+1)(p+1)(s-p)}\,\alpha_{1,\,p+1}\alpha_{0,\,p-1} = -\frac{s(s-1)}{2}\,.$$

Hence

$$\sum_{p=0}^{s} (s-2p) \sqrt{p(s-p+1)(p+1)(s-p)}\,\alpha_{0,\,p+1}\alpha_{0,\,p-1} = 0$$

and

$$-\frac{\varepsilon}{2}\cos(2\tau) \sum_{p=0}^{s} H^{(1)}_{p,\,p-1} \sqrt{(p+1)(s-p)}\,\alpha_{0,\,p+1}\alpha_{0,\,p-1} = \frac{\varepsilon s(s-1)}{32}\cos(2\tau). \tag{D.14}$$

For the third and the fourth terms in (D.7) we have

$$\frac{\varepsilon}{2}\cos(2\tau)\left\{\sum_{p=0}^{s} H^{(1)}_{p,\,p-1}\sqrt{p(s-p+1)}\,\alpha^2_{0p} - \right.$$

$$\left. - \sum_{p=0}^{s} H^{(1)}_{p,\,p-1}\sqrt{p(s-p+1)}\,\alpha^2_{0,\,p-1}\right\} =$$

$$= \frac{\varepsilon}{4}\cos(2\tau)\sum_{p=0}^{s}(s-2p+1)\,p\,(s-p+1)\,\alpha^2_{0,\,p-1} = -\frac{\varepsilon s(s-1)}{16}\cos(2\tau). \tag{D.15}$$

Finally, for the last term in (D.7) we get

$$\frac{\varepsilon}{2}\cos(2\tau)\sum_{p=0}^{s} H^{(1)}_{p,\,p+1}\sqrt{p(s-p+1)}\,\alpha_{0,\,p+1}\alpha_{0,\,p-1} =$$

$$= -\frac{\varepsilon}{8}\cos(2\tau)\sum_{p=0}^{s}(s-2p-1)\times$$

$$\times\sqrt{p(s-p+1)(p+1)(s-p)}\,\alpha_{0,\,p+1}\alpha_{0,\,p-1} = \frac{\varepsilon s(s-1)}{32}\cos(2\tau). \tag{D.16}$$

Now we can easily complete the calculation of sums in (D.6):

$$-\varepsilon s\cos(2\tau)\sum_{p=0}^{s} H^{(1)}_{p,\,p-1}\alpha_{0,\,p}\alpha_{0,\,p-1} =$$

$$= \frac{\varepsilon s}{4}\cos(2\tau)\sum_{p=0}^{s}(s-2p+1)\sqrt{p(s-p+1)}\,\alpha_{0p}\alpha_{0,\,p-1} = 0, \tag{D.17}$$

$$\varepsilon\cos(4\tau)\sum_{p=0}^{s} H^{(1)}_{p,\,p-1}\sqrt{(p+1)(s-p)}\,\alpha_{0,\,p+1}\alpha_{0,\,p-1} =$$

$$= -\frac{\varepsilon}{4}\cos(4\tau)\sum_{p=0}^{s}(s-2p+1)\times$$

$$\times\sqrt{p(s-p+1)(p+1)(s-p)}\,\alpha_{0,\,p+1}\alpha_{0,\,p-1} = -\frac{\varepsilon s(s-1)}{16}\cos(4\tau). \tag{D.18}$$

We only have to calculate the last term in (D.1). Using the formulas already obtained one can show that

$$\sum_{p=0}^{s} A^2_{p0}\sum_{m=1}^{s} mA^2_{pm} = \frac{s}{2} - \frac{\varepsilon s(s-1)}{16}. \tag{D.19}$$

Adding (D.8), (D.14) and (D.19) we get

$$E^{(1)}_{\mathrm{a}\tau} = {}^1\!/_2\, s\cos(2\tau) + {}^1\!/_{16}\,\varepsilon s(s-1)(1-\cos(4\tau)).$$

Appendix E

Here we perform the calculations necessary for determining the degree of squeezing in the spontaneous emission by an ensemble of two-level atoms in an ideal resonator. We use the wave function of the system in the zeroth approximation (7.12):

$$|\psi(t)\rangle^{(0)} = \sum_{l=0}^{s} b_l \sum_{k=0}^{l}\sum_{n=0}^{l}\alpha^{(l)}_{0p}\alpha^{(l)}_{np_s}\exp\left(\frac{igt\lambda^{(0)}_{k(l)}}{\sqrt{\varepsilon_l}}\right)|l,n\rangle. \tag{E.1}$$

We have introduced the notation

$$b_k = e^{ik\varphi} \cos^k \theta \cdot \sin^{s-k} \theta \, (s!/k!\,(s-k)!)^{1/2}$$

for the initial amplitudes of the states with different numbers of excited atoms: $\varepsilon_1 = [N-(\iota-1)/2]^{-1}$. According to definition (6.4) of the $|\iota,n\rangle$-basis, the photon operators act on its vectors as follows:

$$a \mid l, \quad n\rangle = \sqrt{n} \mid l-1,\ n-1\rangle, \quad a^+ \mid l,\ n\rangle = \sqrt{n+1} \mid l+1,\ n+1\rangle.$$

Let us begin with calculating the average $\langle\alpha\rangle$. In the zeroth approximation we have

$$\langle a\rangle = {}^{(0)}\langle\psi(t)|a|\psi(t)\rangle^{(0)} = \sum_{l=1}^{s} b^*_{l-1} b_l \sum_{k',k=0}^{s} \exp\left\{\left(\frac{\lambda^{(0)}_{k'(l-1)}}{\sqrt{\varepsilon_{l-1}}} - \frac{\lambda^{(0)}_{k(l)}}{\sqrt{\varepsilon_l}}\right) itg\right\} \times \alpha^{*(l-1)}_{0k'} \alpha^{(l)}_{0k} \sum_{n=1}^{l} \sqrt{n}\, \alpha^{*(l-1)}_{n-1,k} \alpha^{(l)}_{nk}. \tag{E.2}$$

Let us calculate the inner sum over n. For this purpose we use the formula relating matrix elements from different representations of SU(2) [62]

$$\alpha^{(l)}_{kn} = \frac{1}{\sqrt{2n}} \left\{\sqrt{l-k}\, \alpha^{(l-1)}_{k,\,n-1} - \sqrt{k}\, \alpha^{(l-1)}_{k-1,n-1}\right\}.$$

We have

$$\sum_{n=1}^{l} \sqrt{n}\, \alpha^{(l)}_{kn} \alpha^{(l-1)}_{k',\,n-1} = \sum_{n=1}^{l} \frac{1}{\sqrt{2}} \left\{\sqrt{l-k}\, \alpha^{(l-1)}_{k,\,n-1} - k^{1/2} \alpha^{(l-1)}_{k-1,\,n-1}\right\} \alpha^{(l-1)}_{k',\,n-1} =$$

$$= \left(\frac{l-k}{2}\right)^{1/2} \sum_{n=1}^{l} \alpha^{(l-1)}_{k,\,n-1} \alpha^{(l-1)}_{k',\,n-1} - \left(\frac{k}{2}\right)^{1/2} \sum_{n=1}^{l} \alpha^{(l-1)}_{k-1,\,n-1} \alpha^{(l-1)}_{k',\,n-1} =$$

$$= \left(\frac{l-k}{2}\right)^{1/2} \delta_{k,\,k'} - \left(\frac{k}{2}\right)^{1/2} \delta_{k-1,\,k'}.$$

For the double sum over k and k' we now get

$$\sum_{k=0}^{l} \sum_{k'=0}^{l-1} \exp\left\{igt \left(\frac{\lambda^{(0)}_{k'(l-1)}}{\sqrt{\varepsilon_{l-1}}} - \frac{\lambda^{(0)}_{k(l)}}{\sqrt{\varepsilon_l}}\right)\right\} \alpha^{(l-1)}_{0k'} \alpha^{(l)}_{0k} \left[\left(\frac{l-k}{2}\right)^{1/2} \delta_{kk'} - \left(\frac{k}{2}\right)^{1/2} \delta_{k-1,\,k'}\right] =$$

$$= \frac{1}{\sqrt{2}} \sum_{k=0}^{l-1} \exp\left\{itg \left(\frac{\lambda^{(0)}_{k(l-1)}}{\sqrt{\varepsilon_{l-1}}} - \frac{\lambda^{(0)}_{k(l)}}{\sqrt{\varepsilon_l}}\right)\right\} \sqrt{l-k}\, \alpha^{(l)}_{0k} \alpha^{(l-1)}_{0k} -$$

$$- \frac{1}{\sqrt{2}} \sum_{k=0}^{l-1} \exp\left\{itg \left(\frac{\lambda^{(0)}_{k(l-1)}}{\sqrt{\varepsilon_{l-1}}} - \frac{\lambda^{(0)}_{k+1(l)}}{\sqrt{\varepsilon_l}}\right)\right\} \sqrt{k+1}\, \alpha^{(l-1)}_{0k} \alpha^{(l)}_{0,\,k+1}. \tag{E.3}$$

Consider the difference in the argument of the exponential function

$$\frac{\lambda^{(0)}_{k(l-1)}}{\sqrt{\varepsilon_{l-1}}} - \frac{\lambda^{(0)}_{k+1(l)}}{\sqrt{\varepsilon_l}} = \left(N - \frac{l-1}{2}\right)^{1/2} \left[\lambda^{(0)}_{k(l-1)} \left(1 + \frac{\varepsilon_l}{2}\right)^{1/2} - \lambda^{(0)}_{k+1(l)}\right] =$$

$$= \frac{1}{\sqrt{\varepsilon_l}} \left[\lambda^{(0)}_{k(l-1)} - \lambda^{(0)}_{k+1(l)} + \frac{\varepsilon_l}{4} \lambda^{(0)}_{k(l-1)}\right] = \frac{1 + \varepsilon_l\,(l-1-2k)/4}{\sqrt{\varepsilon_l}}$$

and similarly for the second exponential. One can write for (E.3)

$$\frac{1}{\sqrt{2}} \exp\left(-\frac{itg}{\sqrt{\varepsilon_l}}\right) \sum_{k=0}^{l-1} \exp\left(itg \frac{\sqrt{\varepsilon_l}}{4} (l-1-2k)\right) \alpha^{(l)}_{0k} \alpha^{(l-1)}_{0k} -$$

$$- \frac{1}{\sqrt{2}} \exp\left(\frac{itg}{\sqrt{\varepsilon_l}}\right) \sum_{k=0}^{l-1} \exp\left(itg \frac{\sqrt{\varepsilon_l}}{4} (l-1-2k)\right) \alpha^{(l-1)}_{0k} \alpha^{(l)}_{0,\,k+1}.$$

Further we use the relations [62]

$$\alpha^{(l)}_{0k} = \sqrt{l/2\,(l-k)}\, \alpha^{(l-1)}_{0k}, \quad \alpha^{(l)}_{0,\,k+1} = \sqrt{l/2\,(k+1)}\, \alpha^{(l-1)}_{0k}$$

and obtain

$$\exp\left(-\frac{itg}{\sqrt{\varepsilon_l}}\right)\frac{\sqrt{l}}{2}\sum_{k=0}^{l-1}\exp\left(itg\frac{\sqrt{\varepsilon_l}}{4}(l-1-2k)\right)|\alpha_{0k}^{(l-1)}|^2 -$$
$$-\exp\left(\frac{itg}{\sqrt{\varepsilon_l}}\right)\frac{\sqrt{l}}{2}\sum_{k=0}^{l-1}\exp\left(itg\frac{\sqrt{\varepsilon_l}}{4}(l-1-2k)\right)|\alpha_{0k}^{(l-1)}|^2. \tag{E.4}$$

Let us elucidate in what cases the replacement of the exponential functions inside the sums by unity is admissible. It is true for not very large times, when the conditions

$$gt\sqrt{\varepsilon_l}\,(l-1-2k)/4 < gt\sqrt{\varepsilon_l}\,(l-1)/4 \ll 2\pi. \tag{E.5}$$

are satisfied. The characteristic frequencies of oscillations are confined within the interval

$$g\varepsilon_0^{-1/2} = g(N+\frac{1}{2})^{1/2} \leq \omega \leq g\varepsilon_s^{-1/2} = g(N-(s-1)/2)^{1/2}$$

The de-phasing during the time of the first oscillation $t \approx 2\pi\sqrt{\varepsilon}/g$ can be neglected provided (see (E.5))

$$\varepsilon_s(s-1)/4 = [N-(s-1]^{-1}(s-1)/4 << 1$$

This requirement is obviously fulfilled when

$$N \gg s/2. \tag{E.6}$$

Believing the last condition to be valid, we expand the exponentials in (E.4) into Taylor's series and throw aside the terms beginning with t^2. Then terms linear with respect to t give no contribution, and we obtain the following expression for sum (E.4):

$$-i\sqrt{l}\sin(gt/\sqrt{\varepsilon_l});$$

so that

$$\langle a\rangle = -i\sum_{l=1}^{s} b_{l-1}^{*}b_l\sqrt{l}\sin\frac{gt}{\sqrt{\varepsilon_l}}.$$

Within the same accuracy one can take $\sin(gt/\sqrt{\varepsilon_1})$ out of the sum mark and obtain

$$\langle a\rangle = -i\sin\frac{gt}{\sqrt{\varepsilon_l}}\sum_{l=1}^{s}\sqrt{l}\,b_{l-1}^{*}b_l. \tag{E.7}$$

Let us proceed to calculating $<\alpha^2>$. Since

$$<\iota',n'|\alpha^2|\iota,n> = \sqrt{n(n-1)}\,\delta_{\iota',\iota-2}\delta_{n',n-2'}$$

one can write

$$\langle a^2\rangle = {}^{(0)}\langle\psi(t)|a^2|\psi(t)\rangle^{(0)} = \sum_{l=2}^{s} b_{l-2}^{*}b_l\sum_{k',k=0}^{l}\exp\left[itg\left(\frac{\lambda_{k'(l-2)}^{(0)}}{\sqrt{\varepsilon_{l-2}}}-\frac{\lambda_{k(l)}^{(0)}}{\sqrt{\varepsilon_l}}\right)\right]\times$$
$$\times\,\alpha_{0k}^{(l)}\alpha_{0k'}^{(l-2)}\sum_{n=2}^{s}\sqrt{n(n-1)}\,\alpha_{kn}^{(l)}\alpha_{k',n-2}^{(l-2)}. \tag{E.8}$$

To calculate the sum over n we use the relation [62]

$$\alpha_{kn}^{(l)} = \frac{1}{2}\left(\frac{(l-k)(l-k-1)}{n(n-1)}\right)^{1/2} \alpha_{k,n-2}^{(l-2)} + \frac{1}{2}\left(\frac{k(k-1)}{n(n-1)}\right)^{1/2} \alpha_{k-2,n-2}^{(l-2)} - \\ - \left(\frac{k(l-k)}{n(n-1)}\right)^{1/2} \alpha_{k-1,n-2}^{(l-2)}$$

and obtain

$$\sum_{n=2}^{l} \sqrt{n(n-1)}\, \alpha_{kn}^{(l)} \alpha_{k,n-2}^{(l-2)} = \sum_{n=2}^{l} \left\{ \frac{1}{2}\sqrt{(l-k)(l-k-1)}\, \alpha_{n,n-2}^{(l-2)} + \right. \\ \left. + \frac{1}{2}\sqrt{k(k-1)}\, \alpha_{k-2,n-2}^{(l-2)} - \sqrt{k(l-k)}\, \alpha_{k-1,n-2}^{(l-2)} \right\} \alpha_{k'n-2}^{(l-2)} = \\ = \frac{1}{2}\sqrt{(l-k)(l-k-1)}\, \delta_{kk'} + \frac{1}{2}\sqrt{k(k-1)}\, \delta_{k-2,k'} - \sqrt{k(l-k)}\, \delta_{k-1,k'}.$$

For the inner sum over k and k' in (E.8) we get

$$\frac{1}{2}\sum_{k=0}^{l-2} \exp\left[igt\left(\frac{\lambda_{k(l-2)}^{(0)}}{\sqrt{\varepsilon_{l-2}}} - \frac{\lambda_{k(l)}^{(0)}}{\sqrt{\varepsilon_l}}\right)\right] \sqrt{(l-k)(l-k-1)}\, \alpha_{0k}^{(l)} \alpha_{0k}^{(l-2)} + \\ + \frac{1}{2}\sum_{k=0}^{l} \exp\left[igt\left(\frac{\lambda_{k-2(l-2)}^{(0)}}{\sqrt{\varepsilon_{l-2}}} - \frac{\lambda_{k(l)}^{(0)}}{\sqrt{\varepsilon_l}}\right)\right] \sqrt{k(k-1)}\, \alpha_{0k}^{(l)} \alpha_{0,k-2}^{(l-2)} - \\ - \sum_{k=1}^{l-1} \exp\left[igt\left(\frac{\lambda_{k-1(l-2)}^{(0)}}{\sqrt{\varepsilon_{l-2}}} - \frac{\lambda_{k(l)}^{(0)}}{\sqrt{\varepsilon_l}}\right)\right] \sqrt{k(l-k)}\, \alpha_{0k}^{(l)} \alpha_{0,k-1}^{(l-2)}. \tag{E.9}$$

As before, we replace the exponential functions inside sums by unity. The satisfactory approximation during the first several periods of oscillations, i.e., for $t \approx 2\pi\sqrt{\varepsilon_s}/g$, is then ensured by condition (E.6). After this expression (E.9) is reduced to

$$\frac{1}{2} e^{-i2gt/\sqrt{\varepsilon_l}} \sum_{k=0}^{l-2} \sqrt{(l-k)(l-k-1)}\, \alpha_{0k}^{(l)} \alpha_{0k}^{(l-2)} + \\ + \frac{1}{2} e^{i2gt/\sqrt{\varepsilon_l}} \sum_{k=2}^{l} \sqrt{k(k-1)}\, \alpha_{0k}^{(l)} \alpha_{0,k-2}^{(l-2)} - \sum_{k=1}^{l-1} \sqrt{k(l-k)}\, \alpha_{0k}^{(l)} \alpha_{0,k-1}^{(l-2)}. \tag{E.10}$$

To calculate the remaining sums we use the relations [62]

$$\alpha_{0k}^{(l)} = \frac{1}{2}\left(l(l-1)/(l-k)(l-k-1)\right)^{1/2} \alpha_{0k}^{(l-2)},$$

$$\alpha_{0k}^{(l)} = \frac{1}{2}\left(l(l-1)/k(k-1)\right)^{1/2} \alpha_{0,k-2}^{(l-2)}, \qquad \alpha_{0k}^{(l)} = \frac{1}{2}\left(l(l-1)/k(l-k)\right)^{1/2} \alpha_{0,k-1}^{(l-2)}$$

and obtain the following expressions:

$$\sum_{k=0}^{l-2} \sqrt{(l-k)(l-k-1)}\, \alpha_{0k}^{(l)} \alpha_{0k}^{(l-2)} = \sum_{k=2}^{l} \sqrt{k(k-1)}\, \alpha_{0k}^{(l)} \alpha_{0,k-2}^{(l-2)} = \\ = \sum_{k=1}^{l-1} \sqrt{k(l-k)}\, \alpha_{0k}^{(l)} \alpha_{0,k-1}^{(l-2)} = \frac{1}{2}\sqrt{l(l-1)}.$$

Finally, substituting (E.10) into (E.8) and taking into account condition (E.6) enabling us to take the exponential functions out of the mark of summation over ι, we have

$$\langle a^2 \rangle = -\sin^2 \frac{gt}{\sqrt{\varepsilon_s}} \sum_{l=2}^{s} \sqrt{l(l-1)}\, b_{l-2}^{*} b_l.$$

The calculation of $\langle \alpha^{+}\alpha \rangle$ is performed analogously. From (E.7) and (E.8) we get

$$\langle a_1 \rangle = \frac{1}{2} \langle a + a^+ \rangle = \sin \frac{gt}{\sqrt{\varepsilon_s}} \sum_{l=1}^{s} \sqrt{l} \sin (\varphi_{l+1} - \varphi_l) \, | b_{l-1}^* b_l |,$$

$$\langle a_2 \rangle = \frac{1}{2i} \langle a - a^+ \rangle = - \sin \frac{gt}{\sqrt{\varepsilon_s}} \sum_{l=1}^{s} \sqrt{l} \cos (\varphi_{l+1} - \varphi_l) \, | b_{l-2}^* b_l |,$$

$$\frac{1}{2} \langle a^2 + a^{+2} \rangle = - \sin^2 \frac{gt}{\sqrt{\varepsilon_s}} \sum_{l=2}^{s} \sqrt{l(l-1)} \cos (\varphi_{l+1} - \varphi_{l-1}) \, | b_{l-2}^* b_l |,$$

$$b_l = | b_l | e^{i\varphi_l}.$$

Substituting the expressions obtained into the formulas for the quadrature components variances

$$(\Delta a_1)^2 = \langle a_1^2 \rangle - \langle a_1 \rangle^2 = {}^1/_4 + {}^1/_2 \langle a^+ a \rangle + {}^1/_4 \langle a^2 + a^{+2} \rangle - \langle a_1 \rangle^2,$$

$$(\Delta a_2)^2 = \langle a_2^2 \rangle - \langle a_2 \rangle^2 = {}^1/_4 + {}^1/_2 \langle a^+ a \rangle + {}^1/_4 \langle a^2 + a^{+2} \rangle - \langle a_2 \rangle^2,$$

we arrive at relations (7.14) and (7.15).

Received in December, 1988

REFERENCES

1. Dodonov, V.V., Kurmyshev, E.V., Man'ko, V.I. "Generalized uncertainty relation and correlated coherent states" *Phys. Lett. A*, 1980, vol. 79, no. 2/3, pp. 150-152.
2. Stoler, D. "Equivalence classes of minimum-uncertainty packets" *Phys. Rev. D*, 1970, vol. 1, no. 12, pp. 3217-3219.
3. Stoler, D. "Equivalence classes of minimum-uncertainty packets. II" *Phys. Rev. D*, 1971, vol. 4, no. 6, pp. 1925-1926.
4. Yuen, H.P. "Two-photon coherent states of the radiation field" *Phys. Rev. A*, 1976, vol. 13, no. 6, pp. 2226-2243.
5. Lu, E.Y.C. "New coherent states of the electromagnetic field" *Lett. Nuovo Cim.*, 1971, vol. 2, no. 24, pp. 1241-1244.
6. Walls, D.F. "Squeezed states of light" *Nature*, 1983, vol. 306, no. 10, pp. 141-146.
7. Dodonov, V.V., Kurmyshev, E.V., Man'ko, V.I. "Correlated coherent states" Proc. Lebedev Phys. Inst., Moscow: Nauka, 1986, vol. 176, pp. 128-150.
8. Dodonov, V.V., Man'ko, V.I. "Invariants and correlated states of nonstationary quantum systems" Proc. Lebedev Phys. Inst., Moscow: Nauka, 1987, vol. 183, pp. 71-181.
9. Schrodinger, E. "Der stetige Ubergang von den Mikro-zur Mikromechanik" *Naturwissenschaften*, 1926, Bd. 14, S. 664-666.
10. Steiner, F. "Schrodinger's discovery of coherent states" *Prepr. DESY 87-142*, 1987, pp. 1-8.
11. Klauder, J.R. "Continuous-representation theory. I: Postulates of continuous-representation theory" *J. Math. Phys.*, 1963, vol. 4, no. 8, pp. 1055-1058.
12. Klauder, J.R. "Continuous-representation theory. II: Generalized relation between quantum and classical dynamics" *J. Math. Phys.*, 1963, vol. 4, no. 8, pp. 1058-1073.
13. Klauder, J.R., Skaugerstam, B.-S. "Coherent states: Applications in physics and mathematical physics" Singapore: World Sci. Publ., 1985, p. 327.
14. Glauber, R.J. "Coherent and incoherent states of the radiation field" *Phys. Rev.*, 1963, vol. 131, no. 6, pp. 2766-2788.
15. Yuen, H.P., Shapiro, J.H. "Optical communication with two-photon coherent states. Pt 1: Quantum-state propagation and quantum noise reduction" *IEEE Trans. Inform. Theory*, 1978, vol. IT-24, no. 6, pp. 657-668.
16. Caves, C.M., Thorne, K.S., Drever, R.W.P., Sandberg, V.D., Zimmermann, M. "On the measurement of a weak classical force coupled to a quantum-mechanical oscillator. I: Issues of principle" *Rev. Mod. Phys.*, 1980, vol. 52, no. 2, pt 1, pp. 341-392.
17. Kimble, H.J., Dagenais, M., Mandel, L. "Photon antibunching in resonance fluorescence" *Phys. Rev. Lett.*, 1977, vol. 39, no. 11, pp. 691-695.
18. Kask, P., Piksarv, P., Mets, U. "Fluorescence correlation spectroscopy in the nanosecond time range: Photon antibunching in dye fluorescence" *Eur. Biophys. J.*, 1985, vol. 12, no. 3, pp. 163-166.
19. Wagner, J., Kurowski, P., Martienssen, W. "Simulation of photon antibunching" *Ztschr. Phys. B*, 1979, Bd. 33, S. 391-402.

20. Short, R., Mandel, L. "Observation of sub-Poissonian photon statistics" *Phys. Rev. Lett.*, 1983, vol. 51, no. 5, pp. 384-387.
21. Hong, C.K., Ou, Z.Y., Mandel, L. "Measurement of subpicosend time intervals between two photons by interference" *Phys. Rev. Lett.*, 1986, vol. 59, no. 18, pp. 2044-2046.
22. Smirnov, D.F., Troshin, A.C. "New phenomena in quantum optics: antibunching and sub-Poissonian photon statistics; squeezed states" *UFN*, 1987, vol. 153, pp. 233-272.
23. Bogolubov, N.N. (Jn.), Kozierowski, M., Quang Than, Shumovsky, A.S. "New effects in quantum electrodynamics" Fizika Elementarnych Chastits i Atomnogo Yadra *[Physics of Elementary Particles and Atom Nuclei]*, 1988, vol. 19, no. 4, pp. 831-863.
24. Loudon, R., Knight, P.L. "Squeezed light" *J. Mod. Opt.*, 1987, vol. 34, no. 1/7, pp. 709-759.
25. Jannusis, A., Bartzis, V. "Generalized properties of the squeezed states" *Nuovo cim. B*, 1987, vol. 100, no. 5, pp. 633-650.
26. Slusher, R.E., Hollberg, L.W., Yurke, B., Mertz, J.C., Valley, J.F. Observation of squeezed states generated by four-wave mixing in an optical cavity" *Phys. Rev. Lett.*, 1985, vol. 55, no. 22, pp. 2409-2412.
27. Shelby, R.M., Levenson, M.D., Perlmutter, S.H., de Voe, R.G., Walls, D.F. "Broad-band parametric deamplification of quantum noise in an optical fiber" *Phys. Rev. Lett.*, 1986, vol. 57, no. 6, pp. 691-694.
28. Wu, L., Kimble, H.J., Hall, L.J., Wu, H. "Generation of squeezed states by parametric down conversion" *Phys. Rev. Lett.*, 1986, vol. 57, no. 20, pp. 2520-2523.
29. Xiao, M., Wu, L., Kimble, H.J. "Precision measurement beyond the shot-noise limit" *Phys. Rev. Lett.*, 1987, vol. 59, no. 3, pp. 278-281.
30. Milburn, G.J., "Interaction of a two-level atom with squeezed light" *Opt. acta.*, 1984, vol. 31, no. 6, pp. 671-679.
31. Abdel-Hafez, A.M., Obada, A.S.F., Ahmad, M.M.A. "N-level atom and N-1 modes: Statistical aspects and interaction with squeezed light" *Phys. Rev.*, 1987, vol. 35, no. 4, pp. 1634-1647.
32. Gariner, C.W. "Inhibition of atomic-phase decays by squeezed light: A direct effect of squeezing" *Phys. Rev. Lett.*, 1986, vol. 56, no. 18, pp. 1917-1920.
33. Carmichael, H.J., Lane, A.S., Walls, D.F. "Resonance fluorescence from an atom in a squeezed vacuum" *Phys. Rev. Lett.*, 1987, vol. 58, no. 24, pp. 2539-2542.
34. Quang, T., Kozierowski, M., Lan, L.H. "Collective resonance fluorescence in a squeezed vacuum" *Phys. Rev. A*, 1989, vol. 39, no. 2, pp. 644-646.
35. Kozierowski, M., Tanas, R., Quantum fluctuations in second-harmonic generation of light" *Opt. Communs.* 1977, vol. 21, no. 3, pp. 229-321.
36. Mandel, L. "Squeezing and photon antibunching in harmonic generation" *Opt. Communs.*, 1982, vol. 42, no. 6, pp. 437-439.
37. Kozierowski, M., Man'ko, V.I. "Second-harmonic generation as a source of correlated coherent states" *Opt. Communs.*, 1988, vol. 69, pp. 71-74.
38. Walls, D.F., Tindke, C.T. "Nonlinear quantum effects in optics" *J. Phys. A.*, 1972, vol. 5, no. 4, pp. 534-545.

39. Mostowski, J., Rzazewski, K. "Photon bunching and antibunching in second harmonics generation" *Phys. Lett. A*, 1978, vol. 66, no. 4, pp. 275-278.
40. Chmela, P. "Theory of light fluctuation filtering in two-photon absorption process by multiple second harmonic generation. I: Basic equations. Classical description" *J. Opt. Quant. Electron.*, 1984, vol. 16, no. 5, pp. 445-454.
41. Chmela, P. "Filtering of light fluctuations by multiple second-harmonic generation" *Opt. Communs.*, 1982, vol. 42, no. 3, pp. 201-204.
42. Chmela, P., Kozierowski, M., Kielich, S. "Squeezing by multiple higher-harmonic generation" *Czech. J. Phys.*, 1987, vol. 337, pp. 846-849.
43. Hong, C.K., Mandel, L. "Generation of higher-order squeezing of quantum electromagnetic fields" *Phys. Rev. A*, 1985, vol. 32, p. 974-982.
44. Kozierowski, M. "Higher-order squeezing in k-th harmonic generation" *Phys. Rev. A*, 1986, vol. 34, no. 4, pp. 3474-3477.
45. Kien Fam Le, Kozierowski, M., Quang, T. "Fourth-order squeezing in the multiphoton Jaynes-Cummings model" *Phys. Rev. A*, 1988, vol. 38, no. 1, pp. 263-266.
46. Meystre, P., Zubairy, M.S. "Squeezed states in the Jaynes-Cummings model" *Phys. Lett. A*, 1982, vol. 89, no. 8, pp. 390-392.
47. Wodkiewicz, K., Knight, P.L., Buckle, S.J., Barnett, S.M. "Squeezing and superposition states" *Phys. Rev. A*, 1987, vol. 35, no. 6, pp. 2567-2577.
48. Shumovsky, A.S., Kien Fam Le, Aliskenderov, E.L. "Squeezing in the multiphoton Jaynes-Cummings model" *Phys. Lett. A*, 1987, vol. 124, no. 6/7, pp. 351-354.
49. Zhu, S.Y., Liu, Z.D., Li, X.S. "Squeezing in a three-level Jaynes-Cummings model" *Phys. Lett. A*, 1988, vol. 128, no. 1/2, pp. 89-96.
50. Kozierowski, M., Man'ko, V.I., Chumakov, S.M. "Multilevels systems as a source of squeezed and correlated light" Abstracts of XIII Conf. "Coherent and nonlinear optics," Minsk: 1988, pt. 1, Section I-VIII, p. 111.
51. Wallz, D.F., Barakat, R. "Quantum-mechanical amplification and frequency conversion with a trilinear Hamiltonian" *Phys. Rev. A*, 1970, vol. 1, no. 2, pp. 446-453.
52. Banifacio, R., Preparata, G., "Coherent spontaneous emission" *Phys. Rev. A*, 1970, vol. 2, no. 2, pp. 336-347.
53. Kumar, S., Mehta, C.L. "Theory of the interaction of a single-mode resonant field with N two-level atoms" *Phys. Rev. A*, 1980, vol. 21, no. 5, pp. 1573-1588.
54. Kumar, S., Mehta, C.L. "Theory of the interaction of a single-mode radiation field with N two-level atoms. II: Time evolution of the statistics of the system" *Phys. Rev. A*, 1981, vol. 24, no. 3, pp. 1460-1468.
55. Heidmann, A., Raimond, J.M., Reynaud, S. "Squeezing in a Rydberg atom maser" *Phys. Rev. Lett.*, 1985, vol. 54, no. 4, pp. 326-328.
56. Dicke, R.H. "Coherence in spontaneous radiation processes" *Phys. Rev.*, 1954, vol. 93, pp. 99-110.
57. Chumakov, S.M., Kozierowski, M. "Exactly solvable Dicke models and radiation capture" Proc. Lebedev Phys. Inst., Moscow: Nauka, 1989, vol. 191, pp. 150-170.

58. Leonardi, C., Persico, F., Verti, G. "Dicke model and the theory of driven and spontaneous emission" *Riv. Nuovo. Cim.*, 1986, vol. 9, no. 4, pp. 1-85.
59. Gross, M., Haroche, S. "Superradiance: An essay on the theory of collective spontaneous emission" *Phys. Rept.*, 1982, vol. 93, pp. 1-85.
60. Louisell, W.H. "Radiation and noise in quantum electronics" New York San Francisco, Toronto, London: McGraw-Hill Book Co., 1964.
61. Allen, L., Eberly, I.E. "Opticheskii rezonans i dvuchurovnevye atomy" [Optical resonance and two level atoms] Moscow: Mir, 1978, p. 222.
62. Vilenkin, N.Ya, "Spetsial'nye funktsii i teoriya predstavleniy grupp" [Special functions and group representation theory] Moscow: Nauka, 1965, p. 588.
63. Knight, P.L. "Quantum fluctuations and squeezing in the interaction of an atom with a single field mode" *Phys. Scr.*, 1986, vol. 12, pp. 51-55.
64. Barnett, S.M., Dupertuis, M.-A. "Multation squeezed states, a new class of collective atomic states" *J. Opt. Soc. Amer. B*, 1987, vol. 4, no. 4, pp. 505-511.

CORRELATED STATES OF STRING AND GRAVITATIONAL WAVEGUIDE

V.V. Dodonov, V.I. Man'ko, O.V. Man'ko

Abstract: The theory of coherent and correlated states of quantum resonant circuit and continuous string is constructed. The excitation of the mentioned systems due to time-dependence of their parameters is considered. The hypothesis of the existence of the gravitational waveguide in the Universe and its possible role are discussed.

1. INTRODUCTION

Recently, the theory of nonstationary and stationary quantum systems was developed both due to problems of the quantum mechanics and in connection with important applications (see, for example, the reviews [1-3]). The quantum systems are called the quadratic ones if they have the Hamiltonians which are arbitrary nonstationary quadratic forms in coordinates and momenta. The open systems are called quadratic ones if the Fokker-Planck equations for these systems are reduced to the equations which formally coincide with the Schrodinger equations for the systems with quadratic Hamiltonians. The number of degrees of freedom may be both finite and infinite. The theory of nonstationary quantum systems with one or finite number of degrees of freedom is well studied. In the case of nonstationary quadratic systems, the infinite number of degrees of freedom is interesting from the point of view of finding the explicit formulae for the transition amplitudes due to parametric excitations of the system and due to such application as, for example, a chain of atoms-oscillators (the model of phonons and the parametric excitation of phonons), the string without damping and the string with damping, the string excited by an external force which is distributed along the length of the string and the elasticity of the string being time-dependent. For such systems the integrals of the motion, the adiabatic invariants, the transition probabilities, the connection of the evolution operator with the group of infinite dimensional symplectic transformations, the limit transitions from the discrete chain of the atoms-oscillators to the continuous string must be studied separately.

In addition to considering the examples of such quadratic systems, it is interesting to consider some applications of these systems. In this study we will consider such applications as resonant circuit, Josephson junction, gravitational waveguide. It is worthy of note that all the discussed systems may be considered with the help of the theory of quadratic systems [1-3] using the linear time-dependent integrals of the motion. But all these systems possess their own properties that demands the separate consideration for obtaining the explicit formulae and, especially, for discussing the physical results which are absolutely different for these different physical systems unified from the point of view of the mathematical schemes. The coherent states [4] of nonstationary quantum systems [1-3] may be constructed both for the chain of atoms and the string. It is also important for these systems to take into account the Schrodinger-Robertson uncertainty relation [5,6] minimization which gave the possibility to introduce [7] the notion of the correlated states, the partial case of this state being the squeezed states, (see [8,9]). Taking into account the damping in the Caldirola-Kanai model [12-14] also may be applied in the theory of the string. The universal invariants for the systems of oscillators and for a charge in the magnetic field [12-14] may be constructed in the theory of quantum resonant circuit and Josephson junction. The linear adiabatic invariants of the quadratic systems [15] and the Berry phase of the oscillator [16] as well as the change of the adiabatic invariants [17,18] may also be calculated for the Josephson junction and for the waveguides propagating the waves of different nature. The Berry phase [19 and 20] and the correlated states [21] are very interesting to study in connection with the quadratic systems possessing the linear adiabatic invariants with the problem of nonstationary classical harmonic oscillator (see [22]). The path integral method of quantum mechanics [23] gives the possibility to see the Feynmann integral and study the resonant circuit and the Josephson junction and also to apply it in the theory of the paraxial beams of different waves propagating along a waveguide. The aim of this paper is to give a review of the properties of quadratic quantum systems and to apply the theory of these systems to the quantum string, a resonant circuit, a Josephson junction and a gravitational waveguide, as well as to discuss the physical consequences of the models under consideration.

1. RESONANT CIRCUIT WITHOUT DAMPING

Following studies [1 and 2] let us consider the quantum resonant circuit with the inductance L and the capacitance C and with high Q-factor as an interesting example for application of the developed methods. In order to consider the resonant circuit as the quantum one, it is necessary for the energy of this circuit to be comparable with the energy of one quantum, i.e., the following inequality holds

$$\hbar\omega \gtrsim kT \tag{a}$$

Hence at wavelength ~1 cm and the temperature less than 1.4 K the resonant circuit must be examined as the quantum one. Besides, the model of the quantum resonant circuit is useful for considering the Josephson junction in some approximation, and we will discuss it in detail in further sections.

Let us write down the Hamiltonian of the resonant circuit

$$\hat{H} = \frac{1}{2}\left(\hat{Q}^2/C + L\hat{\mathfrak{J}}^2\right), \tag{1.1}$$

where $\hat{Q}$ is the operator of the charge in the capacitance and $\hat{\mathfrak{J}}$ is the operator of the current in the circuit. These variables satisfy the commutation relation

$$[\hat{\mathfrak{J}},\ \hat{Q}] = i\hbar/L. \tag{1.2}$$

Let us introduce the voltage operator in the capacitance $\hat{V} = \hat{Q}/C$. The commutation relation of the current operator and the voltage operator has the form

$$[\hat{\mathfrak{J}},\ \hat{V}] = i\hbar\omega^2, \tag{1.3}$$

where $\omega=(LC)^{-1/2}$ is the plasma frequency of the circuit.

The Heisenberg equations

$$\frac{\partial \hat{\mathfrak{J}}_H(t)}{\partial t} - \frac{i}{\hbar}[\hat{H}, \hat{\mathfrak{J}}_H(t)] = 0, \qquad \frac{\partial \hat{V}_H(t)}{\partial t} - \frac{i}{\hbar}[\hat{H}, \hat{V}_H(t)] = 0 \tag{b}$$

for the Hamiltonian (1.1) and the commutation relations (1.3) coincide with the classical equations of the motion of the resonant circuit

$$\partial\mathfrak{J}/\partial t = -V/L, \qquad \partial V/\partial t = \mathfrak{J}/C, \tag{c}$$

which lead to the usual equation for the charge vibrations

$$\ddot{Q} + \omega^2 Q = 0. \tag{d}$$

The solutions to the Heisenberg equations have the form

$$\begin{aligned} \hat{\mathfrak{J}}_H(t) &= \hat{\mathfrak{J}}_H(0)\cos\omega t + \left(\frac{i_0^2}{\hbar\omega^2}\right)\hat{V}_H(0)\sin\omega t, \\ \hat{V}_H(t) &= -\left(\frac{\hbar\omega^2}{i_0^2}\right)\hat{\mathfrak{J}}_H(0)\sin\omega t + \hat{V}_H(0)\cos\omega t. \end{aligned} \tag{e}$$

Following the general scheme [1] let us construct the "creation" operator and its hermitian conjugate "annihilation" operator (annihilating the vibrational quantum), these operators being linear in $\hat{\mathfrak{J}}$ and $\hat{V}$

$$\hat{A}(t) = \frac{e^{i\omega t}}{\sqrt{2}}\left(\frac{\hat{\mathfrak{J}}}{i_0} + \frac{i i_0}{\hbar\omega^2}\hat{V}\right), \tag{1.4}$$

where $i_0=(\hbar\omega/L)^{1/2}$ is the amplitude of the ground state current fluctuations.

The operators $\hat{A}(t)$ and $A^+(t)$ satisfy the commutation relations for boson creation and annihilation operators

$$[\hat{A}(t), \hat{A}^+(t)] = 1 \tag{1.5}$$

and they satisfy the equations for determining the integrals of the motion

$$\partial\hat{A}/\partial t = -(i/\hbar)[\hat{H}, \hat{A}]. \tag{1.6}$$

The Hamiltonian (1.1) may be rewritten in terms of the creation and annihilation operators in the form

$$\hat{H} = \hbar\omega\,[\hat{A}^+(t)\hat{A}(t) + {}^1/_2]. \tag{1.7}$$

Let us construct, in the Schrodinger representation, two Hermitian operators $\hat{V}_0(t)$ and $\hat{\mathfrak{I}}_0(t)$ which are linear forms in $\hat{A}(t)$ and $A^+(t)$

$$\hat{\mathfrak{I}}_0(t) = \frac{i_0}{\sqrt{2}}(\hat{A}(t) + \hat{A}^+(t)), \quad \hat{V}_0(t) = -\frac{i\omega^2\hbar}{i_0\sqrt{2}}(\hat{A}(t) - \hat{A}^+(t)). \tag{1.8}$$

If one rewrites these expressions in terms of the current and voltage operators, it is possible to obtain

$$\hat{\mathfrak{I}}_0(t) = \hat{\mathfrak{I}}\cos\omega t - \frac{i_0^2}{\hbar\omega^2}\hat{V}\sin\omega t, \quad \hat{V}_0(t) = \frac{\hbar\omega^2}{i_0^2}\sin\omega t\hat{\mathfrak{I}} + \hat{V}\cos\omega t. \tag{1.9}$$

The physical variables $\hat{\mathfrak{I}}_0(t)$ and $\hat{V}_0(t)$ are the explicitly time-dependent integrals of the motion and they determine the initial point at the voltage-current curve of the resonant circuit.

2. STATIONARY STATES OF QUANTUM RESONANT CIRCUIT

Let us find the ground state of the resonant circuit following to the general scheme [3]. This scheme satisfies the equations

$$\hat{A}\Psi_0 = 0, \quad i\hbar\dot{\Psi}_0 = \hat{H}\Psi_0, \quad \langle\Psi_0 \mid \Psi_0\rangle = 1, \tag{2.1}$$

where the operator $\hat{A}$ is the "annihilation" operator (1.4). We obtain

$$\Psi_0(\mathcal{J}, t) = e^{-i\omega t/2}(\pi i_0^2)^{-1/4}\exp[-\mathcal{J}^2/2i_0^2]. \tag{2.2}$$

The stationary states may be found from the condition $\hat{H}\Psi_n = E_n\Psi_n$ where E_n is the energy of the state, $E_n=\hbar\omega(n+1/2)$, n=0,1,2,...; n-th state may be obtained acting n times on the ground state by the "creation" operator, i.e.,

$$\Psi_n(\mathcal{J}, t) = (\hat{A}^+)^n(n!)^{-1/2}\Psi_0(\mathcal{J}, t). \tag{2.3}$$

Let us write down the explicit expression for the states with the fixed energy $\Psi_n(\mathfrak{I},t)$ in the current representation

$$\Psi_n(\mathcal{J}, t) = i_0^{-1/2}\pi^{-1/4}e^{-i\omega t(n+1/2)}(2^n n!)^{-1/2}e^{-\mathcal{J}^2/2i_0^2}H_n(\mathcal{J}/i_0), \tag{2.4}$$

where $H_n(x)$ is the Hermite polynomial. In the voltage representation this state with fixed energy has the form

$$\Psi_n(V,t)=\left(\frac{\hbar\omega^2}{i_0}\right)^{-1/2}\pi^{-1/4}\exp\left[-i\left(\omega t+\frac{\pi}{2}\right)\left(n+\frac{1}{2}\right)-\frac{V^2 i_0^2}{2\omega^4\hbar^2}\right]\times$$
$$\times\frac{1}{(2^n n!)^{1/2}}H_n\left(\frac{Vi_0}{\hbar\omega^2}\right). \tag{2.5}$$

The quantum fluctuations of the current and the voltage in the stationary state have the form

$$\sigma^2_{V,n}=\frac{\hbar^2\omega^4}{i_0^2}\left(n+\frac{1}{2}\right),\qquad \sigma^2_{\mathcal{I},n}=i_0^2\left(n+\frac{1}{2}\right). \tag{2.6}$$

3. GREEN FUNCTION OF QUANTUM RESONANT CIRCUIT

Let us introduce the Green function of the resonant circuit where $\hat{U}(t)$ is the evolution operator.

Solving the eigenvalue problem for the operators-integrals of the motion (1.9)

$$\hat{\mathcal{I}}_0 G(\mathcal{I},\mathcal{I}',t)=\mathcal{I}_0' G(\mathcal{I},\mathcal{I}',t),\qquad \hat{V}_0 G(\mathcal{I},\mathcal{I}',t)=i\partial G(\mathcal{I},\mathcal{I}',t)/\partial\mathcal{I}_0', \tag{3.1}$$

we obtain the explicit expresssion for the Green function in the current representation

$$G(\mathcal{I},\mathcal{I}',t)=(2\pi i i_0^2\sin\omega t)^{-1/2}\exp\left[\frac{i}{2}\operatorname{ctg}\omega t\,\frac{\mathcal{I}^2+\mathcal{I}'^2}{i_0^2}-\frac{i}{i_0^2}\frac{\mathcal{I}\mathcal{I}'}{\sin\omega t}\right]. \tag{3.2}$$

In the voltage representation the Green function has the form

$$G(V,V',t)=\frac{i_0}{\hbar\omega^2}(2\pi i\sin\omega t)^{-1/2}\exp\left[\frac{ii_0^2}{\hbar^2\omega^4}\left(\frac{1}{2}\operatorname{ctg}\omega t\,(V^2+V'^2)-\frac{VV'}{\sin\omega t}\right)\right]. \tag{3.3}$$

The Green function (3.2) and (3.3) are related with the help of the Fourier transform.

4. QUANTUM RESONANT CIRCUIT BEING SUBJECT TO EXTERNAL CURRENT

Let us apply an external current to the resonant circuit with high Q-factor. The Hamiltonians of the quantum resonant circuit being subject to the external current has the form

$$\hat{H}={}^1\!/_2\,(L\hat{\mathcal{I}}^2+C\hat{V}^2)-LI(t)\,\hat{\mathcal{I}}. \tag{4.1}$$

The third term in the Hamiltonian is chosen in such a form in order to apply the model of the quantum resonant circuit to Josephson effect which will be considered in further sections. We construct the "annihilation" operator

$$\hat{A}_\delta(t)=\hat{A}(t)-\frac{ii_0L}{\hbar\sqrt{2}}\int_0^t I(\tau)\,e^{i\omega\tau}\,d\tau. \tag{4.2}$$

where $\hat{A}(t)$ is the "annihilation" operator for the vibrations in the quantum resonant circuit without the external current described by the formula (1.4). It may be checked that this operator (4.2) is the integral of the motion for the resonant circuit with the external current. Let us write down the time-dependent integrals of the motion (the real and imaginary parts of the invariant (4.2))

$$\hat{\mathcal{I}}_{0,\delta}(t)=\frac{i_0}{\sqrt{2}}(\hat{A}_\delta(t)+\hat{A}_\delta^+(t)),\qquad \hat{V}_{0,\delta}(t)=-\frac{i\hbar\omega^2}{i_0\sqrt{2}}(\hat{A}_\delta(t)-\hat{A}_\delta^+(t)). \tag{4.3}$$

Substituting the dependence of the invariant (4.2) on the current and voltage operators given by the formula (1.4) into the relation (4.3), we obtain the integrals of the motion in the explicit form

$$\hat{\mathcal{I}}_{0,\delta}(t) = \hat{\mathcal{I}} \cos \omega t - \frac{i_0^2}{\hbar\omega^2} \hat{V} \sin \omega t - \frac{L i_0^2}{\hbar} \int_0^t I(\tau) \sin \omega\tau \, d\tau,$$

$$\hat{V}_{0,\delta}(t) = \hat{V} \cos \omega t + \hat{\mathcal{I}} \frac{\hbar\omega^2}{i_0^2} \sin \omega t - \omega^2 L \int_0^t I(\tau) \cos \omega\tau \, d\tau. \tag{4.4}$$

Let us write out the Green function in the current representation for the resonant circuit acted by the external current

$$G_\delta(\mathcal{I}, \mathcal{I}', t) = G(\mathcal{I}, \mathcal{I}', t) \exp\Big[\frac{iL}{\hbar}\Big(-\frac{\mathcal{I}'}{\sin \omega t} \int_0^t I(\tau) \sin \omega\tau \, d\tau + {}$$
$$+ \mathcal{I} \int_0^t I(\tau) \cos \omega\tau \, d\tau + \mathcal{I} \operatorname{ctg} \omega t \int_0^t I(\tau) \sin \omega\tau \, d\tau\Big) + {}$$
$$+ \frac{i i_0^2 L^2}{2\hbar^2}\Big(\operatorname{ctg} \omega t \int_0^t I(\tau) \sin \omega\tau \, d\tau \int_0^t I(\xi) \sin \xi\omega \, d\xi + {}$$
$$+ 2 \int_0^t d\tau I(\tau) \cos \omega\tau \int_0^\tau I(\tau') \sin \omega\tau' \, d\tau'\Big)\Big], \tag{f}$$

where the function G(ℑ, ℑ', t) is given by Equation (3.2).

In the voltage representation the Green function of the resonant circuit acted by the external current has the form

$$G_\delta(V, V', t) = G(V, V', t) \exp\Big[-\frac{i i_0^2 L}{\omega^2\hbar^2}\Big(\frac{V'}{\sin \omega t} \int_0^t I(\tau) \cos \omega\tau \, d\tau + {}$$
$$+ V \int_0^t I(\tau) \sin \omega\tau \, d\tau - V \operatorname{ctg} \omega t \int_0^t I(\tau) \cos \omega\tau \, d\tau\Big) + {}$$
$$+ \frac{i i_0^2 L^2}{2\hbar^2}\Big(\operatorname{ctg} \omega t \int_0^t I(\tau) \cos \omega\tau \, d\tau \int_0^t I(\xi) \cos \omega\xi \, d\xi - {}$$
$$- \int_0^t I(\tau) \cos \omega\tau \, d\tau \int_0^\tau \sin \omega\tau' I(\tau') \, d\tau'\Big)\Big], \tag{g}$$

where the Green function G(V, V',t) is given by the formula (3.3).

5. COHERENT STATES OF QUANTUM RESONANT CIRCUIT

Following study [1], let us find the states of the quantum-resonant circuit acted by an external classical current which minimize the Heisenberg uncertainty relation

$$\delta\mathcal{I}\delta V \geqslant \hbar\omega^2/2, \tag{5.1}$$

and satisfy the Schrodinger equation with the Hamiltonian (4.1). Here δℑ, δV are the dispersions of the current and voltage in the circuit. These states are called the coherent states [4] if the dimensionless dispersions of the current and voltage are equal

To find these states we obtain the ground state from the condition $\hat{A}_\delta(t)\Psi_{0,\delta}=0$, where $\hat{A}_\delta(t)$ is determined by the relation (4.2). Then we

$$\Psi_{0,\delta}(t) = i_0^{-1/2}\pi^{-1/4}\exp\left[-\frac{i\omega t}{2} - \frac{\mathcal{J}^2}{2i_0^2} - \frac{\sqrt{2}\,\mathcal{J}}{i_0}\delta(t)\,e^{-i\omega t} + i\omega\int_0^t d\tau\delta^2(\tau)\,e^{-2i\omega\tau}\right],$$

$$\delta(t) = -\frac{ii_0L}{\hbar\sqrt{2}}\int_0^t I(\tau)\,e^{i\omega\tau}\,d\tau. \tag{5.2}$$

In the absence of the external current $\mathfrak{J}(t)$ the state (5.2) coincides with the ground state of the quantum resonant circuit (2.2). We introduce the shift operator

$$\hat{D}(\alpha) = \exp\{\alpha\hat{A}_\delta^+(t) - \alpha^*\hat{A}_\delta(t)\}, \tag{5.3}$$

where α is a complex number. It is worthy of note that the operator $\hat{D}(\alpha)$ is unitary operator and it is the integral of the motion. Acting by the shift operator (5.3) on the ground state (5.2)

$$\hat{D}(\alpha)\,\Psi_{0,\delta}(\mathcal{J},t) = \Psi_{\alpha,\delta}(\mathcal{J},t), \tag{h}$$

we obtain the function

$$\Psi_{\alpha,\delta}(\mathcal{J},t) = i_0^{-1/2}\pi^{-1/4}\exp\left[-\frac{i\omega t}{2} + i\omega\int_0^t d\tau\delta^2(\tau)\,e^{-2i\omega\tau} + \frac{\sqrt{2}\,\mathcal{J}}{i_0}e^{-i\omega t}(\alpha-\delta) - \right.$$
$$\left. - \frac{\mathcal{J}^2}{2i_0^2} - \frac{\alpha^2}{2}e^{-2i\omega t} - \frac{|\alpha|^2}{2} + \alpha e^{-i\omega t}(\delta e^{-i\omega t} + \delta^* e^{i\omega t})\right]. \tag{5.4}$$

The states $\Psi_{\alpha,\delta}(\mathfrak{J},t)$ are the eigenstates of the "annihilation" operator

$$\hat{A}_\delta(t)\,\Psi_{\alpha,\delta}(\mathcal{J},t) = \alpha\Psi_{\alpha,\delta}(\mathcal{J},t), \tag{5.5}$$

They satisfy the Schrodinger equation with the Hamiltonian (4.1)

$$i\hbar\dot{\Psi}_{\alpha,\delta}(\mathcal{J},t) = \hat{H}\Psi_{\alpha,\delta}(\mathcal{J},t) \tag{5.6}$$

and minimize the Heisenberg uncertainty relation (5.1), i.e., they are coherent states of the quantum resonant circuit.

We consider the case when the external current is equal to zero. Then the coherent states of the resonant circuit in the current representation have the form

$$\Psi_\alpha(\mathcal{J},t) = \frac{e^{-|\alpha|^2/2-i\omega t/2}}{(i_0^2\pi)^{1/4}}\exp\left(-\frac{\mathcal{J}^2}{2i_0^2} + \frac{\sqrt{2}\,e^{-i\omega t}\alpha\mathcal{J}}{i_0} - \frac{\alpha^2 e^{-2i\omega t}}{2}\right). \tag{5.7}$$

The coherent states of the quantum resonant circuit without the external current in the voltage representation are given by the relation

$$\Psi_\alpha(V,t) = \left(\frac{i_0}{\hbar\omega^2}\right)^{-1/2}\pi^{-1/4}\exp\left[-\frac{i}{2}\left(\omega t + \frac{\pi}{2}\right) - \frac{|\alpha|^2}{2} - \right.$$
$$\left. - \frac{i_0^2}{2\hbar^2\omega^4}V^2 + \frac{\alpha^2}{2}e^{-2i\omega t} + \frac{\alpha\sqrt{2}}{\omega^2\hbar}i_0 V e^{-i(\omega t+\pi/2)}\right]. \tag{5.8}$$

The distribution function for current in the coherent state (5.7) $W_\alpha(\mathfrak{J},t)=\Psi^*{}_\alpha(\mathfrak{J},t)\Psi_\alpha(\mathfrak{J},t)$ has the form

$$W_\alpha(\mathcal{J},t) = (i_0\sqrt{\pi})^{-1}\exp\{-[\mathcal{J}/i_0 - |\alpha|\sqrt{2}\cos(\varphi_\alpha - \omega t)]^2\}, \tag{5.9}$$

where $\phi_\alpha=\arg\alpha$. The distribution function for the voltage in the coherent state (5.8) $W_\alpha(V,t)=\Psi^*{}_\alpha(V,t)\Psi_\alpha(V,t)$ has the form

$$W_\alpha(V,t) = \frac{1}{\sqrt{\pi}}\frac{i_0}{\hbar\omega^2}\exp\left[-\left(\frac{Vi_0}{\hbar\omega^2} - \sqrt{2}\,|\alpha|\sin(\varphi_\alpha - \omega t)\right)^2\right]. \qquad (5.10)$$

The mean values of the current and the voltage in the resonant circuit in the coherent state are given by the relations

$$\langle\Psi_\alpha|\hat{\mathcal{J}}|\Psi_\alpha\rangle = \frac{i_0}{\sqrt{2}}(\alpha e^{-i\omega t} + \alpha^* e^{i\omega t}),$$

$$\langle\Psi_\alpha|\hat{V}|\Psi_\alpha\rangle = \frac{\hbar\omega^2}{i_0\sqrt{2}}(\alpha e^{-i\omega t} - \alpha^* e^{i\omega t}). \qquad (5.11)$$

The variances of the current and the voltage have the form

$$\delta\mathcal{J} = i_0/\sqrt{2},\ \delta V = \hbar\omega^2/i_0\sqrt{2}, \qquad (5.12)$$

i.e., the equality is achieved in the Heisenberg uncertainty relation (5.1)

$$\delta\mathcal{J}\delta V = \hbar\omega^2/2. \qquad (5.13)$$

Let us calculate some characteristics of the circuit without the external current using the scheme suggested in study [3]. The mean value of the resonant circuit energy in the coherent state (5.7) is described by the formula

$$\bar{E} = \langle\alpha\,|\,\hat{H}\,|\,\alpha\rangle = \hbar\omega\,\langle\alpha\,|\,\hat{A}^+\hat{A}\,|\,\alpha\rangle + {}^1/_2\hbar\omega\,\langle\alpha\,|\,\alpha\rangle, \qquad (5.14)$$

and it takes the value

$$\bar{E} = \hbar\omega\,(|\,\alpha\,|^2 + {}^1/_2). \qquad (i)$$

The variance of the energy in the coherent state has the form

$$\Delta E = \hbar^2\omega^2\,|\,\alpha\,|^2. \qquad (j)$$

The energy distribution function in the coherent state is described by the Poisson distribution

$$W_n = |\langle n\,|\,\alpha\rangle|^2 = \alpha^{2n}(n!)^{-1}e^{-|\alpha|^2}. \qquad (5.15)$$

The evolution operator of the quantum resonant circuit in the coherent state representation has the form

$$\langle\alpha\,|\,\hat{U}\,(t)|\,\beta\rangle = \exp\,[-i\omega t/2 - |\,\alpha\,|^2/2 + |\,\beta\,|^2/2 - \alpha^*\beta e^{-i\omega t}]. \qquad (k)$$

The Green function of the harmonic oscillator may be expressed as follows:

$$\langle\alpha\,|\,\hat{U}\,(t)\,|\,\alpha\rangle = e^{-|\alpha|^2}\exp\left(\alpha^*\alpha e^{-i\omega t} - \frac{i\omega t}{2}\right). \qquad (l)$$

In n-α – representation the Green function has the form

$$\langle n\,|\,\hat{U}\,(t)|\alpha\rangle = \exp\,(-i\omega t/2 - |\,\alpha\,|^2/2)\alpha^n\,(n!)^{-1/2}e^{-i\omega t n}. \qquad (m)$$

Let us consider the level population for the quantum resonant circuit. The population of the ground state of the circuit has the form

$$\rho_{00}(t)\,|_{t=0} = 2\,\mathrm{sh}\left(\frac{\hbar\omega}{2T}\right)e^{-\hbar\omega/2T}, \qquad (5.16)$$

where T is the temperature. The population of the stationary states is given by the formula

$$\rho_{nn}(t)\,|_{t=0} = 2\,\mathrm{sh}\left(\frac{\hbar\omega}{2T}\right)\exp\left[-\frac{\hbar\omega}{2T}\left(n + \frac{1}{2}\right)\right]. \qquad (5.17)$$

If the external current acts on the equilibrium state of the resonant circuit which has at moment t=0 the temperature $T=\beta^{-1}$, the new state with ground populations and excited states is created

$$\rho_{nn}(t) = \rho_{00}(t)\, e^{-\hbar\omega\beta n} L_n\left(-\left(2\,\mathrm{sh}\left(\frac{\hbar\omega\beta}{2}\right)|\delta|\right)^2\right), \tag{5.18}$$

where L_n is Lagerre polynomial.

The results found correspond to the results obtained for the harmonic oscillator in study [1].

6. QUANTUM RESONANT CIRCUIT WITH VARYING FREQUENCY

Let us consider the resonant circuit with the time-dependent frequency that may be accomplished, for example, by a specified time-variation of the inductance. On the example of a Josephson junction we will suggest the concrete method of the parametric excitation of the resonant circuit.

The Hamiltonian of the quantum resonant circuit with varying frequency has the form

$$\hat{H} = {}^1/_2\,(C\hat{V}^2 + L\,(t)\,\hat{\mathcal{J}}^2), \tag{6.1}$$

i.e.

$$\omega\,(t) = (CL\,(t))^{-1/2}, \quad \omega\,(0) = \omega_0, \quad L\,(0) = L. \tag{n}$$

According to general scheme [1] we find out the "annihilation" operator for the parametric resonant circuit. We have

$$\hat{A}_\omega(t) = \frac{i}{\sqrt{2}}\left(\frac{i_0}{\omega_0^2\hbar}\,\varepsilon\hat{V} - \frac{1}{\omega_0 i_0}\,\dot{\varepsilon}\hat{\mathcal{J}}\right), \tag{6.2}$$

where the function $\varepsilon(t)$ satisfies the equation

$$\ddot{\varepsilon}\,(t) + \omega^2\,(t)\varepsilon\,(t) = 0 \tag{6.3}$$

with the additional condition

$$\dot{\varepsilon}\varepsilon^* - \dot{\varepsilon}^*\varepsilon = 2i\omega_0. \tag{6.4}$$

Let us write down the time-dependent integrals of the motion (real and imaginary parts of the "creation" and "annihilation" operators (6.2))

$$\hat{\mathcal{J}}_{0,\omega}(t) = \frac{i_0}{\sqrt{2}}\,(\hat{A}_\omega(t) + \hat{A}_\omega^+(t)),$$

$$\hat{V}_{0,\omega}(t) = -\frac{i\hbar\omega_0^2}{i_0\sqrt{2}}\,(\hat{A}_\omega(t) - \hat{A}_\omega^+(t)). \tag{6.5}$$

If one substitutes the functional dependence of the operators (6.2) on the current and voltage operators into Equation (5.5), the integrals of the motion may be obtained explicitly

$$\hat{\mathcal{J}}_{0,\omega}(t) = \frac{i}{2}\left[\frac{i_0^2}{\hbar\omega_0^2}\,\hat{V}\,(\varepsilon - \varepsilon^*) - \frac{\hat{\mathcal{J}}}{\omega_0}\,(\dot{\varepsilon} - \dot{\varepsilon}^*)\right],$$

$$\hat{V}_{0,\omega}(t) = \frac{1}{2}\left[(\varepsilon + \varepsilon^*)\,\hat{V} - \frac{\hbar\omega_0{}^2}{i_0^2}\,\hat{\mathcal{J}}\,(\dot{\varepsilon} + \dot{\varepsilon}^*)\right]. \tag{6.6}$$

Solving the eigenvalues problem for the operator integrals of the motion, we obtain the Green function of the parametric resonant circuit in the current representation

$$G_\omega(\mathcal{J}, \mathcal{J}', t) = (-2\pi i i_0^2 \varepsilon_2)^{-1/2} \exp\left[-\frac{i}{2i_0^2\varepsilon_2}\left(\frac{\dot\varepsilon_2}{\omega_0}\mathcal{J}'^2 + \varepsilon_1 \mathcal{J}^2 - 2\mathcal{J}\mathcal{J}'\right)\right],$$

$$\varepsilon_1 = \mathrm{Re}\,\varepsilon, \quad \varepsilon_2 = \mathrm{Im}\,\varepsilon; \tag{6.7}$$

and in the voltage representation

$$G_\omega(V, V', t) = \left(\frac{i_0^2\omega_0}{2\pi i\hbar^2\omega_0^4\dot\varepsilon_1}\right)^{1/2} \exp\left[\frac{i i_0^2}{2\hbar^2\omega_0^4}\left(V^2\frac{\dot\varepsilon_2}{\dot\varepsilon_1} + V'^2\frac{\varepsilon_1\omega_0}{\dot\varepsilon_1} - 2VV'\frac{\omega_0}{\dot\varepsilon_1}\right)\right]. \tag{6.8}$$

The functions (6.7) and (6.8) are related by the Fourier transtorm.

7. CORRELATED STATES OF THE QUANTUM RESONANT CIRCUIT

Let us find the states of the quantum resonant circuit with varying frequency which are the solutions to Schrodinger equation with the Hamiltonian (6.1) minimizing the uncertainty relation by Schrodinger-Robertson (5.6)

$$\delta\mathcal{J}\delta V \geqslant \hbar\omega^2/2\sqrt{1 - r^2}, \tag{7.1}$$

where

$$r = (\delta\mathcal{J}\delta V)^{-1}\left(\frac{1}{2}\langle\hat{\mathcal{J}}\hat{V} + \hat{V}\hat{\mathcal{J}}\rangle - \langle\hat{\mathcal{J}}\rangle\langle\hat{V}\rangle\right) \tag{o}$$

is the coefficient of correlation of the current and voltage. These states are called the correlated states [7].

The ground state may be found from the condition $\hat{A}_\omega \Psi_{\omega,0}(\mathcal{J},t) = 0$, where $\hat{A}_\omega$ is given by the formula (6.2). We then have

$$\Psi_{0,\omega}(\mathcal{J}, t) = \pi^{-1/4}(i_0\varepsilon)^{-1/2} \exp\left[\frac{i\dot\varepsilon}{\varepsilon\omega_0}\frac{\mathcal{J}^2}{2i_0^2}\right]. \tag{7.2}$$

The wave function of the ground state in the voltage representation has the form

$$\Psi_{0,\omega}(V, t) = \pi^{-1/4}e^{i\pi/4}\left(\frac{i_0}{\hbar\omega_0\dot\varepsilon}\right)^{1/2} \exp\left[-\frac{i i_0^2}{\hbar^2\omega_0^4}V^2\frac{\omega_0\varepsilon}{2\dot\varepsilon}\right]. \tag{7.3}$$

The probability density for current and voltage in the resonant circuit with varying frequency has the form

$$W_{0,\omega}(\mathcal{J}, t) = \pi^{-1/2}(|\varepsilon| i_0)^{-1} \exp\left[-\frac{\mathcal{J}^2}{i_0^2|\varepsilon|^2}\right],$$

$$W_{0,\omega}(V, t) = \pi^{-1/2}\frac{\omega_0}{|\dot\varepsilon|}\frac{i_0}{\hbar\omega_0^2} \exp\left[-\frac{i_0^2\omega_0^2}{\hbar^2\omega_0^4}\frac{V^2}{|\dot\varepsilon|^2}\right]. \tag{7.4}$$

The mean values of the current and voltage in the resonant circuit in the state (7.2) and (7.3) are equal to zero, and the standard deviations of the voltage and the current take the values

$$\delta\mathcal{J}_\omega = \frac{i_0}{\sqrt{2}}|\varepsilon|, \quad \delta V_\omega = \frac{\hbar\omega_0^2}{i_0\sqrt{2}}|\dot\varepsilon|. \tag{7.5}$$

Let us construct the shift operator

$$\hat{D}(\alpha) = \exp[\alpha\hat{A}_\omega^+(t) - \alpha^*\hat{A}_\omega(t)], \tag{7.6}$$

where α is a complex number. Acting by this operator on the ground state (7.2) $\hat{D}(\alpha)\Psi_0=\Psi_\alpha$, we obtain

$$\Psi_{\alpha,\omega}(\mathcal{J},t)=\Psi_{0,\omega}\exp\left[-\frac{|\alpha|^2}{2}-\frac{\alpha^2\varepsilon^*}{2\varepsilon}+\frac{\sqrt{2}\,\alpha\mathcal{J}}{\varepsilon i_0}\right]. \tag{7.7}$$

The state $\Psi_{\alpha,\omega}(\mathcal{J},t)$ is the eigenstate of the "annihilation" operator (6.2)

$$\hat{A}_\omega\Psi_{\alpha,\omega}(\mathcal{J},t)=\alpha\Psi_{\alpha,\omega}(\mathcal{J},t), \tag{p}$$

and it satisfies the Schrodinger equation with the Hamiltonian (6.1) minimizing the Schrodinger-Robertson uncertainty relation (7.1) with the correlation coefficient

$$r=|\varepsilon\dot{\varepsilon}|^{-1}(|\varepsilon\dot{\varepsilon}|^2-1)^{1/2} \tag{7.8}$$

and the squeezing coefficient

$$K=\frac{\delta\mathcal{J}_\omega}{i_0^2}\frac{\hbar\omega_0^2}{\delta V_\omega}=\left|\frac{\varepsilon}{\dot{\varepsilon}}\right|\omega_0, \tag{7.9}$$

i.e., these states are the correlated states of the quantum resonant circuit. At $|\varepsilon\dot{\varepsilon}|=1$ these states are the squeezed state of the circuit [8]. The wave function of the correlated states in the voltage representation has the form

$$\Psi_{\alpha,\omega}(V,t)=e^{i\pi/4}\pi^{-1/4}\left(\frac{i_0}{\hbar\omega_0\dot{\varepsilon}}\right)^{1/2}\exp\left[-\frac{|\alpha|^2}{2}-\frac{\alpha^2}{2}\frac{\dot{\varepsilon}^*}{\dot{\varepsilon}}+\right.$$
$$\left.+\frac{\sqrt{2}\,\alpha i_0 V}{\omega_0\hbar\dot{\varepsilon}}-\frac{i i_0^2}{\hbar^2\omega_0^4}\frac{\varepsilon\omega_0}{2\dot{\varepsilon}}V^2\right]. \tag{7.10}$$

According to known scheme [1] the n-th excited state of the circuit with varying frequency may be obtained acting n times by the "creation" operator (6.2) on the ground state (7.2), i.e.,

$$\Psi_n(\mathcal{J},t)=(\hat{A}_\omega^+)^n(n!)^{-1/2}\Psi_{0,\omega}(\mathcal{J},t). \tag{7.11}$$

Let us demonstrate the explicit expression of the wave function of the excited states $\Psi_n(\mathcal{J},t)$ for the resonant circuit with varying frequency in the current representation

$$\Psi_n(\mathcal{J},t)=\Psi_{0,\omega}(\mathcal{J},t)(\varepsilon^*/2\varepsilon)^{n/2}(n!)^{-1/2}H_n(\mathcal{J}/i_0|\varepsilon|). \tag{7.12}$$

In the voltage representation the wave function of the Fock state has the form

$$\Psi_n(V,t)=\Psi_{0,\omega}(V,t)(\dot{\varepsilon}^*/2\dot{\varepsilon})^{n/2}(n!)^{-1/2}H_n(i_0V/\hbar\omega_0|\dot{\varepsilon}|), \tag{7.13}$$

where $H_n(x)$ is the Hermite polynomial.

8. PARAMETRIC QUANTUM RESONANT CIRCUIT ACTED BY EXTERNAL CURRENT

Let us consider the resonant circuit acted by both parametric excitation and external current by analogy with the harmonic oscillator discussed in study [1].

The Hamiltonian of the parametric quantum resonant circuit with the external current has the form

$$\hat{H}=\tfrac{1}{2}(C\hat{V}^2+L(t)\hat{\mathcal{J}}^2)-LI(t)\hat{\mathcal{J}}, \tag{8.1}$$

where I(t) is the external classical current.

In this case the "annihilation" operator takes the form

$$\hat{a}(t) = \frac{\omega_0 c - i\dot{\varepsilon}}{2\omega_0\sqrt{2}}\left(\frac{\hat{\mathcal{I}}}{i_0} + \frac{i i_0}{\hbar\omega_0^2}\hat{V}\right) - \frac{\omega_0\varepsilon - i\dot{\varepsilon}}{2\sqrt{2}\,\omega_0}\left(\frac{\hat{\mathcal{I}}}{i_0} - \frac{i i_0}{\omega_0^2\hbar}\hat{V}\right) + \delta(t), \tag{8.2}$$

where

$$\delta(t) = -\frac{i i_0 L}{\hbar\sqrt{2}}\int_0^t I(\tau)\,\varepsilon(\tau)\,d\tau, \tag{q}$$

and the function $\varepsilon(t)$ satisfies the equation (6.3) with the additional condition (6.4). One can check that the "annihilation" operator $\hat{\alpha}$ and its hermitian conjugate "creation" operator $\hat{\alpha}^+$ are the integrals of the motion.

Let us write out the time-dependent integrals of the motion (real and imaginary parts of the "creation" operator)

$$\hat{\mathcal{I}}_{0,\,\omega,\,\delta}(t) = \frac{i_0}{\sqrt{2}}(\hat{a}(t) + \hat{a}^+(t)),$$

$$\hat{V}_{0,\,\omega,\,\delta}(t) = -\frac{i\hbar\omega_0^2}{i_0\sqrt{2}}(\hat{a}(t) - \hat{a}^+(t)). \tag{8.3}$$

Substituting Equation (8.2) into Equation (8.3) we obtain the integrals of the motion in explicit form

$$\hat{\mathcal{I}}_{0,\,\omega,\,\delta}(t) = -\frac{i\hat{\mathcal{I}}}{2\omega_0}(\dot{\varepsilon} - \dot{\varepsilon}^*) + \frac{i i_0^2}{\hbar\omega_0^2}\hat{V}\,\frac{(\varepsilon - \varepsilon^*)}{2} + \frac{i_0}{\sqrt{2}}(\delta + \delta^*),$$

$$\hat{V}_{0,\,\omega,\,\delta}(t) = -\frac{\hbar\omega_0}{2i_0^2}(\dot{\varepsilon} + \dot{\varepsilon}^*)\hat{\mathcal{I}} + \frac{\varepsilon + \varepsilon^*}{2}\hat{V} - \frac{i\hbar\omega_0^2}{i_0\sqrt{2}}(\delta - \delta^*). \tag{8.4}$$

Solving the eigenvalue problem for the operators-integrals of the motion (8.4) we obtain the Green function in the current representation

$$\begin{aligned} G_{\omega,\,\delta}(\mathcal{I}, \mathcal{I}', t) &= (2\pi i i_0^2\varepsilon_2)^{-1/2}\exp\left[\frac{i}{2i_0^2}\left(\mathcal{I}'^2\frac{\dot{\varepsilon}_2}{\omega_0\varepsilon_2} + \mathcal{I}^2\frac{\dot{\varepsilon}_1}{\varepsilon_2} - \frac{\mathcal{I}\mathcal{I}'}{\varepsilon_2}\right) + \right.\\ &+ \frac{iL}{\hbar}\left(\mathcal{I}'\int_0^t I(\tau)\frac{\varepsilon_2(\tau)}{\varepsilon_2(t)}\,d\tau + \mathcal{I}\int_0^t I(\tau)\,\varepsilon_1(\tau)\,d\tau - \int_0^t \frac{I(\tau)\,\varepsilon_2(\tau)\,\varepsilon_1(t)}{\varepsilon_2(t)}\,d\tau\right) + \\ &+ \frac{i i_0^2 L^2}{\hbar^2}\left(\int_0^t \frac{I(\tau)\,\varepsilon_2(\tau)\,\varepsilon_1(t)}{\varepsilon_2(t)}\,d\tau\int_0^t I(\xi)\,\varepsilon_2(\xi)\,d\xi - \right.\\ &\left.\left. - \int_0^t I(\tau)\,\varepsilon_1(\tau)\,d\tau\int_0^\tau \varepsilon_2(\tau')\,I(\tau')\,d\tau'\right)\right] \end{aligned} \tag{8.5}$$

and in the voltage representation

$$\begin{aligned} G_{\omega,\,\delta}(V, V', t) &= \left(\frac{i_0^2}{2\pi i\hbar^2\omega_0^3\varepsilon_1}\right)^{1/2}\exp\left[-\frac{i i_0^2}{2\hbar^2\omega_0^4}\left(V'^2\frac{\varepsilon_1\omega_0}{\varepsilon_1} + 2VV'\frac{\omega_0}{\varepsilon_1} + \right.\right.\\ &\left. + V^2\frac{\dot{\varepsilon}_2}{\varepsilon_1}\right) + \frac{i i_0^2 L}{\hbar^2\omega_0^2}\left(V'\int_0^t \frac{I(\tau)\,\varepsilon_1(\tau)\,\omega_0}{\varepsilon_1(t)}\,d\tau + V\int_0^t I(\tau)\,\varepsilon_2(\tau)\,d\tau - \right.\\ &\left. - V\int_0^t \frac{I(\tau)\,\varepsilon_1(\tau)}{\dot{\varepsilon}_1(t)}\,\dot{\varepsilon}_2(t)\,d\tau\right) - \frac{i i_0^2 L^2}{2\hbar^2\omega_0^2}\int_0^t \frac{\dot{\varepsilon}_2(t)}{\dot{\varepsilon}_1(t)}\,I(\tau)\,\varepsilon_1(\tau)\,d\tau\int_0^t I(\tau')\,\varepsilon_1(\tau')\,d\tau' - \\ &\left. - \frac{i i_0^2 L^2}{2\hbar^2\omega_0^2}\int_0^t \varepsilon_2(\tau)\,I(\tau)\,d\tau\int_0^\tau I(\tau')\,\varepsilon_1(\tau')\,d\tau'\right]. \end{aligned} \tag{8.6}$$

According to the usual scheme [1] we find the ground state of the parametric resonant circuit acted by the external current using the condition $\hat{\alpha}\Psi_{0,\delta,\omega}=0$. We obtain the function in the current representation

$$\Psi_{0,\omega,\delta}(\mathcal{I},t)=\pi^{-1/4}(\varepsilon i_0)^{-1/2}\exp\left[\frac{i\dot{\varepsilon}}{2\omega_0\varepsilon}\frac{\mathcal{I}^2}{i_0^2}-\frac{\sqrt{2}\,\delta\mathcal{I}}{i_0\varepsilon}-\frac{\delta^2\varepsilon^*}{2\varepsilon}-\frac{|\delta|^2}{2}+\right.$$

$$\left.+\frac{1}{2}\int_0^t(\delta\dot{\delta}^*-\dot{\delta}^*\delta)\,d\tau\right]. \tag{8.7}$$

In the voltage representation the wave function of the ground state has the form

$$\Psi_{0,\omega,\delta}(V,t)=\pi^{-1/4}e^{i\pi/4}\left(\frac{i_0}{\varepsilon\hbar\omega_0}\right)^{1/2}\exp\left[-\frac{i\varepsilon i_0^2}{2\varepsilon\omega_0^3\hbar^2}V^2-\sqrt{2}\frac{\delta i_0}{\dot{\varepsilon}\hbar\omega_0}V-\frac{\delta^2\dot{\varepsilon}^*}{2\dot{\varepsilon}}-\right.$$

$$\left.-\frac{|\delta|^2}{2}+\frac{1}{2}\int_0^t(\dot{\delta}\delta^*-\delta\dot{\delta}^*)\,d\tau\right]. \tag{8.8}$$

Let us construct the shift operator

$$\hat{D}(\alpha)=\exp\left[\alpha\hat{a}^+(t)-\alpha^*\hat{a}(t)\right], \tag{r}$$

where α is a complex number. Acting on the ground state $\Psi_{0,\omega,\delta}(\mathcal{I},t)$ by the shift operator, we obtain all the family of the correlated states for the quantum parametric resonant circuit with the external current in the current representation

$$\Psi_{\alpha,\omega,\delta}(\mathcal{I},t)=\Psi_{0,\omega\delta}(\mathcal{I},t)\exp\left[-\frac{|\alpha|^2}{2}+\frac{\sqrt{2}\,\alpha\mathcal{I}}{\varepsilon i_0}+\frac{\alpha(\delta\varepsilon^*+\delta^*\varepsilon)}{\varepsilon}-\frac{\alpha^2\varepsilon^*}{2\varepsilon}\right] \tag{8.9}$$

and in the voltage representation

$$\Psi_{\alpha,\omega,\delta}(V,t)=\Psi_{0,\omega,\delta}(V,t)\exp\left[-\frac{|\alpha|^2}{2}-\frac{\alpha^2\dot{\varepsilon}^*}{2\dot{\varepsilon}}+\right.$$

$$\left.+\frac{\sqrt{2}\,\alpha V i_0}{\omega_0\hbar\dot{\varepsilon}}+\frac{\alpha}{\dot{\varepsilon}}(\delta\dot{\varepsilon}^*+\delta^*\dot{\varepsilon})\right]. \tag{8.10}$$

The standard deviations, the coefficients of correlation and squeezing are not changed when the external current I(t) is applied to the parametric circuit and these observables are described by the formulae (7.5), (7.8) and (7.9); the mean values of the current and the voltage are changed when the external current is applied to the circuit and they take the form

$$\langle\alpha\,|\,\hat{V}\,|\,\alpha\rangle=\sqrt{2}\,(\hbar\omega_0/i_0)\mathrm{Re}\,((\alpha-\delta)\dot{\varepsilon}^*),$$
$$\langle\alpha\,|\,\hat{\mathcal{I}}\,|\,\alpha\rangle=\sqrt{2}i_0\,\mathrm{Re}\,((\alpha-\delta)\varepsilon^*). \tag{8.11}$$

In the Fock-representation the wave function is

$$\Psi_n(\mathcal{I},t)=\Psi_{0,\omega,\delta}(\mathcal{I},t)\left(\frac{\varepsilon^*}{2\varepsilon}\right)^{n/2}(n!)^{-1/2}H_n\left(\frac{\mathcal{I}}{i_0|\varepsilon|}+\frac{\delta\varepsilon^*+\delta^*\varepsilon}{\sqrt{2}|\varepsilon|}\right), \tag{8.12}$$

where $H_n(x)$ is the Hermite polynomial.

9. UNIVERSAL INVARIANTS FOR CIRCUIT WITH VARYING FREQUENCY ACTED BY EXTERNAL VARYING CURRENT

In the study [9], it was shown that for nonstationary Hamiltonians which are quadratic forms in operators having the commutators equal to c-numers, there exist the universal invariants depending on the initial states of the system and the form of the commutation relations. But these universal invariants do not at all depend on the coefficients of the corresponding quadratic or linear forms.

Let the operators $\hat{Q}_\alpha$ obey the boson commutation relations

$$\hat{Q}_\alpha \hat{Q}_\beta - \hat{Q}_\beta \hat{Q}_\alpha = \Sigma_{\alpha\beta} = -\Sigma_{\beta\alpha}, \quad \alpha, \beta = 1, 2, \ldots, \tag{9.1}$$

where $\Sigma_{\alpha\beta}$ are arbitrary complex numbers and the Hamiltonian is the quadratic form in operators $\hat{Q}_\alpha$

$$\hat{H} = {}^1/_2 \hat{Q}_\alpha B_{\alpha\beta}(t) \hat{Q}_\beta + C_\alpha(t) \hat{Q}_\alpha. \tag{9.2}$$

Then, in the Heisenberg representation the operators $\hat{Q}_\alpha(t)$ are expressed through the initial values of the operators $\hat{Q}_\alpha(0)$ as follows:

$$\hat{Q}_\alpha(t) = \Lambda_{\alpha\beta}(t) \hat{Q}_\beta(t) + \delta_\alpha(t). \tag{9.3}$$

The mean values $<\hat{Q}_\alpha(t)>$ satisfy the same relations (9.2) where the term $<\ldots>$ is understood as quantum-mechanical averages. Consequently, we obtain

$$Q_{\alpha\beta}(t) = \Lambda_{\alpha\mu}(t) Q_{\mu\nu}(0) \Lambda_{\beta\nu}(t), \tag{9.4}$$

where

$$Q_{\alpha\beta} = {}^1/_2 \langle (\hat{Q}_\alpha - Q_\alpha)(\hat{Q}_\beta - Q_\beta) - (\hat{Q}_\beta - Q_\beta)(\hat{Q}_\alpha - Q_\alpha) \rangle. \tag{s}$$

Let us rewrite Equation (9.4) in the matrix form

$$Q(t) = \| Q_{\alpha\beta}(t) \| = \Lambda(t) Q(0) \tilde{\Lambda}(t), \tag{9.5}$$

where tilda means the transposed matrix.

In the study [9] it is shown that for aribtrary matrices $B = \|\beta_{\alpha\beta}\|$ and $\Sigma = \|\Sigma_{\alpha\beta}\| = -\tilde{\Sigma}$, the matrix $\Lambda(t)$ has the unity determinant and satisfies the identity $\Lambda(t)\Sigma\tilde{\Lambda}(t) = \Sigma$. Consequently, for any parameter μ we have

$$D(\mu) = \det [Q(t) - \mu\Sigma] = \sum_{m=0}^{N} D_m \mu^m = \text{const}. \tag{9.6}$$

The expansion coefficients D_m were called the universal invariants. It was also shown in study [9] that these invariants are quantum-mechanical analogs of the universal Poincare-Cartan invariants in classical Hamiltonian mechanics.

Let us find the universal invariant (9.5) of the quantum resonant circuit with a varying frequency acted by an external current using the suggested approach [9]. The Hamiltonian of the parametric circuit with an external current has the form (8.1), i.e. $\hat{Q}_1 = V$, $\hat{Q}_2 = \mathfrak{I}$, and the commutator of the operators of the current and the voltage is given by the formula (1.3), consequently, the matrix Σ is equal

$$\Sigma = \begin{Vmatrix} 0 & -i\hbar\omega_0^2 \\ i\hbar\omega_0^2 & 0 \end{Vmatrix}. \tag{t}$$

For the quantum resonant circuit the universal invariant (9.5) has the form

$$D_0 = \delta_{\mathfrak{I}}^2 \delta_V^2 - \delta_{\mathfrak{I}V}^2 = \text{const}, \tag{9.7}$$

where $\delta_{\mathfrak{J}}$, δ_V are the standard deviations of the current and the voltage in the circuit, and $\delta_{\mathfrak{J}V}$ is the correlator of the current and the voltage in the circuit. If at the initial moment of time the quantum resonant circuit was either in coherent or in correlated state, the expression (9.7) does not depend on time and it is equal to

$$D_0 = \hbar^2\omega_0^4/4, \tag{u}$$

i.e., it is the universal invariant. This statement takes place in spite of the dependence of the mean values of the current and the voltage (8.11), (5.9) on the external current applied to the circuit, and in spite of the dependence of the standard deviations of the current and the voltage in the circuit on the varying frequency given by the formula (7.5).

10. QUANTUM RESONANT CIRCUIT WITH CONSTANT DAMPING

Each resonant circuit has an active resistance R. The energy of the circuit is gradually wasted in this resistance heating it, and due to this heat, the free oscillations damp.

Let us write down the classical equation of the oscillation in the resonant circuit with damping

$$\ddot{\mathfrak{J}} + (R/L)\,\dot{\mathfrak{J}} + \omega_0^2\,\mathfrak{J} = 0. \tag{10.1}$$

We consider the quantum analog of the classical resonant circuit with damping in the frame of the model by Caldirola [10 and 11]. In this model the Hamiltonian of the quantum circuit with a constant damping has the form

$$\hat{H}(t) = {}^1/_2\,[C\hat{V}^2e^{-Rt/L} + L\hat{\mathfrak{J}}^2e^{Rt/L}]. \tag{10.2}$$

The Hamiltonian (10.2) and the commutation relations (1.3) follow the equations of the motion

$$\dot{\hat{\mathfrak{J}}} = (\hat{V}/L)\,e^{-Rt/L}, \qquad \dot{\hat{V}} = -(\hat{\mathfrak{J}}/C)\,e^{Rt/L}. \tag{10.3}$$

The equations (10.3) coincide with the classical equations of the oscillations (10.1).

Using the general scheme [1] we find the integrals of the motion $\hat{A}(t)$ satisfying the equation

$$[i\hbar\partial/\partial t - \hat{H},\ \hat{A}(t)] = 0. \tag{v}$$

We look for the integrals of the motion in the form of a linear combination in the operators of the current and the voltage

$$\hat{A}(t) = a(t)\frac{\hat{\mathfrak{J}}}{i_0} + b(t)\frac{i_0}{\hbar\omega_0^2}\hat{V}, \qquad i_0 = \left(\frac{\hbar\omega_0}{L}\right)^{1/2}. \tag{w}$$

Then we obtain the equations for the coefficients $\alpha(t)$, b(t)

$$\ddot{b} + (R/L)\,\dot{b} + (1/LC)\,b = 0, \qquad a = -Le^{Rt/L}\dot{b}. \tag{x}$$

Let us choose the coefficient b(t) in the form $b(t)=i2^{-1/2}\varepsilon(t)$ where the complex function $\varepsilon(t)$ satisfies the equation

$$\ddot{\varepsilon}(t) + 2\Gamma\dot{\varepsilon}(t) + \omega_0^2\varepsilon(t) = 0. \tag{10.4}$$

where r=R/2L.

In order for the operators $\hat{A}(t)$ and $\hat{A}^{+}(t)$ to satisfy the commutation relations of the creation and the annihilation operators of the boson, we demand that the function $\varepsilon(t)$ obeys the additional condition

$$e^{2\Gamma t}(\dot{\varepsilon}\varepsilon^{*}-\varepsilon\dot{\varepsilon}^{*})=2i. \tag{10.5}$$

In the case of weak damping, i.e., when inequality condition $\omega^2{}_0>\Gamma^2$ holds, the equation (10.4) has two solutions

$$\varepsilon=\exp[-\Gamma t+i(\omega_0^2-\Gamma^2)^{1/2}t],\quad \varepsilon'=\exp[-\Gamma t-i(\omega_0^2-\Gamma^2)^{1/2}t], \tag{y}$$

and using the initial conditions we choose the solution

$$\varepsilon=\exp[-\Gamma t+i\Omega t],\quad \Omega=(\omega_0^2-\Gamma^2)^{1/2}. \tag{10.6}$$

We obtain the annihilation operator $\hat{A}(t)$ of the quantum resonant circuit with the finite quality

$$\hat{A}(t)=\left(\frac{\omega_0}{2\Omega}\right)^{1/2}e^{i\Omega t}\left[\frac{\Omega+i\Gamma}{\omega_0 i_0}e^{\Gamma t}\hat{\mathfrak{J}}+\frac{i i_0}{\hbar\omega_0^2}e^{-\Gamma t}\hat{\Gamma}\right]. \tag{10.7}$$

The ground state of the resonant circuit with nonzero resistance may be found from the condition $\hat{A}(t)\Psi_0(\mathfrak{J},t)=0$

$$\Psi_0(\mathcal{J},t)=\left(\frac{\Omega}{\pi i_0^2\omega_0}\right)^{1/4}\exp\left[-\frac{i(\Omega+i\Gamma)t}{2}-\frac{\mathcal{J}^2}{2i_0^2\omega_0}e^{2\Gamma t}(\Omega+i\Gamma)\right]. \tag{10.8}$$

In the ground state the current distribution function has the form

$$W_0(\mathcal{J},t)=\left(\frac{\Omega}{\pi\omega_0 i_0^2}\right)^{1/2}\exp\left[-\frac{\mathcal{J}^2}{2i_0^2}e^{2\Gamma t}\frac{2\Omega}{\omega_0}\right]. \tag{10.9}$$

The standard deviations of the current and voltage in the quantum resonant circuit with the finite quality in the ground state

$$\delta_{\mathcal{J}}=i_0 e^{-\Gamma t}(\omega_0/2\Omega)^{1/2},\qquad \delta_V=e^{\Gamma t}(\hbar/i_0)(\omega_0^5/2\Omega)^{1/2} \tag{10.10}$$

satisfy the Heisenberg uncertainty relation

$$\delta_{\mathcal{J}}^2\delta_V^2=\frac{\hbar^2}{4}\frac{\omega_0^2}{\Omega^2}\omega_0^4\geqslant\frac{\hbar^2\omega_0^4}{4}, \tag{10.11}$$

and the mean values of the current and voltage are equal to zero. The coherent states of the damped circuit may be found from the equation $\hat{A}(t)\Psi_\alpha=\alpha\Psi_\alpha$; we obtain

$$\Psi_\alpha(\mathcal{J},t)=\Psi_0(\mathcal{J},t)\exp\left[-\frac{|\alpha|^2}{2}-\frac{\alpha^2}{2}e^{-2i\Omega t}+\left(\frac{2\Omega}{\omega_0}\right)^{1/2}\frac{\alpha}{i_0}\mathfrak{J}e^{(\Gamma-i\Omega)t}\right]. \tag{10.12}$$

If the quantum resonant circuit is subject to an external varying current, then the Hamiltonian takes the form

$$\hat{H}(t)=\tfrac{1}{2}(C\hat{V}^2e^{-2\Gamma t}+L\hat{\mathfrak{J}}^2e^{2\Gamma t})+L\hat{\mathfrak{J}}I(t)e^{2\Gamma t}. \tag{10.13}$$

The equations of the motion follow from Equations (10.13) and (1.3)

$$\dot{\hat{\mathfrak{J}}}_H(t)=(\hat{V}_H/L)e^{-2\Gamma t},\qquad \dot{\hat{V}}_H(t)=-\hat{\mathfrak{J}}C^{-1}e^{2\Gamma t}+I(t)C^{-1}e^{2\Gamma t}. \tag{10.14}$$

Accomplishing the calculations analogous to the calculations in the beginning of the Section, according to method [1] we obtain the "annihilation" operator for the oscillations in the resonant circuit with finite quality which is excited by the external classical current

$$\hat{A}_\delta(t)=\hat{A}(t)+\delta(t), \tag{10.15}$$

where $\hat{A}(t)$ is given by the formula (10.7) and

$$\delta(t) = -\frac{i}{(2\hbar C\Omega)^{1/2}} \int_0^t I(\tau)\, e^{(\Gamma + i\Omega)\tau}\, d\tau. \tag{z}$$

The ground state of the circuit with the external current has the form

$$\Psi_{0,\delta}(\mathcal{J}, t) = \Psi_0(\mathcal{J}, t) \exp\left[-\left(\frac{\Omega}{\omega}\right)^{1/2} \frac{\mathcal{J}}{i_0} e^{(\Gamma - i\Omega)t}\, \delta(t) - \right.$$
$$\left. - \frac{\Omega}{2} [\mathrm{Re}\,(\Omega^{-1/2}\delta(t)\, e^{-(\Gamma + i\Omega)t}]^2 + \frac{i\Omega}{2} \mathrm{Re}\left(\int_0^t e^{-2i\Omega\tau}\delta^2(\tau)\, d\tau\right)\right], \tag{10.16}$$

where $\Psi_0(\mathcal{J},t)$ is the wavefunction of the ground state of the damped circuit without the external current (10.8).

The set of coherent states of the resonant circuit with constant damping which is excited by the external varying current has the form

$$\Psi_{\alpha,\delta}(\mathcal{J}, t) = \Psi_{0,\delta}(\mathcal{J}, t) \exp\left[-\frac{\alpha^2}{2} e^{-2i\Omega t} - \frac{|\alpha|^2}{2} + \left(\frac{2\Omega}{\omega}\right)^{1/2} \frac{\alpha}{i_0} \mathcal{J} e^{(\Gamma - i\Omega)t} + \right.$$
$$\left. + \alpha (2\Omega)^{1/2} e^{(\Gamma - i\Omega)t}\, \mathrm{Re}\,[\Omega^{-1/2}\delta(t)\, e^{-(\Gamma - i\Omega)t}]\right]. \tag{10.17}$$

In these states the mean values of the current and the voltage are given by the formulae

$$\overline{\mathfrak{J}} = \frac{i_0\omega_0}{\Omega} e^{-(\Gamma + i\Omega)t}\left(\frac{\alpha}{\sqrt{2}} - \frac{\delta}{2}\right) + \frac{i_0\omega_0}{\Omega} e^{-(\Gamma - i\Omega)t}\left(\frac{\alpha^*}{\sqrt{2}} - \frac{\delta^*}{2}\right),$$
$$\overline{V} = \frac{\hbar\omega_0^2}{i_0} e^{2\Gamma t}\left[-\sqrt{2}\,\mathrm{Re}\left(\frac{\Gamma + i\Omega}{\Omega}\, \alpha e^{-(\Gamma + i\Omega)t}\right) - \frac{d}{dt}\mathrm{Re}\left[\frac{\delta}{\Omega} e^{-(\Gamma + i\Omega)t}\right]\right], \tag{10.18}$$

and the standard deviations of the current and voltage do not depend on the external current being described by the relations (10.10).

Let us consider the case when the resistance depends on time and the inductance also varies with time. In this case the Hamiltonian takes the form

$$\hat{H}(t) = {}^1/_2 L(t)\, \hat{\mathfrak{J}}^2 e^{2\Gamma(t)} + {}^1/_2 C\hat{V}^2 e^{-2\Gamma(t)} - L\hat{\mathfrak{J}} e^{2\Gamma(t)} I(t), \qquad \Gamma(t) = R(t)/2L(t). \tag{10.19}$$

From this Hamiltonian and the commutation relations follow the equations of the motion

$$\ddot{\hat{\mathfrak{J}}}_H + \frac{\dot{R}L(t) - R\dot{L}(t)}{L^2(t)}\, \dot{\hat{\mathfrak{J}}} + \omega^2(t)\, \hat{\mathfrak{J}} = LI(t). \tag{A}$$

The "annihilation" operator has the form

$$\hat{A}_{\varepsilon,\delta}(t) = \frac{i}{\sqrt{2}}\left(\frac{i_0\varepsilon}{\hbar\omega_0^2}\hat{V} - \frac{\dot{\varepsilon}}{i_0\omega_0} e^{2\Gamma(t)}\hat{\mathfrak{J}}\right) + \frac{\delta(t)}{\sqrt{2}}, \tag{10.20}$$

where

$$\delta(t) = -\frac{i}{(\hbar C)^{1/2}} \int_0^t \varepsilon(\tau)\, e^{2\Gamma(\tau)} I(\tau)\, d\tau, \tag{B}$$

and the complex function $\varepsilon(t)$ satisfies the equation (10.4) with $r = \frac{d}{dt}\left(\frac{R(t)}{2L(t)}\right)$ and the additional condition (10.5). The eigenfunctions of the operator (10.20) are the coherent states of the parametric circuit with the finite time-independent quality. They have the form

$$\Psi_\alpha(\mathcal{J},t) = (\pi i_0^2\omega_0\varepsilon^2)^{-1/4}\exp\left[i\frac{\dot{\varepsilon}\mathcal{J}^2}{2\varepsilon\omega_0 i_0^2}e^{2\Gamma(t)} + \frac{\sqrt{2}}{\varepsilon}\frac{\alpha\mathcal{J}}{i_0\sqrt{\omega_0}} - \frac{\varepsilon^*}{2\varepsilon}\alpha^2 - \right.$$
$$\left. - \frac{|\alpha|^2}{2} - \frac{\mathcal{J}\delta}{i_0\varepsilon\sqrt{\omega_0}} - \frac{\varepsilon^*\delta^2}{4\varepsilon} - \frac{|\delta|^2}{4} + \frac{\alpha}{\sqrt{2}}\left(\delta^* + \frac{\varepsilon^*\delta}{\varepsilon}\right) - \frac{i}{2}\int_0^t \mathrm{Im}(\dot{\delta}^*\delta)\,d\tau\right]. \tag{10.21}$$

The mean values of the current and voltage in the circuit in the states (10.21) have the form

$$\bar{\mathcal{J}} = i_0[\sqrt{2}\,\mathrm{Re}(\alpha\varepsilon^*) - \mathrm{Re}(\varepsilon^*\delta)]\,\omega_0^{1/2},$$
$$\bar{V} = \frac{\hbar\omega_0^2}{i_0}e^{2\Gamma(t)}\left[\sqrt{2}\,\mathrm{Re}\left(\frac{\alpha\dot{\varepsilon}^*}{\sqrt{\omega_0}}\right) - \frac{d}{dt}\mathrm{Re}\left(\frac{\varepsilon^*\delta}{\sqrt{\omega_0}}\right)\right], \tag{10.22}$$

and the standard deviations of the current and voltage are given by the formulae

$$(\delta\mathcal{J})^2 = \frac{i_0^2\omega_0}{2}|\varepsilon|^2, \qquad (\delta V)^2 = \frac{e^{4\Gamma(t)}}{2\omega_0}\frac{\hbar^2\omega_0^4}{i_0^2}|\dot{\varepsilon}|^2. \tag{10.23}$$

Consequently, the Schrodinger-Robertson uncertainty relation holds

$$(\delta\mathcal{J}\delta V)^2 = {}^1/_4 e^{4\Gamma(t)}\,|\varepsilon\dot{\varepsilon}|^2\hbar^2\omega_0^4 \geqslant {}^1/_4\hbar^2\omega_0^4\,(1 - r^2)^{-1/2}. \tag{10.24}$$

The wave functions of the pseudo stationary states of the resonant circuit with the time dependent parameters are the eigenfunctions of the operator $\hat{A}^+{}_\varepsilon$, $\delta\hat{A}_\varepsilon$, $\delta\Psi_n$=$E\Psi_n$ and they have the form

$$\Psi_n(\mathcal{J},t) = \frac{1}{\sqrt{n!}}\left(\frac{\varepsilon^*}{2\varepsilon}\right)^{n/2}\Psi_0(\mathcal{J},t)\,H_n\left(\frac{\mathcal{J}}{i_0|\varepsilon|\sqrt{\omega_0}} + \frac{\mathrm{Re}(\varepsilon^*\delta)}{|\varepsilon|}\right), \tag{10.25}$$

where $H_n(x)$ is a Hermite polynomial that corresponds to the results obtained in study [1] for the damped harmonic oscillator.

11. RESONANS OF QUANTUM RESONANT CIRCUIT

Let us consider the the damped quantum resonant circuit being subject to an external classical voltage or an external classical current. The dependence on time for this external interaction is assumed to be the harmonic oscillation with the frequency ω. The operator of the charge on the plates of the capitance in the Heisenberg representation obeys the following equation:

$$\ddot{\hat{Q}} + \Gamma\dot{\hat{Q}} + \omega_0^2\hat{Q} = f\cos\omega t. \tag{11.1}$$

The charge operator is connected with the voltage operator $\hat{V}=\hat{Q}/C$ where C is the capacitance. Averaging the relation (11.1) over the coherent states of the damped circuit, we obtain that the mean value of the charge $<\hat{Q}>=x$ satisfies the same equation

$$\ddot{x} + \Gamma\dot{x} + \omega_0^2 x = f\cos\omega t. \tag{11.2}$$

The average current obeys the equation ($y=\dot{x}$)

$$\ddot{y} + \Gamma\dot{y} + \omega_0^2 y = -f\omega\sin\omega t. \tag{11.3}$$

In the quantum resonant circuit the resonance phenomenon for the current and the voltage takes place. The average charge is described by the formula

$$x = f\,[(\omega_0^2 - \omega^2)^2 + \omega^2\Gamma^2]^{-1/2} \cos\left(\omega t - \operatorname{arctg}\frac{\omega\Gamma}{\omega_0^2 - \omega^2}\right), \tag{11.4}$$

and the average current $y=\dot{x}$. Finding the maximum amplitude of the forced oscillations of the average charge in the quantum resonant circuit, we obtain the following resonant value of the frequency

$$\omega_p = (\omega_0^2 - \Gamma^2/2)^{1/2}. \tag{11.5}$$

The resonant frequency for the average current y is determined by the condition of the maximum forced oscillations of the current. It coincides with ω_0. It is worthy of note that the classical parametric resonant circuit may serve as the electrical model for measuring the reflection and transmission coefficients of a quantum particle by the potential barrier. Measuring the electrical energy transmitted to the resonant circuit by the external source of the voltage, we can connect this energy with the reflection and transmission coefficients of the quantum particle which is the potential barrier determined by the time dependence of the frequency of the classical resonant circuit.

12. DENSITY MATRIX OF RESONANT CIRCUIT

If the temperature is not equal to zero the resonant circuit will be in a mixed state. Let us consider non-normalized density matrix of the quantum resonant circuit according to the scheme discussed in study [3]. The density matrix in the mixed representation $\rho_W = e^{-\beta\hat{H}}$ has the form

$$\rho_W(\mathcal{J}, V, \beta) = \left(\operatorname{ch}\left(\frac{\hbar\omega}{2T}\right)\right)^{-1} \exp\left[-\operatorname{th}\frac{\hbar\omega}{2T}\left(\frac{\mathcal{J}^2}{i_0^2} + V^2\frac{i_0^2}{\hbar^2\omega^4}\right)\right], \tag{12.1}$$

where $\beta=T^{-1}$ is the inverse temperature.

We write the kernel of the current operator in the current representation

$$\langle \mathcal{J}' | \hat{\mathcal{J}} | \mathcal{J} \rangle = \mathcal{J}\delta(\mathcal{J} - \mathcal{J}'). \tag{12.2}$$

In the mixed representation the symbol of the current operator is given by the relation

$$\int_{-\infty}^{\infty} \left(\mathcal{J} + \frac{\eta}{2}\right)\delta(\eta)\exp\left(-\frac{iV\eta}{\hbar\omega^2}\right)d\eta = \mathcal{J}. \tag{12.3}$$

The symbol of the voltage operator $\hat{V}$ is also c-number V.

Let us find out mean values of the powers of the current and voltage in the state (12.1). We introduce the partition function

$$Z(T) = [2\,\operatorname{sh}(\hbar\omega/2T)]^{-1} \tag{C}$$

and the characteristic functions $x = \langle e^{z\hat{\mathcal{J}}} \rangle$ and $x' = \langle e^{z\hat{V}} \rangle$. The characteristic function x(z) takes the form

$$\chi(z) = \frac{\iint_{-\infty}^{\infty} d\mathcal{J}\, dV \rho_V(\mathcal{J}, V, T)\, e^{z\mathcal{J}}}{\iint_{-\infty}^{\infty} \rho_V(\mathcal{J}, V, T)\, d\mathcal{J}\, dV}, \tag{12.4}$$

and we have

$$\chi(z) = \exp\left[(\hbar z^2/4i_0^2)\operatorname{ctg}(\hbar\omega/2T)\right]. \tag{12.5}$$

Developing Equation (12.5) into the power series in z, we obtain that the mean values of the even powers of the current are

$$\langle \hat{\mathcal{J}}^{2n} \rangle = 2^{-n}(2n-1)!!\,(i_0)^{2n}\operatorname{cth}^n(\hbar\omega/2T). \tag{12.6}$$

Analogously, for the even powers of the voltage one can obtain the expression

$$\langle \hat{V}^{2n} \rangle = 2^{-n}(2n-1)!!\,(\hbar\omega^2/i_0)^{2n}\operatorname{cth}^n(\hbar\omega/2T), \tag{12.7}$$

and the odd powers of the current and voltage are equal to zero.

Let us write down the density matrix of the resonant circuit in the coherent state representation

$$\rho(\beta^*, \alpha, T) = \exp\left[-|\alpha|^2/2 - |\beta|^2/2 + \beta^*\alpha e^{-\hbar\omega/T} - \hbar\omega/2T\right], \tag{12.8}$$

and we obtain from this expression that the mean values of the current and voltage powers in the coherent state (12.8) have the form

$$\begin{aligned} \langle\alpha|\hat{\mathfrak{I}}^n|\alpha\rangle &= (i\sigma_{\mathcal{J}}/\sqrt{2})^n H_n(\bar{\mathfrak{I}}/i\sqrt{2}\,\sigma_{\mathcal{J}}); \quad \sigma_{\mathcal{J}} = i_0/\sqrt{2}, \\ \langle\alpha|\hat{V}^n|\alpha\rangle &= (i\sigma_V/\sqrt{2})^n H_n(\bar{V}/i\sqrt{2}\,\sigma_V), \quad \sigma_V = \hbar\omega^2/i_0\sqrt{2}, \end{aligned} \tag{12.9}$$

where

$$\begin{aligned} \bar{\mathfrak{I}}(t) &= \sqrt{2}\, i_0\, |\alpha| \cos(\varphi_\alpha - \omega t), \quad \varphi_\alpha = \arg\alpha. \\ \bar{V}(t) &= \sqrt{2}\,(\hbar\omega^2/i_0)|\alpha| \sin(\varphi_\alpha - \omega t). \end{aligned} \tag{D}$$

Let us consider the Green function of the stationary resonant circuit. By analogy with the oscillator problem [1], it may be written down in the form

$$\begin{aligned} G(\beta^*, \alpha, E) &= \left(\frac{\hbar\omega}{2} - E\right)^{-1} \exp\left[-\frac{|\alpha|^2}{2} - \frac{|\beta|^2}{2}\right] \times \\ &\times \Phi\left(\frac{1}{2} - \frac{E}{\hbar\omega};\, \frac{3}{2} - \frac{E}{\hbar\omega};\, \alpha\beta^*\right), \end{aligned} \tag{12.10}$$

where Φ(a,b,c) is the degenerate hypergeometric function.

In the Fock basis, the matrix elements of the current and voltage powers are given by the expressions

$$\langle m|\hat{\mathfrak{I}}^k|n\rangle = i_0^{2k}(m!\,n!)^{-1/2} H^{(C_{ij})}_{kmn}(0,0,0), \quad \langle m|\hat{V}^k|n\rangle = (\hbar\omega^2/i_0)^{2k}(m!\,n!)^{-1/2} H^{(C'_{ij})}_{kmn}(0,0,0),$$

$$C_{ij} = -\frac{1}{\sqrt{2}}\begin{Vmatrix} 1/\sqrt{2} & 1 & 1 \\ 1 & 0 & \sqrt{2} \\ 1 & \sqrt{2} & 0 \end{Vmatrix}, \quad C'_{ij} = -\frac{1}{\sqrt{2}}\begin{Vmatrix} 1/\sqrt{2} & i & -i \\ i & 0 & \sqrt{2} \\ -i & \sqrt{2} & 0 \end{Vmatrix}, \tag{12.11}$$

where $H^{(C)}_{mnk}$ (0,0,0) is the Hermite polynomial of three variables and it is calculated at zero.

13. WIGNER FUNCTION OF QUANTUM RESONANT CIRCUIT WITH FINITE QUALITY

Let us consider the resonant circuit with finite quality as the open system interacting with the thermostat. In this case, the resonant circuit must be described by the density matrix obeying the equation which is reduced to the classical equation for oscillations of the mean values of the current in the damped resonant circuit

$$\ddot{\mathcal{J}} + 2\Gamma\dot{\mathcal{J}} + \omega_0^2\mathcal{J} = 0, \quad \Gamma = R/2L, \tag{13.1}$$

for arbitrary coefficients Γ and ω_0. In the Josephson junction, the coefficient Γ coincides with the characteristic transition frequency of the junction and ω_0 coincides with plasma frequency. Using the approach suggested in study [12], one can obtain that the most general equation for the density matrix $\hat{\rho}$ of the damped resonant circuit satisfying the uncertainty relation and leading to the equation (12.1) for the mean values of the current has the form

$$\dot{\hat{\rho}} = -i\hbar^{-1}[\hat{H}\hat{\rho}] + 2\hat{\Phi}\hat{\rho}\hat{\Phi}^{+} - \hat{\Phi}^{+}\hat{\Phi}\hat{\rho} - \hat{\rho}\hat{\Phi}^{+}\hat{\Phi}. \tag{13.2}$$

Let us introduce the Wigner function of the resonant circuit

$$W(\mathcal{J}, V, t) = \int \rho(\mathcal{J} + \xi/2, \mathcal{J} - \xi/2, t) \exp(-iV\xi/\hbar\omega_0^2)\, d\xi. \tag{13.3}$$

This Wigner function satisfies the equation

$$\frac{\partial W}{\partial t} + \frac{\partial}{\partial q_\alpha}[A_{\alpha\beta} q_\beta W] = D_{\alpha\beta}\frac{\partial^2 W}{\partial q_\alpha \partial q_\beta}, \tag{13.4}$$

where $\beta, \alpha=1,2$, $q_1=\mathcal{J}$, $q_2=V$, $D_{\alpha\beta}$ are the diffusion coefficients.

Using the approach suggested in studies [13 and 14] we choose as the initial system of equations those which describe the relaxation to the equilibrium state

$$\hat{\rho}_{eq} = \exp(-\beta\hat{H})[\mathrm{Tr}\exp(-\beta\hat{H})]^{-1}, \tag{E}$$

the following system

$$\dot{\mathcal{J}} = -\tau\mathcal{J} + V/\omega_0^2, \quad \dot{V} = -g\omega_0^2\mathcal{J} - \xi V. \tag{13.5}$$

From Equations (13.5) they follow the second order equations

$$\ddot{\mathcal{J}} = -2\gamma\dot{\mathcal{J}} - \omega_0^2\mathcal{J}, \quad \tau + \xi = 2\gamma, \quad \tau\xi + g = \omega_0^2, \tag{F}$$

and the model under discussion describing the circuit with finite quality provides the following diffusion coefficients

$$D_{\mathcal{J}\mathcal{J}} = (\hbar\gamma/2\omega)\mathrm{tg}\,(\hbar\omega\beta/4), \quad D_{\mathcal{J}V} = -(\gamma/\omega_0^2)D_{\mathcal{J}\mathcal{J}},$$
$$D_{VV} = (\hbar\gamma/2\omega)\omega_0^{-4}\,[\omega^2\,\mathrm{ctg}\,(\hbar\omega\beta/4) + \gamma^2\,\mathrm{th}\,(\hbar\omega\beta/4)], \tag{G}$$

where $\omega=(q^2+\gamma^2)^{1/2}$, $\beta=1/T$ is the inverse temperature. In the equation for the statistical operator (13.2) the operator $\hat{\Phi}$ has the form

$$\hat{\Phi} = \left[\frac{\gamma}{2\hbar\omega}\,\mathrm{th}\left(\frac{\hbar\omega\beta}{4}\right)\right]^{1/2}\left\{\frac{\hat{V}}{\omega_0^2} + \hat{\mathcal{J}}\left(\gamma - i\omega_0\,\mathrm{ctg}\left(\frac{\hbar\omega\beta}{4}\right)\right)\right\}, \tag{H}$$

and the Hamiltonian is

$$\hat{H} = {}^1/_2\,(\hat{V}^2/\omega_0^4 + g\hat{\mathcal{J}}^2), \tag{I}$$

For simplicity the product L^2C is taken to be equal to unity.

14. ADIABATIC INVARIANTS OF THE RESONANT CIRCUIT

Let us consider in this Section the invariants of the parametric resonant circuit with varying inductance and let the dependence of the inductance on time be slow. In this case, one can construct adiabatic invariants that are linear in current and voltage invariants and quadratic ones. The theory of the adiabatic invariants for the quantum oscillator (linear in momentum and position operators) has been developed in studies [15 and 16]. Using the above analogy of the quantum resonant circuit and the quantum oscillator, we construct the linear and quadratic adiabatic invariants for the parametric resonant circuit. The Hamiltonian for parametric resonant circuit has the form given by the formula (6.1). The integrals of the motion which are linear forms in the current and voltage operators are described by Equation (6.6). The invariant – "annihilation" operator $\hat{A}_\omega$ is given by the formula (6.2). According to the theorem about the functions of integrals of motion [1] an arbitrary function of integrals of the motion of the resonant circuit is an integral of the motion too. Because of this, the operator $\hat{N}=\hat{A}^+_\omega\hat{A}_\omega$ which has the form

$$\hat{N}=\frac{1}{2}\left(\frac{i_0^2|\varepsilon|^2}{\hbar^2\omega_0^4}\hat{V}^2+\frac{\hat{\mathfrak{J}}^2|\dot{\varepsilon}|^2}{\omega_0^2 i_0^2}+\hat{\mathfrak{J}}\hat{V}\frac{\dot{\varepsilon}^*\varepsilon}{\omega_0^3\hbar}+\hat{V}\hat{\mathfrak{J}}\frac{\varepsilon\dot{\varepsilon}^*}{\omega_0^3\hbar}\right), \tag{14.1}$$

is the exact integral of the motion. It is a quadratic form in the current and voltage operators. The operators-integrals of the motion $\hat{\mathfrak{J}}_0$, $\hat{V}_0$ (6.1) $\hat{A}_\omega$, $\hat{A}^+_\omega$ (6.2) and the operator $\hat{N}$ are determined by the classical trajectory $\varepsilon(t)$ satisfying the equation (6.3).

The idea of constructing the adiabatic invariants of the resonant circuit is the following one. Let us construct an approximate solution $\varepsilon_{ad}(t)$ to a classical equation of motion (6.6), assuming that the frequency $\omega(t)$ in this equation varies slowly with time, i.e., the following inequality holds $\dot{\omega}/\omega^2<<1$. After this, we substitute the solution $\varepsilon_{ad}(t)$ into the formulae for the exact integrals of the motion of the resonant circuit and we obtain the expressions for the adiabatic invariants of the resonant circuit. The approximate solution to the equation (6.3) has the form

$$\varepsilon_{ad}(t)=\left(\frac{\omega_0}{\omega(t)}\right)^{1/2}\exp\left\{i\int_0^t\omega(\tau)\,d\tau\right\}. \tag{14.2}$$

In this case we have the following expressions for the integrals of the motion (6.5)

$$\hat{\mathfrak{J}}^{ad}_{0,\omega}(t)=\left[\frac{ii_0^2}{2\hbar\omega_0^2}\hat{V}(\varepsilon_{ad}-\overset{*}{\varepsilon}_{ad})+\frac{i}{2\omega_0}\hat{\mathfrak{J}}(\dot{\varepsilon}_{ad}-\dot{\overset{*}{\varepsilon}}_{ad})\right],$$

$$\hat{V}^{ad}_{0,\omega}(t)=\frac{1}{2}\left[\hat{V}(\varepsilon_{ad}+\overset{*}{\varepsilon}_{ad})+\frac{\hbar\omega_0}{i_0^2}\hat{\mathfrak{J}}(\dot{\varepsilon}_{ad}+\dot{\overset{*}{\varepsilon}}_{ad})\right]. \tag{14.3}$$

The adiabatic invariants $\hat{\mathfrak{J}}^{ad}{}_{0,\omega}$, $\hat{V}^{ad}{}_{0,\omega}$ take the following explicit form:

$$\hat{\mathfrak{J}}^{ad}_{0,\omega}(t)=-\frac{i_0^2}{\hbar\omega_0^2}\hat{V}\sin\left(\int_0^t\omega(\tau)\,d\tau\right)\left(\frac{\omega_0}{\omega(t)}\right)^{1/2}+\hat{\mathfrak{J}}\left(\frac{1}{\omega_0\omega(t)}\right)^{1/2}\left\{-\omega(t)\times\right.$$

$$\left.\times\cos\left(\int_0^t\omega(\tau)\,d\tau\right)+\frac{\dot{\omega}(t)}{2\omega(t)}\sin\left(\int_0^t\omega(\tau)\,d\tau\right)\right\},$$

$$\hat{V}^{ad}_{0,\omega}(t)=\left(\frac{\omega_0}{\omega(t)}\right)^{1/2}\hat{V}\cos\left(\int_0^t\omega(\tau)\,d\tau\right)+\frac{\hbar\omega_0^2}{i_0^2}\hat{\mathfrak{J}}\left(\frac{\omega_0}{\omega(t)}\right)^{1/2}\times$$

$$\times\left\{-\omega(t)\cos\left(\int_0^t\omega(\tau)\,d\tau\right)+\frac{\dot{\omega}(t)}{2\omega(t)}\sin\left(\int_0^t\omega(\tau)\,d\tau\right)\right\}. \tag{14.4}$$

The operators of the "creation" and "annihilation" of the oscillation quanta are as follows:

$$\hat{A}_{\omega}^{\mathrm{ad}}(t) = \frac{i}{\sqrt{2}}\left\{\frac{i_0}{\omega_0^2\hbar}\hat{\mathcal{V}}\left(\frac{\omega_0}{\omega(t)}\right)^{1/2}\exp\left[i\int_0^t \omega(\tau)\,d\tau\right] + \right.$$

$$\left. + \frac{\hat{\mathcal{J}}}{\omega_0 i_0}\left(\frac{\omega_0}{\omega(t)}\right)^{1/2}\exp\left[i\int_0^t \omega(\tau)\,d\tau\right]\left(i\omega(t) - \frac{\dot{\omega}(t)}{2\omega(t)}\right)\right\},$$

$$\hat{A}_{\omega}^{+\mathrm{ad}}(t) = -\frac{i}{\sqrt{2}}\left\{\frac{i_0}{\omega_0^2\hbar}\hat{\mathcal{V}}\left(\frac{\omega_0}{\omega(t)}\right)^{1/2}\exp\left[-i\int_0^t \omega(\tau)\,d\tau\right] + \right.$$

$$\left. + \frac{\hat{\mathcal{J}}}{\omega_0 i_0}\left(\frac{\omega_0}{\omega(t)}\right)^{1/2}\exp\left[-i\int_0^t \omega(\tau)\,d\tau\right]\left(-i\omega(t) - \frac{\dot{\omega}(t)}{2\omega(t)}\right)\right\}. \qquad (14.5)$$

The adiabatic invariant which is a quadratic form in the current and voltage operators is

$$\hat{N}_{\mathrm{ad}} = \frac{1}{2}\left[\frac{i_0^2\hat{\mathcal{V}}^2}{\hbar^2\omega_0^3\omega(t)} + \frac{\hat{\mathcal{J}}^2}{\omega_0\omega(t)\,i_0^2}\left(\omega^2(t) - \frac{\dot{\omega}^2(t)}{4\omega^2(t)}\right) + \right.$$

$$\left. + \hat{\mathcal{J}}\hat{\mathcal{V}}\frac{\omega_0}{\omega(t)}\left(-i\omega(t) - \frac{\dot{\omega}(t)}{2\omega(t)}\right) + \hat{\mathcal{V}}\hat{\mathcal{J}}\frac{\omega_0}{\omega(t)}\left(i\omega(t) - \frac{\dot{\omega}(t)}{2\omega(t)}\right)\right]. \qquad (14.6)$$

15. CORRELATED STATES AND BERRY PHASE OF THE RESONANT CIRCUIT

In study [16] the connection of the Berry phase with the correlated states and the linear adiabatic invariants of the quantum resonant circuit was analyzed. In this Section we consider the quantum resonant circuit for which there exists an extra term in the expression for its energy due to an interaction of the current with the voltage. In principal, such term may appear if one takes into account the nonlinearity of the resonant circuit. The linearization of the equation for the current and voltage may provide such a term. If this term exists, and the parameters of the circuit depend on time, the nonzero Berry phase appears. In this Section we describe the results of study [16] considering the example of the quantum resonant circuit. This resonant circuit may be considered as the pure phenomenological model and the validity of such approximation needs to be considered in more details.

Using the approach of study [16] in which the connection of the Berry phase with the correlated states of the oscillator has been considered, we construct the following Hamiltonian for the quantum resonant circuit in dimensionless variables

$$\hat{H}(t) = {}^1/_2\left[\mu(t)\hat{\mathcal{V}}^2 + \nu(t)\,\hat{\mathcal{J}}^2 + \rho(t)(\hat{\mathcal{J}}\hat{\mathcal{V}} + \hat{\mathcal{V}}\hat{\mathcal{J}})\right]. \qquad (15.1)$$

According to studies [17 and 18] in the adiabatic limit, one can construct using the method of studies [18-22] the adiabatic invariant which is linear in the current and voltage operators

$$\hat{A}(t) = [\mu\hat{\mathcal{V}} - i(\omega + i\rho)\,\hat{\mathcal{J}}]\,e^{i\Phi(t)}(2\omega\mu)^{-1/2}, \qquad (15.2)$$

where ω(t) is determined by the parameters of the Hamiltonian of the parametric circuit

$$\omega^2 = \mu\nu - \rho^2. \qquad (15.3)$$

The phase Φ(t) is given by the formula

$$\Phi(t) = \int_0^t \omega(t')\,dt' + \int_C \frac{1}{2\omega}\left(\rho\,\frac{d\mu}{\mu} - d\rho\right), \tag{15.4}$$

where the first term is the standard variation of the phase due to the time-dependence of the frequency. The second term is determined by the contour integral where the contour C is the curve in the space of the parameters of the Hamiltonian describing the quantum resonant circuit (the capacitance, the inductance and the coefficient of interaction of the current and the voltage).

The last phenomenological term may appear if we linearize equations of motion of the nonlinear problem of the electrical oscillations in the resonant circuit. However, the physical meaning of the quantum resonant circuit model demands additional clarification. So, the second term in Equation (15.4) does not depend on the concrete form of the time-dependence of the resonant circuit parameters. It depends only on the form of the contour in Equation (15.4). It was demonstrated in study [21] that at $\rho \neq 0$ the correlated state of the oscillator is created. The correlation coefficient in the correlated state of the quantum resonant circuit is determined by the parameters of the Hamiltonian (15.1) analogously to the oscillator problem. It has the form

$$r = -\rho/\sqrt{\mu\nu}. \tag{15.5}$$

If the correlation coefficient is equal to zero, that may be at $\rho=0$, the phase $\Phi(t)$ does not differ from the first term in the formula (15.4), i.e., the Berry phase for the quantum resonant circuit and for Josephson junction is equal to zero in this case. It is worthy of note that the integrals of the motion which are linear forms in the coordinate and momentum operators for the parametric oscillator are analogous to the integral of the motion (15.2) for the quantum resonant circuit. Josephson junction are important in the problem of the sensitive measurements at detecting the gravitational waves by the gravitational detector. These integrals of the motion (quadrature components) are most suitable for the useful signals on the background of the quantum noise.

Thus, the quantum resonant circuit (and the Josephson junction) may serve as the model for studying the quantum measurements of the physical observables which are important in the theory of the gravitational detector and, in particular, in the problem of finding the Berry phase. The correlated states of the resonant circuit may be excited with help of quick variations of the frequency of the resonant circuit. When time-dependence of the frequency is a nonadiabatic one, the formula (15.5) for the correlation coefficient does not hold. But for the adiabatic variations of the parameters of the quantum resonant circuit, the creation of the state with the nonzero correlated coefficient of the current and the voltage turns out to be closely connected with the nonzero Berry phase The Berry phase itself is determined by the phase factor in the expression for the linear adiabatic invariants – the "annihilation" operator of the quantum resonant circuit.

16. PATH INTEGRALS FOR QUANTUM RESONANT CIRCUIT

The expression for the propagator of the resonant circuit and the Josephson junction G may be obtained if the quantization of the classical resonant circuit and Josephson junction is accomplished with the help of the Feynmann integral or path integral method. According to the quantization scheme, with the help of the path integral the transition probability amplitude (Green function or propagator) of the resonant circuit (Josephson junction) from the state with the fixed current $\mathfrak{J}$ into another state with the current $\mathfrak{J}'$ may be written in the form

$$G(\mathcal{J}, \mathcal{J}^{\bullet}, t) = \int_1^2 D[\mathcal{J}(t)] \exp\left\{\frac{i}{\hbar} S[\mathcal{J}(t)]\right\}. \tag{16.1}$$

Here $D[\mathfrak{I}(t)]$ is the measure in the Feynmann path integral. S is the classical action for the resonant circuit (Josephson junction)

$$S = \int_{t_1}^{t_2} \mathcal{L}(\mathcal{J}, \dot{\mathcal{J}}, t)\, dt, \tag{16.2}$$

L is the Lagrangian determining the equation for the current $L=(CL^2\mathfrak{I}^2-2\mathfrak{I}^2)/2$. Generally speaking, the canonical quantization of the Josephson junction accomplished above is equivalent to the quantization with the help of the path integral. The analysis of the quantum-mechanical equations has shown [23] that in the cases when it is necessary to use approximations, the quantization with the help of the path integral has the same advantages. Because of this, in the theory of the quantum resonant circuit (taking into account, for example, a nonlinearity) the expression for the propagator of the resonant circuit (16.1) may be useful. Hence, this expression gives the possibility to easily derive the quasi-classical approximation for the Green function of the quantum resonant circuit acted by thermostat degrees of freedom. By means of the path integral method the functional influence may be easily derived for the resonant circuit analogously to the problem of quantum oscillator [23].

17. QUANTUM RESONANT CIRCUIT AS A MODEL OF JOSEPHSON JUNCTION

Let us consider, as an example of using the model of quantum resonant circuit, the problem of Josephson junction, and transfer to this case the results obtained in the previous sections.

The feasibility of using Josephson junction (simulated by a quantum resonant circuit) to generate squeezed electromagnetic radiation was considered in [24]. In this section, using the analogy between a Josephson junction and a quantum resonant circuit, we demonstrate the possibility to create a new state of a Josephson junction – correlated state – with the aid of some parametric action on the junction which can lead effectively to temporal variations of its parameters, say, the critical current of the junction.

We carry out the analysis within the context of the Hamiltonian [25 and 26]

$$\hat{H} = \frac{\hat{Q}^2}{2C} + \frac{\hbar I_c}{2e}(1 - \cos\hat{\varphi}) - \frac{\hbar}{2e} I(t)\,\hat{\varphi}. \tag{17.1}$$

Here C is the capacity and I_c the critical current of the junction, e the electron charge, $\hbar$–Planck's constant, $\hat{Q}$ the charge operator, $\hat{\phi}$–the phase operator, and I(t) the external current feeding the junction. Here the problem is investigated in the domain of small values of the phase ϕ, when the term $(1\text{-}\cos\hat{\phi})$ in Hamiltonian (17.1) can be replaced by the quadratic expression $\hat{\phi}^2/2$. The necessary conditions justifying this replacement were discussed in [27]. Hamiltonian (17.1) is thus reduced to that of a quantum resonant circuit

$$\hat{H} = \frac{\hat{Q}^2}{2C} + \frac{\hbar I_c}{2e}\frac{\hat{\varphi}^2}{2} - \frac{\hbar}{2e} I(t)\,\hat{\varphi}. \tag{17.2}$$

The commutation relation for the charge and phase operators is of the form [25]

$$[\hat{\varphi}, \hat{Q}] = 2ei, \tag{17.3}$$

which corresponds to commutation relation of the current and voltage operators $\hat{J}=-I_c\hat{\phi}$ and $\hat{V}=\hat{Q}/C$

$$[\hat{\mathcal{J}}, \hat{V}] = i\hbar\omega^2, \tag{17.4}$$

where $w=(2eI_c/\hbar C)^{1/2}$ is the plasma frequency of the junction.

Coherent states for a quantum resonant circuit were considered in [2]. We introduce in similar fashion coherent states for a Josephson junction, starting with the condition of minimizing the Heisenberg uncertainty relations resulting from (17.3) and (17.4)

$$\delta\mathcal{J}\,\delta V \geqslant \hbar\omega^2/2, \tag{17.5}$$

Here δJ and δV are the standard deviations of the current and voltage in the junction. A coherent state of a Josephson junction is distinguished by two properties: $^{i)}$ relation (17.5) turns into an equality, and $^{ii)}$ the energy of quantum fluctuations reaches its minimum.

To solve the problem of exciting a Josephson junction at zero temperature by a classical external current, we construct the "annihilation" operator

$$\hat{A}(t) = \frac{e^{i\omega t}}{\sqrt{2}}\left[\left(\frac{\hbar C\omega}{4e^2}\right)^{1/2}\hat{\varphi} + \frac{i\hat{Q}}{(\hbar C\omega)^{1/2}}\right] - \frac{i}{(2C\hbar\omega)^{1/2}}\int_0^t e^{i\omega\tau} I(\tau)\,d\tau. \tag{17.6}$$

It can be shown that the operator $\hat{A}(t)$ and its Hermitian conjugate $\hat{A}^+(t)$ are integrals of the motion. This means that the mean values $<\hat{A}(t)>$ and $<\hat{A}^+(t)>$ do not change with time. It follows from (17.3) that the operators $\hat{A}$ and $\hat{A}^+$ satisfy the commutation relations for boson creation and annihilation operators: $[\hat{A},\hat{A}^+]=1$. The ground state of the junction can be found from the equation $\hat{A}(t)\psi_0=0$. Its normalized solution is given by

$$\Psi_0(\varphi, t) = \left(\frac{\hbar C\omega}{4e^2\pi}\right)^{1/4}\exp\left[-\frac{i\omega t}{2} - \frac{\hbar C\omega}{4e^2}\frac{\varphi^2}{2} - \right.$$
$$\left. -\sqrt{2}\left(\frac{\hbar C\omega}{4e^2}\right)^{1/2}\varphi\delta(t)\,e^{-i\omega t} + i\omega\int_0^t \delta^2(\tau)\,e^{-2i\omega\tau}\,d\tau\right], \tag{17.7}$$

where

$$\delta(t) = -i\,(2C\hbar\omega)^{-1/2}\int_0^t I(\tau)\,e^{i\omega\tau}d\tau.$$

The entire family of the coherent states of the Josephson junction is determined by the relations

$$\hat{A}\Psi_\alpha = \alpha\Psi_\alpha,$$
$$\Psi_\alpha(\varphi, t) = \Psi_0(\varphi, t)\exp\left[\sqrt{2}\left(\frac{\hbar C\omega}{4e^2}\right)^{1/2}\varphi\alpha e^{-i\omega t} - \right.$$
$$\left. -\frac{1}{\sqrt{2}}\alpha^2 e^{-2i\omega t} + \alpha e^{i\omega t}\left(\delta(t)\,e^{-i\omega t} + \delta^*(t)\,e^{i\omega t}\right) - \frac{|\alpha|^2}{2}\right]. \tag{17.8}$$

These states minimize the uncertainty relation (17.5) and satisfy the Schrodinger equation with Hamiltonian (17.2).

Let us examine the probabilities of transitions between the energy levels of a Josephson junction induced by a classical external current. If the Josephson junction was in the ground state at the initial instant t=0, then we can find by analogy with the resonant circuit [2] that the probability of remaining $W_0(t)$ in the ground state is

$$W_0(t) = \exp[-|\delta(t)|^2]. \tag{17.9}$$

It is possible to express δ(t) in terms of the classicial values of current J(t) and voltage V(t) in a resonant circuit equivalent to the given Josephson junction which is excited by an external current I(t) from the initial state with $J^- = V^- = 0$:

$$|\delta(t)|^2 = E_{cl}/\hbar\omega, \qquad E_{cl} = \hbar\bar{\mathcal{J}}^2/4eI_c + C\bar{V}^2/2. \tag{17.10}$$

The calculation of the transition probability $W_{(n)}(t)$ from the ground state to the n-th energy level yields

$$W_n(t) = (n!)^{-1}|\delta(t)|^{2n} \exp[-|\delta(t)|^2]. \tag{17.11}$$

Consequently, under the influence of an external classical current I(t), a Josephson junction at zero temperature goes from the ground state with energy $\hbar\omega/2$ into a coherent state having a Poisson distribution over the energy levels and constituting a normalized eigenstate of the integral of motion (17.6).

If temperature is nonzero, a normal current I_N is produced, and instead of Hamiltonian (17.2) one should use

$$\hat{H} = \frac{\hat{Q}^2}{2c} + \frac{\hbar I_c}{2e}\frac{\hat{\varphi}^2}{2} - \frac{\hbar}{2e}(I(t) - I_N)\hat{\varphi}. \tag{17.12}$$

We do not take into account the term containing coordinates of the quasiparticles ensemble (thermostate), since it is assumed that the relaxation time is long enough compared with the characteristic time of current variation.

Let us now consider the influence of an external classical current on the populations of stationary states of a Josephson junction. If the initial temperature of the junction is $T \neq 0$, it can be shown that the population of the n-th level changes from the initial value

$$\rho_{nn}(0) = 2\mathrm{sh}(\hbar\omega/2T)\exp\{-(n + {}^1/_2)\hbar\omega/T\} \tag{17.13}$$

to the value

$$\rho_{nn}(t) = \rho_{nn}(0)\exp\{-|\delta(t)|^2(1 - e^{-\hbar\omega/T}\}\, L_n\left(-2\,\mathrm{sh}\left(\frac{\hbar\omega}{2T}\right)|\delta(t)|^2\right), \tag{17.14}$$

where $L_n(x)$ is a Laguerre polynomial, and

$$\delta(t) = -i(2\hbar C\omega)^{-1/2}\int_0^t (I(\tau) - I_N)e^{i\omega\tau}\,d\tau,$$

which corresponds to the solution obtained in [1] for the oscillator-excitation problem. Note that the realization of a coherent state of a Josephson junction is reduced to applying an external current with account taken of the temperature-dependent term in Hamiltonian (17.12).

Correlated states for a quantum resonant circuit were considered in [2]. We introduce analogous correlated states for a Josephson junction by starting from the minimization condition of the Robertson-Schrodinger uncertainty relation (5.6) resulting from Equations (17.3) and (17.4)

$$\delta\mathcal{J}\delta V \geqslant {}^1/_2\hbar\omega^2(1 - r^2)^{-1/2}. \tag{17.15}$$

Here δV and δJ are standard deviations of the voltage and current in the junction, while r is their correlation coefficient

$$r = (\delta\mathcal{J}\,\delta V)^{-1}\left(\tfrac{1}{2}\langle\hat{\mathcal{J}}\hat{V} + \hat{V}\hat{\mathcal{J}}\rangle - \langle\hat{\mathcal{J}}\rangle\langle\hat{V}\rangle\right),$$

where the angle brackets denote averaging over a quantum state. For r=0, relation (17.15) goes over into the Heisenberg uncertainty relation (17.15).

Let us consider a parametrically excited Josephson junction, i.e., a Josephson junction which plasma frequency depends on time. This can be achieved by a specified time variation of the capacitance or of the critical current. Since the critical current is caused by the tunnel effect, i.e., depends exponentially on the junction thickness or on some other junction parameters, it is much easier in practice to obtain noticeable time variations of I_c than of the capacitance. In addition, it was shown recently that it is possible to obtain a manyfold change of the junction current by infrared irradiation. Therefore we henceforth consider a case of constant capacitance, but time-dependent critical current $I_c(t)$, i.e.,

$$\omega(t) = (2eI_c(t)/\hbar C)^{1/2}, \qquad \omega(0) = \omega_0, \qquad I_c(0) = I_c. \tag{17.16}$$

The Hamiltonian of the parametric Josephson junction is given by

$$\hat{H} = \tfrac{1}{2}[\hat{Q}^2/C + (\hbar I_c(t)/2e)\,\hat{\varphi}^2]. \tag{17.17}$$

The "annihilation" operator for a variable-frequency Josephson junction is

$$\hat{A}_\omega(t) = i\,(2\hbar C)^{-1/2}\,[\varepsilon\hat{Q} - (\hbar C/2e)\,\dot{\varepsilon}\hat{\varphi}], \tag{17.18}$$

where function $\varepsilon(t)$ satisfies the equation

$$\ddot{\varepsilon}(t) + \omega^2(t)\,\varepsilon(t) = 0 \tag{17.19}$$

with the additional condition

$$\dot{\varepsilon}\varepsilon^* - \dot{\varepsilon}^*\varepsilon = 2i. \tag{17.20}$$

It can be shown that operator $\hat{A}_\omega(t)$ and its Hermitian conjugate $\hat{A}^+_\omega(t)$ are integrals of the motion and satisfy the commutation relation $[\hat{A}_\omega, \hat{A}^+_\omega]=1$. We obtain the ground state of a variable-frequency Josephson junction from the condition $\hat{A}_\omega(t)\Psi_0=0$:

$$\Psi_0(\varphi, t) = \left(\frac{\hbar C}{\pi}\right)^{1/4}(2e\varepsilon)^{-1/2}\exp\left(\frac{i\hbar C}{4e^2}\frac{\dot{\varepsilon}}{\varepsilon}\frac{\varphi^2}{2}\right). \tag{17.21}$$

It is easy to verify that the state $\Psi_0(\phi,t)$ satisfies the Schrodinger equation with Hamiltonian (17.17) and minimizes the Robertson-Schrodinger uncertainty relation (17.15) with a current and voltage correlation coefficient

$$r = (|\varepsilon\dot{\varepsilon}|^2 - 1)^{1/2}/|\varepsilon\dot{\varepsilon}|, \tag{17.22}$$

i.e., it is a correlated state of the Josephson junction. The entire family of Josephson-junction correlation states is of the form

$$\Psi_\alpha(\varphi, t) = \Psi_0(\varphi, t)\exp\left[-\frac{|\alpha|^2}{2} - \frac{\alpha^2\varepsilon^*}{2\varepsilon} + \frac{\sqrt{2}\,\alpha\varphi}{\varepsilon}\left(\frac{\hbar C}{4e^2}\right)^{1/2}\right]. \tag{17.23}$$

States (17.23) also minimize the uncertainty relation (17.15) with the correlation coefficient (17.22), while the squeezing coefficient for them is

$$k = \frac{\delta\mathcal{J}}{\delta V}\,\frac{\delta V(t=0)}{\delta\mathcal{J}(t=0)} = \frac{|\varepsilon|}{|\dot{\varepsilon}|}\,\frac{I_c(t)}{I_c(0)}\,\omega_0.$$

For $|\varepsilon\dot{\varepsilon}|=1$, meaning r=0, states (17.23) are squeezed [8]. The standard deviations of the current and voltage in a correlated state are

$$\delta \mathcal{J} = (2/\hbar C)^{1/2} e I_c \quad (t) \mid \varepsilon \mid, \quad \delta V = (\hbar/2C)^{1/2} \mid \dot{\varepsilon} \mid. \tag{17.24}$$

Hence, we see that by varying in a definite manner function $I_c(t)$, we can control current and voltage quantum noises of the Josephson junction, i.e., we can decrease the voltage quantum noise at the cost of increasing the current noise, and vice versa.

Let us consider, in analogy with the resonant circuit, the problem of parametric excitation of a Josephson junction. Assume that at the initial instant of time t=0 the Josephson junction was in stationary state n at zero temperature. Suppose the junction parameters began to vary with time at t>0, but at $t \geq t_0$ they regain their initial values. Then the probability of transition into a state m as a result of variation of the parameters is [1]

$$W_n^m = \left(E(t) + \frac{1}{2}\right)^{-1/2} \frac{n!}{m!} \left[P_{(m+n)/2}^{(m-n)/2} \left(E(t) + \frac{1}{2}\right)^{-1/2} \right) \right]^2, \qquad m \geqslant n, t \geqslant t_0, \tag{17.25}$$

where $P^{\beta}_{\alpha}(x)$ is a Legendre polynomial and the parameter E(t) is equal to

$$E(t) = {}^1/_4 \omega_0^{-1} [\omega_0^2 \mid \varepsilon \mid^2 + \mid \dot{\varepsilon} \mid^2], \qquad t \geqslant t_0. \tag{17.26}$$

It can be shown that this parameter is none other than the average energy acquired by the junction by the instant t_0 if it was in the ground state at the initial instant.

Let now a Josephson junction be acted upon both by parametric excitation and by an external current. The Hamiltonian takes the form

$$\hat{H} = \frac{\hat{Q}^2}{2C} + \frac{\hbar I_c(t)}{2e} \frac{\hat{\varphi}^2}{2} - \frac{\hbar}{2e} (I(t) - I_N) \hat{\varphi}. \tag{17.27}$$

If the Josephson junction is at zero temperature, we have $I_N=0$.

We construct the "annihilation" operator

$$\hat{A}_{\omega, I} = \hat{A}_\omega + \delta(t), \qquad \delta(t) = -i (2\hbar C)^{-1/2} \int_0^t I(\tau) \varepsilon(\tau) \, d\tau, \tag{17.28}$$

where $\hat{A}_\omega$ is the "annihilation" operator (17.18) and ε(t) is a function that satisfies Equations (17.19) and (17.20). We obtain the ground state from the condition $\hat{A}_{\omega,I}\Psi_0=0$ and from the normalization condition:

$$\Psi_0(\varphi, t) = \left(\frac{\hbar C}{\pi}\right)^{1/4} (2e\varepsilon)^{-1/2} \exp\left[\frac{i\hbar C}{4e^2} \frac{\dot{\varepsilon}}{\varepsilon} \frac{\varphi^2}{2} - \right.$$
$$\left. - \frac{\sqrt{2}\,\delta\varphi}{\varepsilon} \left(\frac{\hbar C}{4e^2}\right)^{1/2} - \frac{1}{2} \delta^2 \frac{\varepsilon^*}{\varepsilon} - \frac{|\delta|^2}{2} + \frac{1}{2} \int_0^t (\dot{\delta}\delta^* - \delta\dot{\delta}^*) \, d\tau \right]. \tag{17.29}$$

The entire family of correlated states takes the form

$$\Psi_\alpha(\varphi, t) = \Psi_0(\varphi, t) \exp\left[-\frac{|\alpha|^2}{2} + \frac{\sqrt{2}\,\alpha\varphi}{\varepsilon} \left(\frac{\hbar C}{4e^2}\right)^{1/2} + \right.$$
$$\left. + \frac{\alpha(\delta\varepsilon^* + \delta^*\varepsilon)}{\varepsilon} - \frac{\alpha^2 \varepsilon^*}{2\varepsilon} \right]. \tag{17.30}$$

The states $\Psi_0(\phi,t)$, $\Psi_\alpha(\phi,t)$ satisfy the Schrodinger equation with Hamiltonian (17.27) and minimize the Schrodinger-Robertson uncertainty relation (17.15) with the correlation coefficient (17.22). The current and voltage quantum noises are given by (17.24). Therefore, we see that an external current influences the mean values of the current and voltage in the junction, and transforms the ground state and coherent states of a Josephson junction only into coherent ones. To produce a correlated state of a Josephson junction it is necessary to apply a parametric excitation. If the Josephson junction at the initial time was at temperature $T \neq 0$ ($I_N \neq 0$) and in a state of thermodynamic equilibrium described by a density matrix

$$\hat{\rho}(0) = (2\,\mathrm{sh}\,(\hbar\omega\beta/2))^{-1} \exp[-\hbar\omega\beta(\hat{a}^{+}\hat{a} + {}^{1}/_{2})],$$

where

$$\hat{a} = \frac{1}{\sqrt{2}}\left[\left(\frac{\hbar C\omega}{4e^2}\right)^{1/2}\hat{\varphi} + \frac{i\hat{Q}}{(\hbar C\omega)^{1/2}}\right], \quad \beta = \frac{1}{T},$$

then a parametric excitation and an external current will transform it into a state

$$\hat{\rho}(t) = (2\,\mathrm{sh}\,(\hbar\omega\beta/2))^{-1} \exp[-\hbar\omega\beta(\hat{A}^{+}_{\omega,I}\hat{A}_{\omega,I} + {}^{1}/_{2})].$$

It can be shown that the populations of the stationary levels of a Josephson junction are of the form

$$\rho_{nn} = \rho_{00}(n!)^{-1} H_{nn}^{\{D_{ij}\}}(x_1, x_2),$$

where $H^{(D_{ij})}_{nm}(x_1, x_2)$ is a Hermite polynomial of two variables,

$$D_{ij} = D^{-1}\begin{Vmatrix} u^{*}v(1 - e^{-2\hbar\omega\beta}) & -e^{-\hbar\omega\beta} \\ -e^{-\hbar\omega\beta} & uv^{*}(1 - e^{-2\hbar\omega\beta}) \end{Vmatrix}, \quad x_1 = (1 - e^{-\hbar\omega\beta})\frac{\delta^{*}ue^{-\hbar\omega\beta} + \delta v^{*}}{|u|^2 e^{-2\hbar\omega\beta} - |v|^2}, \quad u = \frac{\omega_0\varepsilon - i\dot{\varepsilon}}{2\sqrt{\omega_0}},$$

$$D = |u|^2 - |v|^2 e^{-2\hbar\omega\beta}, \quad x_2 = (1 - e^{-\hbar\omega\beta})\frac{\delta u^{*}e^{-\hbar\omega\beta} + \delta^{*}v}{|u|^2 e^{-2\hbar\omega\beta} - |v|^2}, \quad v = -\frac{\omega_0\varepsilon + i\dot{\varepsilon}}{2\sqrt{\omega_0}},$$

the population of the ground state is

$$\begin{aligned}\rho_{00} = {} & 2D^{-1/2}\exp(-\hbar\omega\beta/2)\,\mathrm{sh}\,(\hbar\omega\beta/2)\exp\{D^{-1}[{}^{1}/_{2}(\delta^{*2}uv + \delta^{2}u^{*}v^{*}) - \\ & - |\delta|^2|v|^2 + [{}^{1}/_{2}(\delta^{*2}uv + \delta^{2}u^{*}v^{*}) - |\delta|^2|v|^2]e^{-2\hbar\omega\beta} + \\ & + e^{-\hbar\omega\beta}[|\delta|^2(|u|^2 + |v|^2) - (\delta^{*2}uv + \delta^{2}u^{*}v^{*})]]\},\end{aligned}$$

which accords with the oscillator-excitation solution obtained in [1].

The results are valid under the conditions $\delta\phi << 1$, $\delta Q >> 2e$. The first ensures the possibility of replacing Hamiltonian (17.1) by Hamiltonian (17.2), and the second means that the charge quantization is not substantial, i.e., that the shot noise is small compared with the quantum fluctuations. For parameters that are constant in time these inequalities lead to the following restriction,

$$I_c C >> 32e^3/\hbar \approx 10^{-21}\,F\cdot A.$$

It was shown in [9] that for certain classes of Hamiltonians there exist quantities that depend on the initial state of the system and on the form of the commutation relations, but do not depend at all on the coefficients of the corresponding quadratic or linear forms. Such quantities were named *universal invariant*, by analogy with the Poincare-Cartan universal invariants of classical mechanics. For Hamiltonian (17.2) with $I(t)=0$, the universal invariant for a Josephson junction takes the form

$$D = (\langle\hat{\varphi}^2\rangle - \langle\hat{\varphi}\rangle^2)(\langle\hat{Q}^2\rangle - \langle\hat{Q}\rangle^2) - ({}^{1}/_{2}\langle\hat{\varphi}\hat{Q} + \hat{Q}\hat{\varphi}\rangle - \langle\hat{\varphi}\rangle\langle\hat{Q}\rangle)^2. \qquad (17.31)$$

Using the connection of the phase and charge with the current and voltage

$$\hat{\varphi} = -\hat{\mathfrak{J}}/I_c(t), \qquad \hat{Q} = C(t)\hat{V},$$

we can rewrite the universal invariant in the form

$$\begin{aligned}D = \frac{C^2(t)}{I_c^2(t)}[& (\langle\hat{\mathfrak{J}}^2\rangle - \langle\hat{\mathfrak{J}}\rangle^2)(\langle\hat{V}^2\rangle - \langle\hat{V}\rangle^2) - \\ & - ({}^{1}/_{2}\langle\hat{\mathfrak{J}}\hat{V} + \hat{V}\hat{\mathfrak{J}}\rangle - \langle\hat{\mathfrak{J}}\rangle\langle\hat{V}\rangle)^2]. \end{aligned} \qquad (17.32)$$

The universal invariant D is conserved in time. If a Josephson junction was initially in a coherent or correlated state, the value of the universal invariant is

$$D = \hbar^2\omega^4/4.$$

If we consider a universal invariant in the range of variable ϕ for which the term 1-$\cos\hat{\phi}$ in Hamiltonian (17.1) cannot be replaced by $\hat{\phi}^2/2$, we find by expanding $\cos\hat{\phi}$ in a Taylor series that the universal invariant becomes time dependent and satisfies the equation

$$\frac{dD}{dt} = \sum_{k=2}^{\infty} \frac{\hbar^2}{4e^2} \frac{(-1)^{k+1}}{(2k-1)!} [\langle\hat{\varphi}^{2k}\rangle \langle\hat{Q}\hat{\varphi} + \hat{\varphi}\hat{Q}\rangle - \langle\hat{\varphi}^2\rangle \langle\hat{Q}\hat{\varphi}^{2k-1} + \hat{\varphi}^{2k-1}\hat{Q}\rangle]. \tag{17.33}$$

In conclusion we note that for a Josephson junction there exists a correlated state described by function (17.23). It can be produced with the aid of some parametric action: by varying, for example, the critical current of the junction (by irradiating the junction by a laser or by a sound source). The parameters of the correlated state produced in this manner depend on the concrete time dependence of the critical current. The mechanism proposed may be able to control the current and voltage noise in a junction, and this may turn out to be important for measurements at sensitivities close to a quantum limit.

18. MATHEMATICAL RELATIONS IN JOSEPHSON JUNCTION THEORY

Let us discuss some mathematical relations used in studying the Josephson junction. So, the expressions (17.25) and (17.26) describe the transition probability between the stationary states of the junction due to the parametric excitation. In this case, the argument of the Legendre polynomial in the formula (17.25) is determined by the probability to be in the ground state W_0 and it has the form

$$W_0 = (E + {}^1/_2)^{-1/2}. \tag{18.1}$$

At the same time, considering the problem of the parametric excitation of the quantum harmonic oscillator, the analogous expression for the probability of the oscillator after the parametric excitation to be in the ground state may be obtained in the form

$$W_0 = \cos(\theta/2), \tag{18.2}$$

where the parameter

$$R = \sin(\theta/2) \tag{18.3}$$

has the meaning of the reflection coefficient by a potential barrier in the following problem. Let us have the oscillator with varying frequency $\Omega(t)$ and for the remote past the frequency $\Omega \to \Omega_{in}$ where Ω_{in} is a constant one. Then, replacing the time t by the coordinate x and constructing the function U(x)–potentially satisfying the condition

$$\Omega^2(x) = \Omega_{in}^2 + 2M[U(-\infty) - U(x)], \tag{18.4}$$

where M is the mass of the particle, one can formulate the following rule. Let the particle with the mass M be scattered by the potential U(x) given by the formula (18.4) and let this particle have the energy

$$\varepsilon = \Omega_{in}^2/2M + U(-\infty). \tag{18.5}$$

Then the parameter R given by the formula (18.3) is the reflection coefficient for the particle with mass M and energy ε by the potential barrier U(x) determined by the varying oscillator frequency in accordance with formula (18.4). On the other hand, the physical sense of parameter E in Equation (18.1) is also known. It is dimensionless energy which the classical oscillator receives due to the parametric excitation, i.e., due to the time-dependence of the function $\Omega(t)$. Hence, these two different problems are related. The reflection coefficient of the quantum-mechanical particle by the potential barrier R and the energy E which is received by the classical parametric oscillator due to the parametric excitation are connected. We have

$$R = (E - 1/2)/(E + 1/2) \tag{18.6}$$

and

$$E = 1/2\,(1 + R)/(1 - R). \tag{18.7}$$

Thus in classical mechanics one can model the calculations of the reflection coefficient of the quantum particle by the potential barrier. To do this, it is necessary to measure the energy which is received by the paramtric oscillator with the frequency which is determined by the form of the potential barrier in accordance with Equation (18.4). Studying the probability of Josephson junction to be in the ground state after the parametric excitation, one can conclude what energy is received by the classical oscillator and what is the reflection coefficient by the correspondong potential barrier. Let us write some formulae reducing the expressions for the exponents of the operators squared to the integrals for the exponents by the first powers of these operators. To do this let us use the known Poisson integral

$$\int_{-\infty}^{\infty} e^{-ax^2+bx}dx = \left(\frac{\pi}{a}\right)^{1/2} e^{b^2/4a}, \tag{18.8}$$

which was applied for normalization of the wave functions of the resonant circuit and Josephson junction. Rewriting it in the form

$$e^{b^2/4a} = \left(\frac{a}{\pi}\right)^{1/2} \int_{-\infty}^{\infty} e^{-ax^2+bx}dx \tag{18.9}$$

and replacing now the variable b by an operator $\hat{b}$ and considering the variable α as the c-number we have

$$e^{\hat{b}^2/4a} = \left(\frac{a}{\pi}\right)^{1/2} \int_{-\infty}^{\infty} e^{-ax^2+\hat{b}x}\,dx. \tag{18.10}$$

The operator $\hat{b}$ may be an arbitrary finite- or infinite-dimensional matrix. For example, if we take as the operator $\hat{b}$ the operator $\Delta=\nabla^2$ and $\alpha=1/4$ the exponent of the Laplacean may be presented in the form of the integral of the shift operator

$$\exp(\Delta) = \frac{1}{2\sqrt{\pi}} \iiint_{-\infty}^{\infty} \exp(-\mathrm{u}^2/4 + \mathrm{u}\nabla)\,du. \tag{18.11}$$

The formulae of the type (18.10) are convenient when one multiplies the exponents of the operators squared. So, one can multiply the following exponents

$$e^{\mu x^2} e^{\nu a^2/ax^2} = \exp\{Q[\mu x^2 + \nu a^2/ax^2 - \mu\nu(xa/ax + a/ax\,x)]\},$$
$$Q = \frac{\ln(1 - 2\mu\nu + 2(\mu\nu(\mu\nu - 1))^{1/2})}{2(\mu\nu(\mu\nu-1))^{1/2}} \tag{18.12}$$

In this representation if operator $\hat{b}$ is a generator of the Lie group representation, the formula of the type (18.10) gives the possibility to express the exponent of such operator squared in terms of the integral of the operator of finite rotation. Let us write the representation of the inverse operator based on the expression for the inverse operator in the form of the integral

$$A^{-1} = \int_0^\infty e^{-\beta A}\, d\beta. \tag{18.13}$$

Of course the written formulae are correct when the integrals have the sense.

19. CLASSICAL STRING

Let us consider in this section a one-dimensional string of lengh L-2π. The classical equation of motion reads

$$c^{-2}\partial^2 u/\partial t^2 = \partial^2 u/\partial x^2. \tag{19.1}$$

u(x,t) is the dimensionless shift of the string from the equilibrium position at the space-time point (x,t), c is the velocity of vibrations propagation along the string. Equation (19.1) can be derived from the extremum condition of the action functional

$$S = \int dt \int dx \,\{{}^1/_2\,[(\partial u/\partial t)^2 - c^2\,(\partial u/\partial x)^2]\}. \tag{19.2}$$

The integration over time is performed from t_1 to t_2, and over space coordinate – from 0 to 2π.

The Lagrangian is given by the integral

$$\mathcal{L} = \int dx\, {}^1/_2\,[(\partial u/\partial t)^2 - c^2\,(\partial u/\partial x)^2]. \tag{19.3}$$

The corresponding Hamiltonian is of the form

$$H = \int dx\, {}^1/_2\,[(\partial u/\partial t)^2 + c^2\,(\partial u/\partial x)^2]. \tag{19.4}$$

If one introduces new variables in equation (19.1), expanding function u(x,t) into Fourier series

$$u\,(x,t) = \sum_k u_k\,(t)\, e^{ikt}, \quad u_k\,(t) = \int \frac{dx}{2\pi}\, e^{-ikx} u\,(x,t), \tag{19.5}$$

where the *integer* wave number k changes from -∞ to +∞, then partial differential equation (19.1) turns into the system of linear ordinary differential equations for complex time-dependent Fourier amplitudes. These equations are the equations of motion of independent oscillators

$$\ddot{u}_k\,(t) + \omega_k^2 u_k\,(t) = 0, \tag{19.6}$$

where the frequency of the k-th oscillator is

$$\omega_k^2 = c^2 k^2. \tag{19.7}$$

Equation (19.6) is derived from the Langrangian

$$\mathcal{L}_k = {}^1/_2\,(\dot{u}_k^2 - \omega_k^2 u_k^2). \tag{19.8}$$

The string Hamiltonian in new variables assumes the form

$$H = \frac{1}{2}\sum_k [|\,p_k\,|^2 + \omega_k^2\,|\,u_k\,|^2], \tag{19.9}$$

where the momentum p_k is connected with the Fourier amplitude as follows:

$$p_k = \dot{u}_k. \tag{19.10}$$

The complex Fourier expansion (19.5) corresponds to the periodical condition u(x+2π, t)=u(x,t). If one considers a string with fixed ends, then the Fourier expansion contains only sine functions. Then one has a single k-oscillator (only the imaginary part of $u_k(t)$) instead of two k-oscillators (the real and imaginary parts of amplitude).

20. THE PARAMETRIC OSCILLATOR CHAIN

It is known that the problem of a string can be considered as a limit of the problem of a chain consisting of atoms-oscillators interacting via a quadratic potential. Having in mind the aim to investigate the string with time-dependent parameters, let us consider the parametrically excited chain. We have N atoms-oscillators with the Hamiltonian

$$H = \sum_{n=1}^{N} \frac{1}{2} [p_n^2 + \Omega^{2(t)} (q_n - q_{n+1})^2], \tag{20.1}$$

which corresponds to the equation of motion

$$\ddot{q}_n = \Omega^2 (t) (q_{n+1} + q_{n-1} - 2q_n), \tag{20.2}$$

The frequency Ω is the function of time, which corresponds to the time-dependent elasticity of the chain. The Lagrangian in this case is

$$\mathcal{L} = \frac{1}{2} \sum_{2n=1}^{N} [\dot{q}_n^2 - \Omega^2 (t) (q_n - q_{n+1})^2], \tag{20.3}$$

and functional action reads

$$S = {}^1\!/_2 \int [\dot{q}_n^2 - \Omega^2 (t) (q_n - q_{n+1})^2]\, dt. \tag{20.4}$$

The momentum coincides with the velocity, q_n is the shift from the equilibrium point of the n-th oscillator (chain's atom). We use the periodical condition, i.e., the chain is supposed to be closed, so that $q_{i+n}=q_i$, i=1,2,...,N. For simplicity we will consider the number N odd.

Let us introduce the normal coordinates (perform a canonical symplectic transformation)

$$q_n = \sum_m q_m^{\text{H}} \frac{1}{\sqrt{N}} \exp\left(imn \frac{2\pi}{N}\right), \quad p_n = \sum_m p_m^{\text{H}} \frac{1}{\sqrt{N}} \exp\left(-imn \frac{2\pi}{N}\right),$$

$$m = -(N-1)/2, \ldots, 0, \ldots, (N-1)/2. \tag{20.5}$$

The inverse transformation is

$$q_m^{\text{H}} = \sum_{n=1}^{N} \frac{q_n}{\sqrt{N}} \exp\left(-imn \frac{2\pi}{N}\right), \quad p_m^{\text{H}} = \sum_{n=1}^{N} \frac{p_n}{\sqrt{N}} \exp\left(imn \frac{2\pi}{N}\right). \tag{20.6}$$

The case m=0 corresponds to the oscillation with zero frequency (the uniform motion of all chain with a constant velocity). Normal coordinates oscillate independently, satisfying the equations of motion

$$\ddot{q}_m^{\text{H}} + \omega_m^2 (t) q_m^{\text{H}} = 0, \qquad \omega_m^2 = 4\Omega^2 (t) \sin^2 (m\pi/N). \tag{20.7}$$

The Hamiltonian expressed in terms of normal coordinates reads

$$H = \sum_m \left[\overset{H}{p}{}_m^{*} \overset{H}{p}_m + \Omega^2(t) \sin^2\left(\frac{m\pi}{N}\right) \overset{H}{q}{}_m^{*} \overset{H}{q}_m \right]. \tag{20.8}$$

The equations of motion can be written in the form of a single equation of motion for 2N-vector (p,q)=X

$$\frac{d\mathbf{X}}{dt} = B\mathbf{X}, \qquad B = \begin{Vmatrix} b_1 & b_2 \\ b_3 & b_4 \end{Vmatrix}. \tag{20.9}$$

In this case, rectangular matrices b_1 and b_4 are equal to zero, the N-dimensional matrix b_3 is equal to unity, and $b_2=-\Omega^2 A$.

The canonical transformation to normal coordinates is a contact transformation (new coordinates are expressed in terms of old ones without any admixture of momenta). From the group theoretical point of view it is nothing but picking out the irreducible representations of the permutation group of oscillators in the chain. This is the group of symmetry of equations of motion (20.2), which is clear from the examples considered below (see, e.g., (21.1) and (21.9). It consists from the following N elements:

$$C_N, C_N^2, \ldots, C_N^N = E. \tag{20.10}$$

Here C_N is the rotation by the angle $2\pi/N$, or the permutation (1,2,...,N-1,N)→(2,3,...,N,1). Other elements are the powers of this transformation. E is the unity element. All the N irreducible representations of the given abelian finite group are unidimensional. They are constructed by the following method. Let us take some root of the N-th degree from the unity, for example the number $k=\exp(2\pi it/N)$. Then all the irreducible representations are described by the set of the characters represented by the matrix

$$K = \frac{1}{\sqrt{N}} \begin{Vmatrix} k & k^2 & \ldots & k^N = 1 \\ k^2 & (k^2)^2 & \ldots & (k^2)^N = 1 \\ \ldots & \ldots & \ldots & \ldots \\ k^m & (k^m)^2 & \ldots & (k^m)^N = 1 \\ \ldots & \ldots & \ldots & \ldots \\ 1 & 1 & \ldots & 1 \end{Vmatrix}. \tag{20.11}$$

The equation of motion (20.2) is split for the variables $q^H{}_s$ which are constructed according to the rule: one takes the set of the variables q_n, multiply them by the coordinates from the s-th row of the character matrix $(k^s)^n$, and forms the sum over index n taking into account the normalization condition

$$\overset{H}{q}_s = \sum_{n=1}^{N} (k^s)^n q_n N^{-1/2}, \qquad s = 1, 2, \ldots, N, \tag{20.12}$$

$$q_n = \sum_{s=1}^{N} (k^{-n})^s \overset{H}{q}_s N^{-1/2}. \tag{20.13}$$

But one can renormalize these normal coordinates by another method. Since there are pairs of the complex conjugate rows in the matrix of the characters corresponding to the powers of the rotation by the angle $(-2\pi/N)$, we can use the index m in formula (20.6) running another set of N integers: from -(N-1)/2 to (N-1)/2 for odd values of N, and from -N/2+1 to N/2-1 for even N. For the sake of convenience, we assume the odd numbers of atoms in the chain. For even numbers N it is necessary to separate another vibrational mode realizing the antisymmetrical representation of the group of cyclic permutations which is absent for the odd number of atoms. The number s in Equation (20.13) is equal to N/2 for this mode. Other modes are described by the formulae

$$Q_m^H = \sum_{n=1}^{N} \frac{q_n}{\sqrt{N}} \exp\left(-imn\frac{2\pi}{N}\right), \quad m = \overline{-\frac{N}{2}+1, \frac{N}{2}-1},$$

$$P_m^H = \sum_{n=1}^{N} \frac{p_n}{\sqrt{N}} \exp\left(imn\frac{2\pi}{N}\right). \tag{20.14}$$

The frequency is equal to zero in the case of a symmetrical oscillation with the coordinate $Q^H{}_N$ describing the uniform motion of the whole system with constant velocity, because $\dot{Q}_N^H = 0$, which is seen from the sysem of equations (20.2). Coordinates $Q^H{}_m$ and $Q^H{}_{-m}$ are complex conjugated

$$Q_m^{H*} = Q_{-m}^H. \tag{20.15}$$

The same is true for momenta

$$P_{-m}^H = P_m^{H*}. \tag{20.16}$$

The solution of classical equations of motion (20.9) in the case of time-independent frequency Ω can be written in the form (N is odd):

$$q_n(t) = \frac{1}{N}\sum_{n'=1}^{N}\sum_{m=-(N-1)/2}^{(N-1)/2}\left\{q_{n'}(0)\cos\left[\frac{2\pi}{N}m(n-n') - \omega_n t\right] + \right.$$

$$\left. + \frac{\dot{q}_{n'}(0)}{\omega_m}\sin\left[\frac{2\pi}{N}m(n-n') - \omega_m t\right]\right\}; \tag{20.17}$$

$$p_n(t) = \dot{q}_n(t). \tag{20.18}$$

Let us rewrite the solution of classical equations of motion in the matrix form for vector

$X(t)=\begin{pmatrix} p(t) \\ q(t) \end{pmatrix}$:

$$\mathbf{X}(t) = \Lambda^{-1}(t)\,\mathbf{X}(0), \tag{20.19}$$

where $\Lambda^{-1}(t)=\Lambda(-t)$, and 2Nx2N matrix $\Lambda(t)$ is divided into N-dimensional blocks

$$\Lambda = \begin{Vmatrix} \lambda_1 & \lambda_2 \\ \lambda_3 & \lambda_4 \end{Vmatrix}. \tag{20.20}$$

Using formulae (20.11), one can write transformation (20.12) in the matrix form

$$\begin{pmatrix} \mathbf{P}^H \\ \mathbf{Q}^H \end{pmatrix} = \begin{Vmatrix} K^{-1} & 0 \\ 0 & K \end{Vmatrix} \begin{pmatrix} \mathbf{p} \\ \mathbf{q} \end{pmatrix}.$$

Since

$$\frac{d}{dt}\begin{pmatrix} \mathbf{p} \\ \mathbf{q} \end{pmatrix} = \begin{pmatrix} -\Omega^2 A\mathbf{q} \\ \mathbf{p} \end{pmatrix} = \begin{Vmatrix} 0 & -\Omega^2 A \\ 1 & 0 \end{Vmatrix} \begin{pmatrix} \mathbf{p} \\ \mathbf{q} \end{pmatrix}, \tag{20.21}$$

multiplying this equation from the left by matrix

$$\begin{vmatrix} k^{-1} & 0 \\ 0 & K \end{vmatrix}$$

and differentiating with respect to time one gets

$$\frac{d}{dt}\begin{pmatrix} P^H(t) \\ Q^H(t) \end{pmatrix} = \begin{Vmatrix} 0 & -\Omega^2 K^{-1} A K^{-1} \\ K^2 & 0 \end{Vmatrix} \begin{pmatrix} P^H \\ Q^H \end{pmatrix} = \chi \begin{pmatrix} P^H \\ Q^H \end{pmatrix}. \tag{20.22}$$

The solution of this equation reads

$$\begin{pmatrix} P^H(t) \\ Q^H(t) \end{pmatrix} = \exp(t\chi) \begin{pmatrix} P^H(0) \\ Q^H(0) \end{pmatrix}, \tag{20.23}$$

where

$$\chi = \begin{Vmatrix} 0 & -K^{-2}\Omega^2 D \\ K^2 & 0 \end{Vmatrix}, \qquad D = KAK^{-1}.$$

Expanding the exponential function in Taylor's series

$$e^\alpha = \sum_{n=0}^{\infty}\left(1 + \alpha + \frac{\alpha^2}{2} + \ldots + \frac{\alpha^n}{n!}\right) = \sum_{n=0}^{\infty}\frac{\alpha^{2n}}{(2n)!} + \sum_{n=0}^{\infty}\frac{\alpha^{2n+1}}{(2n+1)!}$$

and taking into account the similar expansions of functions $\cos\alpha = \sum_{n=0}^{\infty}(-1)^n \frac{\alpha^{2n}}{(2n)!}$ and $\sin\alpha = \sum_{n=0}^{\infty}(-1)^n \frac{\alpha^{2n+1}}{(2n+1)!}$ one can get

$$\begin{pmatrix} P^H(t) \\ Q^H(t) \end{pmatrix} = \begin{Vmatrix} K^{-2}\cos(\Omega\sqrt{D}\,t)\,K^2 & -K^{-2}\Omega\sqrt{D}\sin(\Omega\sqrt{D}\,t) \\ K^2\dfrac{\sin(\Omega\sqrt{D}\,t)}{\Omega\sqrt{D}} & \cos(\Omega\sqrt{D}\,t) \end{Vmatrix} \begin{pmatrix} P^H(0) \\ Q^H(0) \end{pmatrix}, \tag{20.24}$$

Using the transformation inverse to (20.12) and (20.20)

$$\begin{pmatrix} \mathbf{p}(t) \\ \mathbf{q}(t) \end{pmatrix} = \begin{Vmatrix} K & 0 \\ 0 & K^{-1} \end{Vmatrix} \exp(\chi t) \begin{Vmatrix} K^{-1} & 0 \\ 0 & K \end{Vmatrix} \begin{pmatrix} \mathbf{p}(0) \\ \mathbf{q}(0) \end{pmatrix}, \tag{20.25}$$

one gets

$$\lambda_1 = \lambda_4 = K^{-1}\cos(\Omega\sqrt{D}\,t)\,K, \qquad \lambda_2 = K^{-1}\Omega\sqrt{D}\sin(\Omega\sqrt{D}\,t)\,K,$$
$$\lambda_3 = -K(\Omega\sqrt{D})^{-1}\sin(\Omega\sqrt{D}\,t)\,K^{-1}. \tag{20.26}$$

21. THE CHAIN OF TWO AND THREE ATOMS

Let us consider two simplest cases N=2, N=3. In the case of N=2 the equations of motion for two atoms-oscillators with coordinates q_1 and q_2, $q = \begin{pmatrix} q_1 \\ q_2 \end{pmatrix}$ are of the form

$$\ddot{\mathbf{q}} = 2\Omega^2 \begin{Vmatrix} -1 & 1 \\ 1 & -1 \end{Vmatrix} \mathbf{q}. \tag{21.1}$$

The matrix of the canonical transformation operator

$$K = \begin{Vmatrix} k & k^2 \\ k^2 & k^4 \end{Vmatrix}$$

is equal to

$$K = \frac{1}{\sqrt{2}} \begin{Vmatrix} -1 & 1 \\ 1 & 1 \end{Vmatrix}, \qquad K^* = K = K^{-1}, \tag{21.2}$$

because of $k=e^{\pi i}$.

We have two normal coordinates: the antisymmetrical oscillation with coordinate $Q^H{}_1=(q_2-q_1)/2^{1/2}$ and symmetrical oscillation with coordinate $Q^H{}_1=(q_2+q_1)/2^{1/2}$. Canonical coordinates satisfy the equations

$$\ddot{Q}_1^{\text{H}} = -4\Omega^2 Q_1^{\text{H}}, \qquad \ddot{Q}_2^{\text{H}} = 0. \tag{21.3}$$

The solutions of these equations are as follows:

$$Q_1^{\text{H}}(t) = {}^1/_2\Omega^{-1}\sin(2\Omega t)\,\dot{Q}_1^{\text{H}}(0) + Q_1^{\text{H}}\cos(2\Omega t),$$
$$Q_2^{\text{H}}(t) = \dot{Q}_2^{\text{H}}(0)\,t + Q_2^{\text{H}}(0). \tag{21.4}$$

Using the relations between the coordinates and momenta

$$\mathbf{p}(t) = \dot{\mathbf{q}}(t), \quad \mathbf{q}(t) = K^{-1}\mathbf{Q}^{\text{H}}(t), \qquad \mathbf{Q}^{\text{H}}(0) = K\mathbf{q}(0), \tag{21.5}$$

one has for the coordinates

$$q_1(t) = {}^1/_2\,(\dot{q}_2(0)\,t + \dot{q}_1(0)\,t + q_2(0) + q_1(0) -$$
$$- {}^1/_2\Omega^{-1}\sin(2\Omega t)\,(\dot{q}_2(0) - \dot{q}_1(0)) - \cos(2\Omega t)\,(q_2(0) - q_1(0))),$$

$$q_2(t) = {}^1/_2\,(\dot{q}_2(0)\,t + \dot{q}_1(0)\,t + q_2(0) + q_1(0) +$$
$$+ {}^1/_2\Omega^{-1}\sin(2\Omega t)\,(\dot{q}_2(0) - \dot{q}_1(0)) + \cos(2\Omega t)\,(q_2(0) - q_1(0))) \tag{21.6}$$

and for the momenta

$$\dot{q}_1(t) = {}^1/_2\,(\dot{q}_2(0) + \dot{q}_1(0) - \cos(2\Omega t)\,(\dot{q}_2(0) - \dot{q}_1(0)) + 2\Omega\,(q_2(0) -$$
$$- q_1(0))\sin(2\Omega t)),$$
$$\dot{q}_2(t) = {}^1/_2\,(\dot{q}_2(0) + \dot{q}_1(0) + (\dot{q}_2(0) - \dot{q}_1(0))\cos(2\Omega t) - 2\Omega\,(q_2(0) -$$
$$- q_1(0))\sin(2\Omega t)). \tag{21.7}$$

Let us write Λ-matrix for N=2

$$\Lambda = \frac{1}{2}\begin{Vmatrix} 1+\cos(2\Omega t) & 1-\cos(2\Omega t) & 2\Omega\sin(2\Omega t) & -2\Omega\sin(2\Omega t) \\ 1-\cos(2\Omega t) & 1+\cos(2\Omega t) & -2\Omega\sin(2\Omega t) & 2\Omega\sin(2\Omega t) \\ -t-\dfrac{\sin(2\Omega)t}{2\Omega} & -t+\dfrac{\sin(2\Omega t)}{2\Omega} & 1+\cos(2\Omega t) & 1-\cos(2\Omega t) \\ -t+\dfrac{\sin(2\Omega t)}{2\Omega} & -t-\dfrac{\sin(2\Omega t)}{2\Omega} & 1-\cos(2\Omega t) & 1+\cos(2\Omega t) \end{Vmatrix}. \tag{21.8}$$

In the case of N=3 the equations of motion for three atoms-oscillators with coordinates

$q_1, q_2, q_3, \mathbf{q}=\begin{pmatrix} q_1 \\ q_2 \\ q_3 \end{pmatrix}$ assume the form

$$\ddot{\mathbf{q}} = -\Omega^2\begin{Vmatrix} 2 & -1 & -1 \\ -1 & 2 & -1 \\ -1 & -1 & 2 \end{Vmatrix}\mathbf{q}. \tag{21.9}$$

The matrix of canonical transformation operator K is equal to

$$K = \frac{1}{\sqrt{3}}\begin{Vmatrix} e^{2\pi i/3} & e^{4\pi i/3} & 1 \\ e^{4\pi i/3} & e^{2\pi i/3} & 1 \\ 1 & 1 & 1 \end{Vmatrix}. \tag{21.10}$$

Let us perform the canonical transformation to complex normal coordinates $Q^H = Kq$, $P^H = K^*p$. Hamiltonian (20.1) in the case of N=3 rewritten in terms of normal complex coordinates reads

$$H = P_1^H P_2^H + Q_1^H Q_2^H + {}^1/_2 (P_3^H)^2 .$$

Canonical coordinates Q^H satisfy the equations

$$\ddot{Q}_1^H = -3\Omega^2 Q_1^H, \quad \ddot{Q}_2^H = -3\Omega^2 Q_2^H, \quad \ddot{Q}_3^H = 0. \tag{21.11}$$

One can write the solutions of these equations

$$\begin{aligned} Q_1^H(t) &= \exp(i\Omega\sqrt{3}\,t)\left[\frac{Q_1^H(0)}{2} - \frac{i}{2\Omega\sqrt{3}}\dot{Q}_1^H(0)\right] + \\ &+ \exp(-i\Omega\sqrt{3}\,t)\left[\frac{Q_1^H(0)}{2} + \frac{i}{2\Omega\sqrt{3}}\dot{Q}_1^H(0)\right], \\ Q_2^H(t) &= Q_1^{H*}(t), \quad Q_3^H(t) = \dot{Q}_3^H(0)\,t + Q_3^H(0). \end{aligned} \tag{21.12}$$

Using relations (21.5), one can get for the coordinates

$$\begin{aligned} q_1(t) &= {}^1/_3\Big\{(\dot{q}_2(0) + \dot{q}_3(0))\Big(t + \frac{2\sin(\Omega\sqrt{3}\,t)}{\Omega\sqrt{3}}\Big)\cos\frac{2\pi}{3}\Big) + \\ &+ \dot{q}_1(0)\Big(t + \frac{2\sin(\Omega\sqrt{3}\,t)}{\Omega\sqrt{3}}\Big) + q_1(0)(1 + 2\cos(\Omega\sqrt{3}\,t)) + \\ &+ (q_2(0) + q_3(0))(1 + 2\cos(\Omega\sqrt{3}\,t)\cos(2\pi/3))\Big\}, \\ q_2(t) &= {}^1/_3\Big\{(\dot{q}_1(0) + \dot{q}_3(0))\Big(t + 2\cos\frac{2\pi}{3}\,\frac{\sin(\Omega\sqrt{3}\,t)}{\Omega\sqrt{3}}\Big) + \\ &+ \dot{q}_2(0)\Big(t + \frac{2\sin(\Omega\sqrt{3}\,t)}{\Omega\sqrt{3}}\Big) + q_2(0)(1 + 2\cos(\Omega\sqrt{3}\,t)) + \\ &+ (q_1(0) + q_3(0))(1 + 2\cos(\Omega\sqrt{3}t)\cos(2\pi/3))\Big\}, \\ q_3(t) &= {}^1/_3\Big\{(q_1(0) + q_2(0))(1 + 2\cos(\Omega\sqrt{3}\,t)\cos(2\pi/3)) + \\ &+ q_3(0)(1 + 2\cos(\Omega\sqrt{3}\,t)) + (\dot{q}_1(0) + \dot{q}_2(0)) \times \\ &\times \Big(t + 2\frac{\sin(\Omega\sqrt{3}\,t)}{\Omega\sqrt{3}}\cos\frac{2\pi}{3}\Big) + \dot{q}_3(0)\Big(t + \frac{2\sin(\Omega\sqrt{3}\,t)}{\Omega\sqrt{3}}\Big)\Big\} \end{aligned} \tag{21.13}$$

and for the momenta

$$\begin{aligned} \dot{q}_1(t) &= {}^1/_3\{\dot{q}_1(0)(1 + 2\cos(\Omega\sqrt{3}t)) + (\dot{q}_2(0) + \dot{q}_3(0)) \times \\ &\times (1 + 2\cos(2\pi/3)\cos(\Omega\sqrt{3}\,t)) - 2\Omega\sqrt{3}\sin(\Omega\sqrt{3}\,t)\,q_1(0) - \\ &- 2\Omega\sin(\Omega\sqrt{3}\,t)\cos(2\pi/3)(q_3(0) + q_2(0))\}, \end{aligned}$$

$$\begin{aligned} \dot{q}_2(t) &= {}^1/_3\{-2\Omega\sqrt{3}\sin(\sqrt{3}\,\Omega t)\cos(2\pi/3)(q_1(0) + q_2(0)) - \\ &- 2\Omega\sqrt{3}\sin(\Omega\sqrt{3}\,t)\,q_2(0) + (\dot{q}_1(0) + \dot{q}_3(0))(1 + 2\cos(\Omega\sqrt{3}\,t) \times \\ &\times \cos(2\pi/3)) + \dot{q}_2(0)(1 + 2\cos(\Omega\sqrt{3}\,t))\}, \end{aligned}$$

$$\begin{aligned} \dot{q}_3(t) &= {}^1/_3\{-2\Omega\sqrt{3}\sin(\Omega\sqrt{3}\,t)\cos(2\pi/3)(q_1(0) + q_2(0)) - \\ &- 2\Omega\sqrt{3}\sin(\Omega\sqrt{3}\,t)\,q_3(0) + (\dot{q}_1(0) + \dot{q}_2(0))(1 + 2\cos(\Omega\sqrt{3}\,t) \times \\ &\times \cos(2\pi/3)) + \dot{q}_3(0)(1 + 2\cos(\Omega\sqrt{3}\,t))\}. \end{aligned} \tag{21.14}$$

Taking into consideration that $X(t)=\begin{pmatrix} p(t) \\ q(t) \end{pmatrix}$, one gets Λ-matrix

$$\lambda_1 = \lambda_4 = \\ = \frac{1}{3}\begin{Vmatrix} 1+2\cos(\Omega\sqrt{3}t) & 1+2\cos(\Omega\sqrt{3}t)\cos\left(\frac{2\pi}{3}\right) & 1+2\cos(\Omega\sqrt{3}t)\cos\left(\frac{2\pi}{3}\right) \\ 1+2\cos(\Omega\sqrt{3}t)\cos\left(\frac{2\pi}{3}\right) & 1+2\cos(\Omega\sqrt{3}t) & 1+2\cos(\Omega\sqrt{3}t)\cos\left(\frac{2\pi}{3}\right) \\ 1+2\cos(\Omega\sqrt{3}t)\cos\left(\frac{2\pi}{3}\right) & 1+2\cos(\Omega\sqrt{3}t)\cos\left(\frac{2\pi}{3}\right) & 1+2\cos(\Omega\sqrt{3}t) \end{Vmatrix},$$

$$\lambda_3 = \\ = -\frac{1}{3}\begin{Vmatrix} t+\frac{2\sin(\Omega\sqrt{3}t)}{\Omega\sqrt{3}} & t+\frac{2\sin(\Omega\sqrt{3}t)}{\Omega\sqrt{3}}\cos\frac{2\pi}{3} & t+\frac{2\sin(\Omega\sqrt{3}t)}{\Omega\sqrt{3}}\cos\frac{2\pi}{3} \\ t+\frac{2\sin(\Omega\sqrt{3}t)}{\Omega\sqrt{3}}\cos\frac{2\pi}{3} & t+\frac{2\sin(\Omega\sqrt{3}t)}{\Omega\sqrt{3}} & t+\frac{2\sin(\Omega\sqrt{3}t)}{\Omega\sqrt{3}}\cos\frac{2\pi}{3} \\ t+\frac{2\sin(\Omega\sqrt{3}t)}{\Omega\sqrt{3}}\cos\frac{2\pi}{3} & t+\frac{2\sin(\Omega\sqrt{3}t)}{\Omega\sqrt{3}}\cos\frac{2\pi}{3} & t+\frac{2\sin(\Omega\sqrt{3}t)}{\Omega\sqrt{3}} \end{Vmatrix},$$

$$\lambda_2 = \frac{2\Omega\sqrt{3}}{3}\sin(\Omega\sqrt{3}t)\begin{Vmatrix} 1 & \cos(2\pi/3) & \cos(2\pi/3) \\ \cos(2\pi/3) & 1 & \cos(2\pi/3) \\ \cos(2\pi/3) & \cos(2\pi/3) & 1 \end{Vmatrix}. \tag{21.15}$$

22. DISCRETE QUANTUM STRING (OSCILLATOR CHAIN)

Let us proceed to the quantum consideration of an oscillator chain. Consider the set of N quantum one-dimensional oscillators (in the Schrodinger picture) described by Hamiltonian

$$\hat{H} = \sum_{n=1}^{N}\left[\frac{\hat{p}_n^2}{2m} + \frac{m\Omega^2}{2}(\hat{q}_{n+1} - \hat{q}_n)^2\right]; \tag{22.1}$$

where $\hat{p}_n$–momentum operator of n-th oscillator, $\hat{q}_n$–the coordinate operator. Let us introduce the corresponding dimensionless operators

$$\underset{\sim}{\hat{p}}_n = \hat{p}_n(\hbar m\Omega)^{-1/2}, \qquad \underset{\sim}{\hat{q}}_n = \hat{q}_n(\hbar/m\Omega)^{-1/2}. \tag{22.2}$$

Then mass and frequency disappear, so that Hamiltonian assumes the form (we omit the tilde symbol later on for simplicity)

$$\hat{H} = \sum_{n=1}^{N}\left(\frac{\hat{p}_n^2}{2} + \frac{1}{2}[\hat{q}_{n+1} - \hat{q}_n]^2\right). \tag{22.3}$$

The operators of dimensionless coordinates and momenta $\hat{q}_n, \hat{p}_n$ satisfy the commutation relations

$$[\hat{p}_n, \hat{q}_m] = -i\delta_{mn}. \tag{22.4}$$

Let us construct the integrals of motion of discrete string, following the usual scheme [1]. Since these integrals are linear forms of coordinate and momentum operators, they can be represented as follows:

$$\hat{\mathbf{I}} = \begin{pmatrix} \hat{\mathbf{P}}_0(t) \\ \hat{\mathbf{Q}}_0(t) \end{pmatrix} = \Lambda(t)\hat{\mathbf{X}}, \tag{22.5}$$

where $\hat{I}(t), \hat{X}(t)$ are 2N-vectors with the components

$$\hat{\mathbf{I}} = (\hat{P}_{10}(t), \ldots, \hat{P}_{n0}(t), \hat{Q}_{10}(t), \ldots, \hat{Q}_{n0}(t)),$$
$$\hat{\mathbf{X}} = (\hat{p}_1, \ldots, \hat{p}_n, \hat{q}_1, \ldots, \hat{q}_n).$$

Using the unitary symmetrical matrix K of the characters of the cyclic permutations group of N numbers with the matrix elements $K_{mn}=N^{-1/2}k_{mn}$, $n,m=\overline{1,N}$, where $k=\exp(2\pi i/N)$ as the matrix of the linear canonical transformation of the coordinates and momenta, one can reduce the system of equations for the operators – integrals of the motion of the set of N systems for the integrals of the motion of the one-dimensional oscillators with dimensionless frequencies $\omega_s=2\sin(\pi s/N)\Omega(t)$; $s=\overline{1,N}$. In this case matrix $\Lambda(t)$ is described by formulae (20.20) and (20.26).

Let us introduce the Green function of the discrete string $G(Q,Q')$ where $Q(q_1, q_2,\ldots,q_N)$, $Q'=(q'_1, q'_2,\ldots,q'_N)$. Solving the eigenvalue problem

$$Q_{n0}G = Q'_{n0}G, \qquad P_{n0}G = i\frac{\partial}{\partial Q'_{n0}}G, \tag{22.6}$$

we obtain the explicit expression for the Green function of the discrete string in the coordinate representation corresponding to the general expression for the quadratic systems given. For example, in study [1]

$$G(\mathbf{Q}, \mathbf{Q}', t) = \left(\prod_{s=1}^{N}\left(\frac{\omega_s}{2\pi i \sin(\omega_s t)}\right)^{1/2}\right) \times$$
$$\times \exp\left\{\frac{i}{2N}\sum_{n=1}^{N}\sum_{m=1}^{N}\sum_{s=1}^{N}\Big[(q'_n k^{(n-m)s} q'_m + q_n k^{(n-m)s} q_m)\,\omega_s \operatorname{ctg}\omega_s t -\right.$$
$$\left.- 2q_n k^{(n-m)s} q'_m \omega_s/\sin(\omega_s t)\Big]\right\},$$
$$\omega_s = 2\Omega(t)\sin(\pi s/N), \quad s=\overline{1,N}, \quad k=\exp(2\pi i/N). \tag{22.7}$$

We have taken into consideration the relations

$$(\lambda_3^{-1}\lambda_4)_{nm} = (\lambda_1\lambda_3^{-1})_{nm} = -\frac{1}{N}\sum_{s=1}^{N} k^{(n-m)s}\omega_s \operatorname{ctg}(\omega_s t),$$
$$(\lambda_3^{-1})_{nm} = -\frac{1}{N}\sum_{s=1}^{N} k^{(n-m)s}\frac{\omega_s}{\sin(\omega_s t)},$$
$$\det\lambda_3 = \prod_{s=1}^{N}\omega_s^{-1}\sin(\omega_s t). \tag{22.8}$$

In the case of N=2 the Green function reads

$$G(\mathbf{Q}, \mathbf{Q}', t) = \frac{1}{2\pi i}\left(\frac{2\Omega}{t\sin(2\Omega t)}\right)^{1/2}\exp\left\{\frac{i}{4t}(q_1+q_2-q'_1-q'_2)^2 +\right.$$
$$+ {}^1/_2 i\Omega\operatorname{ctg}(2\Omega t)\left[(q_1-q_2)^2-(q'_1-q'_2)^2\right] -$$
$$\left.- i\Omega\sin^{-1}(2\Omega t)(q'_1-q'_2)(q_1-q_2)\right\}.$$

In the case of N=3 we have

$$G(\mathbf{Q},\mathbf{Q}',t)=\frac{\Omega\sqrt{3}}{(2\pi i)^{3/2}\sin(\Omega\sqrt{3}t)\sqrt{t}}\exp\left\{\frac{i}{6t}(q_1'+q_2'+q_3'-q_1-q_2-q_3)^2+\right.$$
$$+{}^1/_6 i\Omega\sqrt{3}\operatorname{ctg}(\Omega\sqrt{3}t)[(q_1-q_2)^2+(q_1'-q_2')^2+(q_1-q_3)^2+(q_1'-q_3')^2+$$
$$+(q_2-q_3)^2+(q_2'-q_3')^2]-{}^1/_3 i\Omega\sqrt{3}\sin^{-1}(\Omega\sqrt{3}t)[(q_1'-q_2')(q_1-q_2)+$$
$$\left.+(q_1'-q_3')(q_1-q_3)+(q_2'-q_3')(q_2-q_3)]\right\}.$$

23. FORCED DISCRETE QUANTUM STRING (OSCILLATOR CHAIN)

In this section we consider in the Schrodinger picture the set of N quantum one-dimensional oscillators, each of them being acted upon by its own external force. The Hamiltonian reads

$$\hat{H}=\sum_{n=1}^{N}\frac{\hat{p}_n^2}{2m}+\frac{m\Omega^2}{2}(\hat{q}_{n+1}-\hat{q}_n)^2-f_n(t)\,\hat{q}_n. \tag{23.1}$$

Let us construct, as in the previous section, the integrals of motion. For this purpose we introduce the dimensionless force

$$F_n(t)=(\hbar/m\omega)^{1/2}f_n(t)\,(1/\hbar\omega). \tag{23.2}$$

Then Hamiltonian (23.1), rewritten in variables (22.2), reads

$$\hat{H}=\sum_{n=1}^{N}\frac{\hat{p}_n^2}{2}+\frac{1}{2}(\hat{q}_{n+1}-\hat{q}_n)^2-F_n(t)\,\hat{q}_n. \tag{23.3}$$

Let us look for the integrals of the motion $\hat{P}^{\delta}{}_{no}(t)$ and $\hat{Q}^{\delta}{}_{no}(t)$ in the form

$$\hat{P}^{\delta}_{n0}(t)=\hat{P}_{n0}(t)+\delta_{1n}(t),\qquad \hat{Q}^{\delta}_{n0}(t)=\hat{Q}_{n0}(t)+\delta_{2n}(t), \tag{23.4}$$

where $\hat{P}_{no}(t)$ and $\hat{Q}_{no}(t)$ are given by formulae (22.5). Introducing 2N-dimension vector $\hat{I}=(\hat{I}_1,\dots,\hat{I}_{2N})$ with the components $(\hat{P}^{\delta}_{10},\dots,\hat{P}^{\delta}_{No},\hat{Q}^{\delta}_{1o},\dots,\hat{Q}^{\delta}_{No})$, vector $\hat{X}=(\hat{X}_1,\hat{X}_2,\dots,\hat{X}_{2N})$ with components $(\hat{p}_1,\dots,\hat{p}_N,\hat{q}_1,\dots,\hat{q}_N)$ and 2N-dimensional vector $\Delta=(\delta_{11},\dots,\delta_{1N},\delta_{21},\dots,\delta_{2N})$, one can rewrite relations (23.4) in the matrix form

$$\hat{\mathbf{I}}=\Lambda(t)\,\hat{\mathbf{X}}+\Delta(t). \tag{23.5}$$

Using the unitary symmetrical matrix K of the characters of the cyclic permutations group of N numbers with the matrix elements

$$K_{nm}=N^{-1/2}k^{mn},\quad n,m=\overline{1,N},\qquad k=\exp(i2\pi/N),$$

as the matrix of the linear canonical transformation of the coordinates and momenta, one can reduce the system of equations for the operators – integrals of the motion to the set of N systems for the integrals of the motion of the one-dimensional forced oscillators with dimensionless frequencies $\omega_s=2\sin(\pi s/N)\Omega(t)$; $s=\overline{1,N}$, and dimensionless forces $\Phi_s(t)=\sum_{m=1}^{N}K^{-1}_{sm}F_m(t)$. In this case matrix $\Lambda(t)$ and components of vector $\Delta(t)$ are of the form

$$\delta_{1n}(t)=\frac{1}{N}\sum_{s=1}^{N}k^{(n-m)s}\int_0^t\cos(\omega_s\tau)\,f_m(\tau)\,d\tau, \tag{23.6}$$

$$\delta_{2n}(t) = -\frac{1}{N}\sum_{s=1}^{N} k^{(n-m)s}\int_0^t \frac{\sin(\omega_s\tau)}{\omega_s} f_m(\tau)\,d\tau,$$

$$(\lambda_1(t))_{nm} = (\lambda_4(t))_{nm} = \frac{1}{N}\sum_{s=1}^{N} k^{(n-m)s}\cos(\omega_s t), \tag{23.7}$$

$$(\lambda_2(t))_{nm} = \frac{1}{N}\sum_{s=1}^{N} k^{(n-m)s}\omega_s \sin(\omega_s t), \tag{23.8}$$

$$(\lambda_3(t))_{nm} = -\frac{1}{N}\sum_{s=1}^{N} k^{(n-m)s}\frac{\sin(\omega_s t)}{\omega_s}, \tag{23.9}$$

which coincides with formulae (20.20) and (20.26) written in matrix form. The inverse matrices $\lambda_1^{-1}, \lambda_2^{-1}, \lambda_3^{-1}, \lambda_4^{-1}$ exist, and the expressions for their matrix elements are the same as determined by formulae (23.7), (23.8) and (23.9), but terms $\cos(\omega_s t)$, $\omega_s \sin(\omega_s t)$, $\omega^{-1}{}_s \sin(\omega_s t)$ in these expressions are replaced by the inverse ones. The sum rules for matrix elements follow from the condition of symplecticity of matrix $\Lambda(t)$

$$\lambda_1\tilde{\lambda}_2 = \lambda_2\tilde{\lambda}_1, \qquad \lambda_3\tilde{\lambda}_4 = \lambda_4\tilde{\lambda}_3, \qquad \tilde{\lambda}_4\lambda_2 = \tilde{\lambda}_2\lambda_4, \qquad \lambda_3\tilde{\lambda}_2 = \lambda_2\tilde{\lambda}_3,$$
$$\lambda_1\tilde{\lambda}_1 - \lambda_3\tilde{\lambda}_2 = \lambda_1\tilde{\lambda}_1 - \lambda_2\tilde{\lambda}_3 = E. \tag{23.10}$$

Solving the eigenvalue problem for the Green function $G_\delta(Q,Q',t)$ with $Q=(q_1,\ldots,q_N)$, $Q'=(q'_1,\ldots,q'_N)$

$$\hat{Q}^{\delta}_{n0}G_\delta = Q'^{\delta}_{n0}G_\delta, \qquad \hat{P}^{\delta}_{n0}G_\delta = i\partial G_\delta/\partial Q'^{\delta}_{n0}, \tag{23.11}$$

we get in agreement with the general formula for quadratic systems (see, for example, [1]) the following explicit expression in the coordinate representation:

$$G_\delta(Q, Q', t) = G(Q, Q', t)\exp\Bigg\{-\frac{i}{2N}\sum_{s=1}^{N} k^{(n-m)s}\Bigg[2q'_n\int_0^t \frac{\sin(\omega_s\tau)}{\sin(\omega_s t)}F_m(\tau)\,d\tau +$$
$$+ 2q_n\int_0^t (\cos\omega_s\tau)F_m(\tau)\,d\tau - 2q_n\int_0^t \operatorname{ctg}(\omega_s t)\sin(\omega_s\tau)F_m(\tau)\,d\tau -$$
$$-\int_0^t \operatorname{ctg}(\omega_s t)\frac{\sin(\omega_s\tau)}{\omega_s}F_m(\tau)\,d\tau\int_0^t \sin(\omega_s\tau')F_n(\tau')\,d\tau' -$$
$$-2\int_0^t d\tau\frac{\cos(\omega_s\tau)}{\omega_s}F_m(\tau)\int_0^\tau \sin(\omega_s\tau')F_m(\tau')\,d\tau'\Bigg]\Bigg\}. \tag{23.12}$$

24. TWO FORCED OSCILLATORS

Let us demonstrate the results of the previous section for the simplest example of two coupled oscillators, each of them being acted on by its own external force. Then the Hamiltonian of the system in dimensionless variables (22.2) and with dimensionless forces

$$F_1(t) = (\hbar/m\omega)^{1/2} f_1(t)/\hbar\omega, \quad F_2(t) = (\hbar/m\omega)^{1/2} f_2(t)/\hbar\omega$$

reads

$$\hat{H} = \hat{p}_1^2/2 + \hat{p}_2^2/2 + [\hat{q}_2 - \hat{q}_1]^2 - F_1(t)\,\hat{q}_1 - F_2(t)\,\hat{q}_2, \tag{24.1}$$

Integrals of the motion are given by

$$\begin{pmatrix} P_{01}^{\delta}(t) \\ P_{02}^{\delta}(t) \\ Q_{10}^{\delta}(t) \\ Q_{02}^{\delta}(t) \end{pmatrix} = \Lambda(t) \begin{pmatrix} \hat{p}_1 \\ \hat{p}_2 \\ \hat{q}_1 \\ \hat{q}_2 \end{pmatrix} + \begin{pmatrix} \delta_{11}(t) \\ \delta_{12}(t) \\ \delta_{21}(t) \\ \delta_{22}(t) \end{pmatrix}, \tag{24.2}$$

where matrix $\Lambda(t)$ is defined by formula (20.29), and components of vector $\Delta(t)$ are of the form

$$\delta_{11}(t) = \frac{1}{2}\int_0^t \cos(2\Omega\tau)(F_1 - F_2)\,d\tau + \frac{1}{2}\int_0^t (F_1 + F_2)\,d\tau,$$

$$\delta_{12}(t) = -\frac{1}{2}\int_0^t \cos(2\Omega\tau)(F_1 - F_2)\,d\tau + \frac{1}{2}\int_0^t (F_1 + F_2)\,d\tau,$$

$$\delta_{21}(t) = -\frac{1}{2}\int_0^t \frac{\sin(2\Omega\tau)}{2\Omega}(F_1 - F_2)\,d\tau - \frac{1}{2}\int_0^t \tau(F_1 + F_2)\,d\tau,$$

$$\delta_{22}(t) = \frac{1}{2}\int_0^t \frac{\sin(2\Omega\tau)}{2\Omega}(F_1 - F_2)\,d\tau - \frac{1}{2}\int_0^t \tau(F_1 + F_2)\,d\tau. \tag{24.3}$$

The Green function is the solution to the system of four equations

$$Q_{10}^{\delta}G = Q_{10}^{\delta'}G, \qquad Q_{20}^{\delta}G = Q_{20}^{\delta'}G,$$
$$P_{10}^{\delta}G = i\,\frac{\partial}{\partial Q_{10}^{\delta'}}\,G, \qquad P_{20}^{\delta}G = i\,\frac{\partial G}{\partial Q_{20}^{\delta'}} \tag{24.4}$$

Its explicit expression is as follows:

$$\begin{aligned} G_{\delta}(\mathbf{Q}, \mathbf{Q}', t) = G(\mathbf{Q}, \mathbf{Q}', t)\exp\Bigg\{-\frac{i}{2}\Bigg[\frac{q_2' - q_1'}{\sin(2\Omega t)}\int_0^t (F_2 - F_1)\sin(2\Omega\tau)\,d\tau + \\ + \frac{q_1' + q_2'}{t}\int_0^t (\tau(F_1 + F_2)\,d\tau + (q_2 - q_1)\int_0^t \cos(2\Omega\tau)(F_2 - F_1)\,d\tau + \\ + (q_2 + q_1)\int_0^t (F_1 + F_2)\,d\tau - \frac{q_1 + q_2}{t}\int_0^t \tau(F_1 + F_2)\,d\tau - \\ - (q_2 - q_1)\operatorname{ctg}(2\Omega t)\int_0^t (F_2 - F_1)\sin(2\Omega\tau)\,d\tau - \\ - \frac{1}{2t}\int_0^t \tau(F_1 + F_2)\,d\tau\int_0^t \tau'(F_1 + F_2)\,d\tau' - \\ - \int_0^t \operatorname{ctg}(2\Omega t)\frac{\sin(2\Omega\tau)}{2\Omega}(F_2 - F_1)\,d\tau\int_0^t \frac{\sin(2\Omega\tau')}{2}(F_2 - F_1)\,d\tau' - \\ - \int_0^t d\tau\cos(2\Omega\tau)(F_2 - F_1)\int_0^{\tau} \frac{\sin(2\Omega\tau')}{2\Omega}(F_2 - F_1)\,d\tau' - \\ - \int_0^t d\tau(F_2 + F_1)\int_0^{\tau} \tau'(F_2 + F_1)\,d\tau'\Bigg]\Bigg\}. \end{aligned} \tag{24.5}$$

25. STATIONARY STATES OF THE CHAIN WITH ODD NUMBER OF ATOMS

Let us consider, for simplicity, the chain consisting of an odd number of atoms N=2p+1, p=(N-1)/2 being an integer. Then, introducing real normal coordinates x_s and y_s according to formulae

$$x_s = \sqrt{2}\,\mathrm{Re}\,Q_s^H, \quad x_N = Q_N^H \equiv Q_N, \quad y_s = -i\sqrt{2}\,\mathrm{Im}\,Q_s^H, \quad s = 1, 2, \ldots, p, \tag{25.1}$$

and real normal momenta

$$p_{xs} = \sqrt{2}\,\mathrm{Re}\,P_s^H, \quad p_{xN} = P_N^H = P_N, \quad p_{ys} = -\sqrt{2}\,\mathrm{Im}\,P_s^H,$$
$$s = 1, 2, \ldots, p, \tag{25.2}$$

or in the explicit form

$$x_s = \left(\frac{2}{N}\right)^{1/2} \sum_{m=1}^{N} \cos\left(\frac{2\pi sm}{N}\right) q_m,$$
$$\tag{25.3}$$
$$y_s = \left(\frac{2}{N}\right)^{1/2} \sum_{m=1}^{N} \sin\left(\frac{2\pi sm}{N}\right) q_m,$$

$$x_N = \left(\frac{1}{N}\right)^{1/2} \sum_{m=1}^{N} q_m, \tag{25.4}$$

one can rewrite the initial Hamiltonian in the standard form: the sum of energies of 2p normal vibrations and the free center of mass motion

$$H = \frac{1}{2} \sum_{s=1}^{p} \left(p_{xs}^2 + p_{ys}^2 + \Omega_s^2 (x_s^2 + y_s^2) + P_N^2\right),$$
$$\Omega_s^2 = 4\Omega^2 \sin^2(\pi s/N). \tag{25.5}$$

Canonical momenta are expressed through the starting ones as follows:

$$p_{xs} = \left(\frac{2}{N}\right)^{1/2} \sum_{m=1}^{N} \cos\left(\frac{2\pi sm}{N}\right) p_m, \tag{25.6}$$

$$p_{ys} = \left(\frac{2}{N}\right)^{1/2} \sum_{m=1}^{N} \sin\left(\frac{2\pi sm}{N}\right) p_m. \tag{25.7}$$

Every normal vibration is doubly degenerate, i.e. the same frequency Ω_s corresponds to coordinates x_s and y_s. So, Hamiltonian (25.5) possesses the symmetry group $U(2) \otimes U(2) \otimes \ldots \otimes U(2)$, where the number of multipliers is equal to p.

After the canonical quantization, Hamiltonian (25.5) becomes the operator. Its explicit form in the coordinate representation is as follows ($\hbar$=1):

$$\hat{H} = \frac{1}{2} \sum_{s=1}^{p} \left(-\frac{\partial^2}{\partial x_s^2} - \frac{\partial^2}{\partial y_s^2}\right) + \frac{\Omega_s^2}{2}(x_s^2 + y_s^2) - \frac{1}{2}\frac{\partial^2}{\partial Q_N^2}. \tag{25.8}$$

The time-dependent Schrodinger equation with Hamiltonian (25.8) admits the separation of variables, so its solution can be written out as the product of well known one-dimensional oscillator wave functions with frequency Ω_s and the free center of mass (coordinate Q_N) wave function

$$\Psi_{\mathbf{n},V}(\mathbf{q},t) = \Psi_V(Q_N,t)\prod_{s=1}^{p}\Psi_{n_{1s}}(x_s,t)\,\Psi_{n_{2s}}(y_s,t), \tag{25.9}$$

where 2p-vector n has projections $n_{11}, n_{12},\ldots,n_{1p}, n_{2p}$, which are integers running meanings from zero to infinity. Suffix V determines the character of the free center of mass motion, and function $\Psi_{n_{1s}}(x_s,t)$ is given by

$$\Psi_{n_{1s}}(x_s\,t) = (l_s\sqrt{\pi})^{-1/2}\,\frac{\exp[-i\Omega_s t\,(n_{1s}+1/2)]}{(2^{n_{1s}}n_{1s}!)^{1/2}}\exp\left(-\frac{x_s^2}{2l_s^2}\right)H_{n_{1s}}\left(\frac{x_s}{l_s}\right). \tag{25.10}$$

The coordinate x_s is determined by formula (25.3), $l_s=\Omega_s^{1/2}$, frequency Ω_s is defined by formula (25.5). The multiplier $\Psi_{n_{2s}}(y_s,t)$ can be obtained from (25.10) with the help of replacements $x\to y$, $n_{1s}\to n_{2s}$. The free motion wave function $\Psi_V(Q_N,t)$ can be chosen, for example, in the form of normalized by δ-function plane wave, corresponding to the free motion with momentum P_N(i.e. $V=P_N$)

$$\Psi_{P_N}(Q_N,t) = (2\pi)^{-1/2}\exp(iP_NQ_N - iP_N^2t/2), \tag{25.11}$$

where $Q_N = \frac{1}{N}\sum_{m=1}^{N} q_m$. The energy of stationary state $\Psi_{\mathbf{n},V}$ reads (when $V=P_N$)

$$E_{\mathbf{n},P_N} = \sum_{s=1}^{p}[(n_{1s}+n_{2s})\Omega_s + \Omega_s] + \frac{P_N^2}{2}. \tag{25.12}$$

In the same way one can construct the wave packets, for example, the coherent state describing the free motion of the chain as the whole: index V will have another meaning. Further we shall consider only that part of wave function, which corresponds to the vibrational motion, omitting the multiplier $\Psi_V(Q_N,t)$. So, the vibrational normalized ground state wave function for the chain with an odd number of particles in q-coordinates reads

$$\Psi_0(\mathbf{q},t) = \pi^{-2p/4}(\Omega_1\Omega_2\ldots\Omega_p)^{1/2}\exp\Big\{-it(\Omega_1+\Omega_2+\ldots+\Omega_p) - \\ -\frac{1}{N}\sum_{s=1}^{p}\Omega_s\Big[\Big(\sum_{m=1}^{N} q_m\cos\frac{2\pi ms}{N}\Big)^2 + \Big(\sum_{m=1}^{N} q_m\sin\frac{2\pi ms}{N}\Big)^2\Big]\Big\}, \tag{25.13}$$
$$\Omega_s^2 = 4\Omega^2\sin^2(\pi s/N).$$

26. INTEGRALS OF MOTION AND COHERENT STATES OF OSCILLATOR CHAIN

One can check that 2p time-dependent operators

$$\hat{A}_s(t) = e^{i\Omega_s t}N^{-1/2}\sum_{m=1}^{N}\cos\frac{2\pi sm}{N}\left(\frac{\hat{q}_m}{l_s} + i\hat{p}_m l_s\right),$$
$$\hat{B}_s(t) = e^{i\Omega_s t}N^{-1/2}\sum_{m=1}^{N}\sin\frac{2\pi ms}{N}\left(\frac{\hat{q}_m}{l_s} + i\hat{p}_m l_s\right), \qquad s=1,2,\ldots,p, \tag{26.1}$$

where $\hat{q}_m$ and $\hat{p}_m$ are the m-th atom coordinate and momentum operators, constitute the set of integrals of motion satisfying the boson commutation relations

$$[\hat{A}_s,\hat{A}_r^+] = \delta_{sr}, \qquad [\hat{B}_s,\hat{B}_r^+] = \delta_{sr}, \qquad [\hat{A}_s,\hat{B}_m] = [\hat{A}_s,\hat{B}_m^+] = 0. \tag{26.2}$$

The ground states (25.13) satisfy the time-dependent Schrodinger equation with Hamiltonain (25.8) (we do not take into consideration the term - $1/2\ \partial^2/\partial Q^2{}_N$) and conditions

$$\hat{A}_s\Psi_0 = \hat{B}_s\Psi_0 = 0,\ s = 1,\ 2,\ \ldots,\ p. \tag{26.3}$$

The stationary state of the chain (25.9) (without taking into consideration the multiplier $\Psi_v(Q_N,t)$ can be obtained from its ground state by acting of operators – integrals of motion

$$\Psi_{\mathbf{n}}(\mathbf{q},t) = \prod_{s=1}^{p} (n_{1s}!)^{-1/2}(n_{2s}!)^{-1/2}\hat{A}_s^{+n_{1s}}(t)\,\hat{B}_s^{+n_{2s}}(t)\,\Psi_0(\mathbf{q},t). \tag{26.4}$$

Using the scheme of constructing coherent states from the ground state with the help of integrals of motion – shift operators [1]

$$\hat{D}(\alpha_s) = \exp(\alpha_s\hat{A}_s^+ - \alpha_s^*\hat{A}_s),$$
$$\hat{D}(\beta_s) = \exp(\beta_s\hat{B}_s^+ - \beta_s^*\hat{B}_s) \tag{26.5}$$

according to formula

$$\Psi_{\boldsymbol{\alpha}}(\mathbf{q},t) = \left[\prod_{s=1}^{p} D(\alpha_s)\,D(\beta_s)\right]\Psi_0, \tag{26.6}$$

where complex 2p-vector α has projections $(\alpha_{11}, \beta_{11}, \alpha_{12}, \beta_{12},\ldots,\alpha_{1p},\beta_{1p})$, one can get the expression for coherent states in normal coordinates of the oscillator chain

$$\Psi_{\boldsymbol{\alpha}}(\mathbf{q},t) = \prod_{s=1}^{p} \Psi_{\alpha_s}(x_s,t)\,\Psi_{\beta_s}(y_s,t). \tag{26.7}$$

Coordinates x_s and y_s are determined by formulae (25.3) and (25.4). The final result reads

$$\Psi_{\alpha_s}(x_s,t) = (\Omega_s/\pi)^{1/4}\exp[-{}^1/_2(|\alpha_s|^2 - i\Omega_s t - \Omega_s^2 x_s^2) +$$
$$+ (2\Omega_s)^{1/2}\exp(-i\Omega_s t)\,\alpha_s x_s - {}^1/_2\alpha_s^2\exp(-2i\Omega_s t)]. \tag{26.8}$$

The multiplier Ψ_{β_s} can be obtained from Ψ_{α_s} by replacements $x\rightarrow y$, $\alpha\rightarrow\beta$. Coherent state of chain (26.7) is the eigenstate for operators – integrals of motion (26.1)

$$\hat{A}_s\Psi_{\boldsymbol{\alpha}} = \alpha_s\Psi_{\boldsymbol{\alpha}},\quad s = 1, 2, \ldots, p,\quad \hat{B}_s\Psi_{\boldsymbol{\alpha}} = \beta_s\Psi_{\boldsymbol{\alpha}}. \tag{26.9}$$

Normal coordinates of vibrations have the following mean values in the constructed coherent state:

$$\bar{x}_s(t) = (2/\Omega_s)^{1/2}|\alpha_s|\cos(\varphi_{\alpha_s} - \Omega_s t),$$
$$\bar{y}_s(t) = (2/\Omega_s)^{1/2}|\beta_s|\cos(\varphi_{\beta_s} - \Omega_s t), \tag{26.10}$$

$\phi_{\alpha s}$, $\phi_{\beta s}$ being the phases of complex numbers α_s and β_s, respectively. The mean values of momenta of normal vibrations are given by

$$\bar{p}_{xs}(t) = (2\Omega_s)^{1/2}|\alpha_s|\sin(\varphi_{\alpha_s} - \Omega_s t),$$
$$\bar{p}_{ys}(t) = (2\Omega_s)^{1/2}|\beta_s|\sin(\varphi_{\beta_s} - \Omega_s t). \tag{26.11}$$

The variances of coordinates and momenta of normal vibrations do not depend on time; they are equal to the values

$$\delta x_s = \delta y_s = (2\Omega_s)^{-1/2}, \qquad \delta p_{xs} = \delta p_{ys} = (\Omega_s/2)^{1/2}, \tag{26.12}$$

which minimize the Heisenberg uncertainty relation. The coherent state of the chain is connected with the stationary states by the standard relation

$$\Psi_\alpha(\mathbf{q},t) = e^{-|\alpha|^2/2}\sum_{n=0}^{\infty}\left(\prod_{s=1}^{p}\frac{\alpha_s^{n_{1s}}\beta_s^{n_{2s}}}{(n_{1s}!\,n_{2s}!)^{1/2}}\right)\Psi_{\mathbf{n}}(\mathbf{q},t). \tag{26.13}$$

Function $\Psi_n(q,t)$ is determined by formulae (25.9) and (25.10).

27. CORRELATED STATES AND PULSATIONS IN THE ATOMIC CHAIN

Coherent states constructed above are coherent with respect to normal coordinates of vibrations. As to the variances of the usual coordinates and momenta relating to the oscillators in the chain, they depend on time, moreover, there are time-dependent correlations of these coordinates and momenta. Really, due to the relation between normal coordinates and initial ones one can write the following matrix formula for N-vector $\mu=(x_1, x_2,\ldots, x_p, y_1, y_2,\ldots, y_p, Q_N)$

$$\mu = T\mathbf{Q}^{H}, \qquad T = \begin{Vmatrix} T_1 & T_2 & 0 \\ T_3 & T_4 & 0 \\ 0 & 0 & 1 \end{Vmatrix}, \tag{27.1}$$

where p-dimensional matrices T_1, T_2, T_3, T_4 are given by

$$iT_3 = T_1 = \frac{1}{\sqrt{2}}E, \qquad T_2 = -iT_4 = \frac{1}{\sqrt{2}}\begin{Vmatrix} 0 & 0 \ldots & 1 \\ & 1 & 0 \\ \vdots & \cdot^{\cdot^{\cdot}} & \vdots \\ 1 & \ldots & 0 \end{Vmatrix},$$

and E is unity p-dimensional matrix. The mean values of coordinates of atoms can be expressed through the mean values of coordinates of normal vibrations follows:

$$\langle \mathbf{q}\rangle = K^*T^{-1}\langle\mu\rangle. \tag{27.2}$$

Matrix T^{-1} is expressed through matrix T

$$T^{-1} = \begin{Vmatrix} E/\sqrt{2} & iE/\sqrt{2} & 0 \\ T_2 & -iT_2 & 0 \\ 0 & 0 & 1 \end{Vmatrix}, \tag{27.3}$$

and matrix K^* is determined by formula (20.11). Let us introduce the designation for the product of matrices K^*T^{-1}

$$K^*T^{-1} = S. \tag{27.4}$$

Then

$$\langle \mathbf{q}\rangle = S\,\langle\mu\rangle. \tag{27.5}$$

We assume that vectors q and μ are operators determined in Heisenberg's representation. The mean values are calculated with respect to an arbitrary initial state. The relation between the momenta vectors is determined by matrix $S^{-1}=TK$

$$\langle \mathbf{p}\rangle = S^{-1}\langle \mathbf{P}_\mu^{H}\rangle, \qquad \mathbf{P}_\mu^{H} \equiv \mathbf{P}_\mu. \tag{27.6}$$

Here (2p+1)-dimensional vector P_μ has the components $P_{x1}, P_{x2}, \ldots, P_{xp}, \ldots, P_{y1}, \ldots, P_{yp}, P_N$. If one introduces 2N-vector with components (p,q)=1 and 2N-vector with the components (P_μ, μ)=L, then relations (27.5) and (27.6) can be rewritten in matrix form

$$\langle \hat{\mathbf{I}} \rangle = \begin{Vmatrix} S^{-1} & 0 \\ 0 & S \end{Vmatrix} \langle \mathbf{L} \rangle = R \langle \mathbf{L} \rangle. \tag{27.7}$$

The mean values for quadratic systems move along classical trajectories according to the Ehrenfest theorem. All quantities $\langle P_{x_s} \rangle$, $\langle P_{y_s} \rangle$, $\langle x_s \rangle$, $\langle y_s \rangle$ describe the classical trajectory of a normal vibration in the coherent states determined by formulae (26.10) and (26.11). The mean values of the usual atomic coordinates and momenta perform the classical vibrations according to (27.7). Let us calculate the variances forming the variance matrix

$$\sigma_{mn} = \langle I_m I_n \rangle - \langle I_m \rangle \langle I_n \rangle, \qquad m, n = 1, 2, \ldots, 2N. \tag{27.8}$$

We have

$$I_m = \sum_{s=0}^{2N} R_{ms} L_s. \tag{27.9}$$

So

$$\sigma_{mn} = \sum_{s,\, q=1}^{2N} R_{ms} R_{nq} (\langle L_s L_q \rangle - \langle L_s \rangle \langle L_q \rangle). \tag{27.10}$$

The expression in brackets is the variance matrix for physical quantities – components of coordinates and momenta of normal vibrations. This matrix is diagonal for vibrational degrees of freedom in the case of coherent states. Moreover, it does not depend on time and the parameters of coherent states. In this case

$$\langle L_s L_q \rangle - \langle L_s \rangle \langle L_q \rangle = \Sigma_{sq},$$

$$\Sigma = \begin{Vmatrix} \Omega & & & \\ & \delta P_N^2 & & \\ & & \Omega^{-1} & \\ & & & \delta Q_N^2 \end{Vmatrix}. \tag{27.11}$$

The 2px2p diagonal matrix Ω reads

$$\Omega = \frac{1}{2} \begin{Vmatrix} \Omega_1 & & & & & 0 \\ & \ddots & & & & \\ & & \Omega_p & & & \\ & & & \Omega_1 & & \\ & & & & \ddots & \\ 0 & & & & & \Omega_p \end{Vmatrix}. \tag{27.12}$$

The quantities δP^2_N and δQ^2_N are the variances corresponding to the free motion of the center of mass. Their explicit forms depend on the concrete wave function describing this motion.

28. THE OSCILLATOR CHAIN WITH TIME DEPENDENT ELASTICITY

Let us now consider the chain with time-dependent elasticity consisting of the odd number of oscillators N=2p+1, each atom of the chain being acted on by its own time-dependent force. The Hamiltonian reads (in dimensionless variables)

$$\hat{H} = \frac{1}{2}\sum_{n=1}^{N}[\hat{p}_n^2 + \Omega^2(t)(\hat{q}_{n+1} - \hat{q}_n)^2 - f_n(t)\hat{q}_n]. \tag{28.1}$$

Let us introduce new coordinates $X=(X_1, X_2,\ldots, X_N)$ and momenta $P_X=(P_{X1}, P_{X2},\ldots, P_{XN})$ using the canonical contact transformation

$$\mathbf{X} = S^{-1}\mathbf{q}, \qquad \mathbf{P}_X = S\mathbf{p}. \tag{28.2}$$

Matrix S is expressed through matrices K and T according to formulae (20.11), (27.3) and (27.4). Hamiltonian (28.1) in new variables assumes the form

$$\hat{H} = \frac{1}{2}\sum_{n=1}^{N}(P_{Xn}^2 + \Omega_n^2(t)X_n^2 - F_n(t)X_n. \tag{28.3}$$

The force F is expressed through the initial force $f=(f_1,\ldots, f_N)$ with the help of the same transformation

$$\mathbf{f}S = \mathbf{F}. \tag{28.4}$$

Hamiltonian (28.3) describes the set of p two-dimensional parametric oscillators with identical frequencies

$$\Omega_n^2(t) = 4\Omega^2(t)\sin^2(\pi n/N), \qquad n = 1,\ 2,\ \ldots.\ p,$$

and free motion of coordinate X_N-the chain center of mass. So, the linear integrals of motion of parametric chain can be expressed through the linear invariants of one-dimensional parametric oscillators. Let us introduce four diagonal matrices λ_i, i=1,2,3,4,

$$\lambda^i = \left\| \begin{matrix} \lambda_1^i & & & & \\ & \lambda_1^i & & & \\ & & \ddots & & \\ & & & \lambda_p^i & \\ & & & & \lambda_p^i \end{matrix} \right\|, \tag{28.5}$$

The N-dimensional integral of motion of the chain $P_0(t)$ having the meaning of initial momentum is expressed through coordinates and momenta q and p as follows:

$$\mathbf{P}_0(t) = \lambda_1\mathbf{p} + \lambda_2\mathbf{q} + \delta_1(t). \tag{28.6}$$

N-vector δ_1 is determined by the external force. For the vector invariant describing initial value of coordinates we get

$$\mathbf{Q}_0(t) = \lambda_3\mathbf{p} + \lambda_4\mathbf{q} + \delta_2(t). \tag{28.7}$$

The N-dimensional matrices $\lambda_i(i=\overline{1,4})$ are expressed through the diagonal matrices λ^i and transformation matrix S:

$$\lambda_1 = S^{-1}\lambda^1 S, \qquad \lambda_2 = S^{-1}\lambda^2 S^{-1}, \qquad \lambda_3 = S\lambda^3 S, \qquad \lambda_4 = S\lambda^4 S^{-1}. \tag{28.8}$$

Let us introduce N-vectors δ^1 and δ^2. Then vectors δ_1 and δ_2 in formulae (28.6) and (28.7) are of the form

$$\delta_1 = S^{-1}\delta^1, \qquad \delta_2 = S\delta^2. \tag{28.9}$$

The sense of transformations considered is as follows. Matrices λ^i and vectors $\delta^{1,2}$ determine well-known integrals of motion of the set of noninteracting parametric one-dimensional oscillators [1]. The required matrices λ_i and vectors $\delta_{1,2}$ determining the invariants of oscillator chain are expressed through these known quantities by means of formulae (28.8) and (28.9). In turn, matrices $\lambda^i{}_{nN}$(i=1,2,3,4) are expressed through the solutions of equation

$$\ddot{\varepsilon}_n + \Omega_n^2(t)\,\varepsilon_n = 0 \qquad (28.10)$$

in the following way:

$$\lambda_n^1 = \varepsilon_{1n}(t), \quad \lambda_n^3 = -\varepsilon_{2n}(t), \quad \lambda_n^2 = -\dot{\varepsilon}_{1n}(t), \quad \lambda_n^4 = \dot{\varepsilon}_{2n}(t). \qquad (28.11)$$

$\varepsilon_{1n}(t)$ and $\varepsilon_{2n}(t)$ are the real solutions of equation (28.10) with the initial conditions

$$\varepsilon_{1n}(0) = \dot{\varepsilon}_{2n}(0) = 1, \quad \varepsilon_{2n}(0) = \dot{\varepsilon}_{1n}(0) = 0. \qquad (28.12)$$

The components of the vector $\delta^{1,2}=(\delta^{1,2}{}_1, \delta^{1,2}{}_2, \ldots, \delta^{1,2}{}_p)$ are expressed through the force F and the solution $\varepsilon^{1,2}{}_n(t)$:

$$\delta_n^{1,2} = \int_0^t \varepsilon_{nk}(\tau)\,F_n(\tau)\,d\tau, \quad k = 1, 2. \qquad (28.13)$$

The Green function of the oscillator chain is expressed in a standard form [1] through the functions $\varepsilon^{1,2}{}_n(t)$ and $\delta^{1,2}{}_n(t)$,which determine the symplectic transformation and set the linear integrals of motion of the chain.

29. QUANTIZED STRING

Let us now consider a continuous string. One can go to the continuous string limit from a discrete chain by decreasing the distances between the atoms to zero and increasing the number of atoms-oscillators to infinity. The quantization of the string is performed canonically, i.e., the shift from the equilibrium point $u(x,t)=u_x(t)$ is replaced by the operator of coordinate $\hat{u}_x(t)$, and the velocity $\dot{u}(x,t)=\dot{u}_x(t)=p_x(t)$ is replaced by the operator of momentum $\hat{p}_x(t)$. The fundamental element in our approach is the symplectic transformation determining linear integrals of motion for the string. To find it, we write the equation of motion (19.1) in the following form:

$$\partial p_x(t)/\partial t = c^2\partial^2 u_x/\partial x^2, \quad \partial u_x(t)/\partial t = p_x, \quad p_x = \dot{u}_x. \qquad (29.1)$$

Rewriting them in matrix form

$$\frac{d}{dt}\begin{pmatrix} p_x \\ u_x \end{pmatrix} = \begin{Vmatrix} 0 & c^2\partial^2/\partial x^2 \\ E & 0 \end{Vmatrix} \begin{pmatrix} p_x \\ u_x \end{pmatrix}, \qquad (29.2)$$

we can formally write the solution

$$\begin{pmatrix} p_x(t) \\ u_x(t) \end{pmatrix} = \exp\left(\begin{Vmatrix} 0 & c^2\partial^2/\partial x^2 \\ 1 & 0 \end{Vmatrix} t \right) \begin{pmatrix} p_x(0) \\ u_x(0) \end{pmatrix}. \qquad (29.3)$$

The exponential function of the operator matrix in (29.3) can be easily calculated, and we have

$$\begin{pmatrix} p(t) \\ u(t) \end{pmatrix} = \begin{Vmatrix} \mathrm{ch}\left(ct\dfrac{\partial}{\partial x}\right) & c\dfrac{\partial}{\partial x}\,\mathrm{sh}\left(tc\dfrac{\partial}{\partial x}\right) \\ \left(c\dfrac{\partial}{\partial x}\right)^{-1}\mathrm{sh}\left(tc\dfrac{\partial}{\partial x}\right) & \mathrm{ch}\left(ct\dfrac{\partial}{\partial x}\right) \end{Vmatrix} \begin{pmatrix} p(0) \\ u(0) \end{pmatrix}. \qquad (29.4)$$

The integrals of motion (which have the sense of initial coordinates in the phase space) are expressed through Λ-matrix of the symplectic transformation. Formulae (29.4) determine the solutions of Heisenberg's equations of motion for the quantized string. As follows from the general relation between the solutions of Heisenberg's equations of motion for stationary quantum systems and integrals of motion (initial coordinates in the phase space in Schrodinger's picture) [1]

$$\begin{pmatrix} p_{0x}(t) \\ u_{0x}(t) \end{pmatrix} = \Lambda(t) \begin{pmatrix} p_x \\ u_x \end{pmatrix}, \qquad \Lambda = \begin{Vmatrix} \lambda_1 & \lambda_2 \\ \lambda_3 & \lambda_4 \end{Vmatrix}, \tag{29.5}$$

the operator matrix Λ has as elements $\lambda_1, \lambda_2, \lambda_3, \lambda_4$ the following operators:

$$\lambda_1 = \lambda_4 = \mathrm{ch}\,(tc\partial/\partial x), \qquad \lambda_2 = -(c\partial/\partial x)^{-1}\,\mathrm{sh}\,(tc\partial/\partial x),$$
$$\lambda_3 = -(c\partial/\partial x)\,\mathrm{sh}\,(tc\partial/\partial x). \tag{29.6}$$

One can notice that this transformation belongs to the infinite-dimensional homogeneous symplectic Lie group. The solutions describing the states of the quantized string realize an irreducible representation of this group. So the infinite-dimensional symplectic group is the dynamical group of symmetry of a continuous string. The element of this group Λ is determined by formulae (29.5) and (29.6). Since the solutions $\hat{p}_H(t)$ and $\hat{u}_H(t)$ of *linear* Heisenberg's equations of motion have the same form as the solutions of corresponding classical equations of motion (29.4), we can express them, taking into account the integral representation

$$\left(c\frac{\partial}{\partial x}\right)^{-1} = \int_0^\infty \exp\left[-\beta c\frac{\partial}{\partial x}\right] d\beta, \tag{29.7}$$

as follows:

$$\hat{p}_H(t) = {}^1/_2\,[\hat{p}_0(ct+x) - \hat{p}_0(ct-x)] -$$
$$- \frac{1}{2}\int_0^\infty d\beta\,[\hat{u}_0(x+ct-\beta c) - \hat{u}_0(x-ct-\beta c)],$$
$$\hat{u}_H(t) = -\frac{c}{2}\left[\frac{\partial}{\partial x}\hat{p}_0(x+ct) - \frac{\partial}{\partial x}\hat{p}_0(x-ct)\right] +$$
$$+ {}^1/_2\hat{u}_0(x+ct) + {}^1/_2\hat{u}_0(x-ct), \tag{29.8}$$

$\hat{p}_0(x)$ and $\hat{u}_0(x)$ being the initial momentum and coordinates operators of the x-th oscillator of the string. The formulae presented are the limit formulae for matrix Λ of the finite oscillator chain. The equations of motion for the latter are nothing but the finite-difference variant of the system of partial differential equations (29.1). The analogous limit transformation in the equations for the Green function of the discrete string reduces to the replacement of the operator matrix kernel K and operator matrix T by infinite-dimensional matrices. In this case the matrix of the characters of the cyclic permutations group $||K_{mn}|| = (N)^{-1/2}||k^{nm}||$ is transformed into the kernel of Fourier transformation $K(x,x') = (2\pi)^{-1/2}\exp(2\pi i x x')$, whereas the kernel of transformation T is expressed through δ-function. The propagator (a function of 2N variables) turns into a functional (a function of an infinite number of variables).

30. THE STRING WITH DAMPING

Let us calculate the integrals of motion for the string with damping. The classical equation of motion for this string is of the form

$$\frac{\partial^2 u}{\partial t^2} + \gamma \frac{\partial u}{\partial t} = c^2 \frac{\partial^2 u}{\partial x^2}\,. \qquad (30.1)$$

One can rewrite it in the matrix form, as for the string without damping, introducing the column consisting of two variables: velocity $\dot{u}$=v and coordinate u

$$\partial v/\partial t = -\gamma v + (c\partial/\partial x)^2 u, \qquad \partial u/\partial t = v. \qquad (30.2)$$

Then the solution of system (30.2) can be written in the matrix form

$$\begin{pmatrix} v\,(t) \\ u\,(t) \end{pmatrix} = \exp\left\{\left\| \begin{matrix} -\gamma & (c\partial/\partial x)^2 \\ 1 & 0 \end{matrix} \right\| t\right\} \begin{pmatrix} v\,(0) \\ u\,(0) \end{pmatrix} = U \begin{pmatrix} v\,(0) \\ u\,(0) \end{pmatrix}. \qquad (30.3)$$

The explicit expression for matrix (30.3) is as follows:

$$U = \left\| \begin{matrix} \operatorname{ch}\tau + \frac{1}{2}\gamma^2 \operatorname{sh}\tau & \left(c\frac{\partial}{\partial x}\right)\left[\left(c\frac{\partial}{\partial x}\right)^2 + \frac{1}{4}\gamma^2\right]^{-1/2} \operatorname{sh}\tau \\ \left[\left(c\frac{\partial}{\partial x}\right)^2 + \frac{1}{4}\gamma^2\right]^{-1/2} \operatorname{sh}\tau & \operatorname{ch}\tau - \frac{1}{2}\gamma^2 \operatorname{sh}\tau \end{matrix} \right\|,$$

$$\tau = t\,[c\partial/\partial x)^2 + \gamma^2/4]^{1/2}. \qquad (30.4)$$

The Lagrangian for equation (30.2) is

$$\mathcal{L} = {}^1/_2 \int dx\,[e^{\gamma t}\,(\dot{u}^2 - c^2\,(\partial u/\partial x)^2)]. \qquad (30.5)$$

The density of momentum p is expressed through the velocity v(x,t) according to the formula

$$p\,(x,\ t) = e^{\gamma t} v. \qquad (30.6)$$

If one quantizes equation (30.1) for the string with damping using the Hamiltonian density (analog of the Caldirola-Kanai Hamiltonian for the one-dimensional oscillator with damping)

$$H = \int dx\,[e^{-\gamma t} p^2 + c^2\,(\partial u/\partial x)^2\,e^{\gamma t}], \qquad (30.7)$$

and introduces canonical commutation relations for operators of coordinate and momentum, then formulae similar to (29.3) and (29.4) give the solutions of Heisenberg's equations for these operators, provided the factor $e^{\gamma t}$ is inserted in relations (29.6) Then the integrals of motion in the Schrodinger picture p_o(x,t) and u_o(x,t) can be expressed through the infinite Λ-matrix

$$\begin{pmatrix} \hat{p}_0\,(x,t) \\ \hat{u}_0\,(x,t) \end{pmatrix} = \Lambda \begin{pmatrix} \hat{p}\,(x) \\ \hat{u}\,(x) \end{pmatrix} = \left\| \begin{matrix} \lambda_1 & \lambda_2 \\ \lambda_3 & \lambda_4 \end{matrix} \right\| \begin{pmatrix} \hat{p}\,(x) \\ \hat{u}\,(x) \end{pmatrix}. \qquad (30.8)$$

In this case

$$\Lambda = U\,(-t) \left\| \begin{matrix} e^{-\gamma t} & 0 \\ 0 & 1 \end{matrix} \right\|, \qquad (30.9)$$

i.e.

$$\begin{aligned} \lambda_1 &= e^{-\gamma t}\,(\operatorname{ch}\tau - {}^1/_2\gamma^2 \operatorname{sh}\tau), \\ \lambda_2 &= -(c\partial/\partial x)\,[(c\partial/\partial x)^2 + \gamma^2/4]^{-1/2} \operatorname{sh}\tau, \\ \lambda_3 &= -e^{-\gamma t}\,[(c\partial/\partial x)^2 + \gamma^2/4]^{-1/2} \operatorname{sh}\tau, \\ \lambda_4 &= \operatorname{ch}\tau + {}^1/_2\gamma^2 \operatorname{sh}\tau, \quad \tau = t\,[(c\partial/\partial x)^2 + \gamma^2/4]^{1/2}. \end{aligned} \qquad (30.10)$$

Formulae (30.10) determine the integral operator in the coordinate representation. The expression for this operator is simplified in the momentum representation, where the expression $(c\partial/\partial x)^2$ in (30.10) is replaced by the expression $-c^2p^2$. Evidently,matrix elements (30.10) turn into (29.6) when $\gamma=0$.

Matrix (30.9) with the matrix elements (30.10) is an element of the infinite symplectic group. It is interesting to note, the states of the string with damping realize a representation of this infinite symplectic group. So, the infinite symplectic group can be used as the dynamical group for the string both with and without damping. It is necessary to note that the group U(N,1) can also be used as the dynamical group of the N-dimensional oscillator. The generators of the group U(N,1) corresponding to subgroup U(N) are realized by operators of the second order with respect to derivatives. In the string limit, these generators are expressed through some functions of functional derivatives. Matrix elements of finite transformations for the string representation of symplectic group are expressed through infinite-dimensional analogs of the Hermite polynomials from many variables. The equations for string with damping considered above are the limit of those for the oscillator chain with damping

$$\ddot{q}_n + \gamma\dot{q}_n + \Omega^2(2q_n - q_{n+1} - q_{n-1}) = 0. \tag{30.11}$$

The Hamiltonian for this chain with damping is written in the form

$$H = \frac{1}{2}\sum_{n=1}^{N}[p_n^2 e^{-\gamma t} + \Omega^2 e^{\gamma t}(q_n - q_{n+1})^2]. \tag{30.12}$$

The momentum p_n is connected with the velocity by the relation $p_n=\dot{q}_n e^{\gamma t}$ in this case. The equations of motion (30.11) can be solved with the help of the same replacement of variables, as in the case of oscillators without damping.

31. OTHER TYPES OF DISCRETE OSCILLATOR LATTICES

Up to now we considered the discrete chain with a single characteristic frequency (both time-dependent and time-independent). The method of integrals of motion permits us to consider the case when there are two or more frequencies. For example, if one introduces into the Hamiltonian additional terms responsible for the vibrations of oscillators which are not connected with the interaction between neighbors, the one arrives at the Hamiltonian model (see, for example [29])

$$\hat{H} = \frac{1}{2}\sum_{n=1}^{N}[\hat{p}_n^2 + \Omega^2(t)(\hat{q}_n - \hat{q}_{n+1})^2 + \Omega_0^2(t)\hat{q}_n^2], \quad N = 2p+1. \tag{31.1}$$

We have taken into account, in comparison with [1], the possibility of time variations of the frequencies $\Omega(t)$ and $\Omega_0(t)$. The cases considered in the previous sections correspond to $\Omega_0=0$. If $\Omega=0$, we have the set of independent parametric oscillators.

The equations of motion for Hamiltonian (31.1) are as follows:

$$\ddot{q}_n + (\Omega_0^2 + 2\Omega^2)q_n - \Omega^2(q_{n+1} + q_{n-1}) = 0. \tag{31.2}$$

If we use the contact transformation to new variables, which has been done earlier, we turn this problem into the problem of the independent parametric normal vibrations with frequencies

$$\omega_s^2(t) = \Omega^2(t)\cdot 4\sin^2(\pi s/N) + \Omega_0^2(t). \tag{31.3}$$

So, we have no zero mode of vibration, which is responsible for the free motion, but we have a nondegenerate vibration with the frequency $\Omega_0(t)$ and doubly degenerate vibrations with frequencies (31.3). The solution of quantum problem with Hamiltonian (31.3) is reduced to the solution of the problem of the set of one-dimensional parametric oscillators. The wave function of the oscillator chain for the discrete set of quantum numbers (Fock's states) is the product of wave functions (Hermite polynomials) of parametric independent oscillator. The wave function of the coherent state of the chain is the product of wave functions of the one-dimensional parametric oscillators (Gaussian exponentials), too. A continuous analog of the chain with Hamiltonian (31.1) is the string with Hamiltonian

$$H = {}^1/_2 \int dx\,[p^2(x,t) + c^2(t)(\partial u/\partial x)^2 + \Omega^2(t)\,u^2(x,t)], \quad p(x,t) = \dot{u}(x,t), \tag{31.4}$$

leading to the equation of motion

$$\ddot{u} = c^2\partial^2 u/\partial x^2 + \Omega_0^2(t)\,u(x,\ t). \tag{31.5}$$

One can take into consideration damping in the oscillator chain proceeding from the Hamiltonian

$$H_f = \frac{1}{2}\sum_{n=1}^{N}[e^{-\gamma t}p_n^2 + \Omega^2(t)\,e^{\gamma t}(u_n - u_{n+1})^2 + \Omega_0^2 e^{\gamma t}u_n^2]. \tag{31.6}$$

The corresponding equations of motion (coinciding with the Heisenberg equations) are given by

$$\ddot{u}_n + \gamma\dot{u}_n + (\Omega_0^2 + 2\Omega^2)\,u_n - \Omega^2(u_{n+1} + u_{n-1}) = 0. \tag{31.7}$$

The similar string with damping is described by the Hamiltonian

$$H_f = {}^1/_2 \int dx\,[e^{-\gamma t}p^2(x,t) + e^{\gamma t}c^2(t)(\partial u/\partial x)^2 + \Omega_0^2(t)\,e^{\gamma t}u^2(x,t)],$$
$$p(x,t) = e^{\gamma t}\dot{u}(x,t). \tag{31.8}$$

Quantum discrete oscillator chain with two frequencies can be considered with the help of the integrals of the motion method in the case when its equations of motion have the form

$$\dot{u}_{2n} = \Omega_1^2(u_{2n+1} + u_{2n-1} - 2u_{2n}),\dot{u}_{2n+1} = \Omega_1^2(u_{2n+2} + u_{2n} - 2u_{2n+1}) \tag{31.9}$$

These equations describe the model of atoms with different masses located in odd and even knots of the lattice. It is nothing but the model (see, for example, [30]) of acoustic and optic branches of the dispersion law in solids. The consideration of this model with two frequencies will be published in detail in other papers.

If one takes into account the possibility of exciting the discrete chain by an external force, then one gets the Hamiltonian

$$H = \frac{1}{2}\sum_{n=1}^{N}[p_n^2 + \Omega^2(t)(u_n - u_{n+1})^2 + \Omega_0^2(t)\,u_n^2 - f_n(t)\,u_n], \tag{31.10}$$

and its continuous analog

$$H = {}^1/_2 \int dx\,[p^2(x,t) + c^2(t)(\partial u/\partial x)^2 + \Omega^2(t)\,u^2(x,t) - f(x,t)\,u]. \tag{31.11}$$

The equations of motion in the discrete case are as follows:

$$\ddot{u}_n + (\Omega_0^2 + 2\Omega^2)\,u_n - \Omega^2(u_{n+1} + u_{n-1}) = f_n(t), \tag{31.12}$$

and for the continuous case we have

$$\ddot{u} = c^2 \partial^2 u/\partial x^2 + \Omega_0^2 (t)\, u + f. \tag{31.13}$$

A result of the parametric excitation in the models considered is the origin of correlated states of discrete and continuous strings, appearing as a set of correlated states of one-dimensional parametric normal modes.

32. GRAVITATIONAL WAVEGUIDE

In previous sections we have discussed the application of the theory of nonstationary and stationary quadratic quantum systems to the following problems: a quantum vibrational circuit, a Josephson circuit, a chain of atoms-oscillators, and a string. Mathematical aspects of these problems have practically the identical properties though their physical features are absolutely different. There is another problem, which is reduced to studying a simple special example of quadratic quantum systems. It is the problem of waves propagation (waves of different nature) through a waveguide. In the case of lightguides, the review of such problems solved on the basis of the theory of quantum integrals of the motion was given, e.g., in Reference [1,32]. But there exists an important problem posed in studies [1,33] and considered in [31] concerning the possibility of a specific waveguide effect for the electromagnetic waves (and also for neutrino beams) due to the presence of the gravitational field potential. From the point of view of mathematical method this problem coincides with that considered in the previous sections, but it is related to a quite different branch of physics.

As is known, the presence of the gravitational field influences the propagation of the electromagnetic radiation. This influence may be demonstrated, for example, in such phenomena as the curvature of light beams propagating near by the sun's surface and also in recently discovered effect of the "gravitational lens" in which it is possible to observe the "twins", i.e. two images of one and the same long distanced object.

The aim of this section is to call attention to the possible existence of still another effect, namely, a "gravitational waveguide" [31,33]. One of its manifestations may be an anomalously high brightness of remote sources whose rays travel to the observer along a waveguide axis and are focused in the gravitational potential produced by the uniformly distributed matter making up the "waveguide."

We first recall the basic equations used in waveguide theory. For our purpose it suffices to restrict ourselves to the paraxial approximation. Then, as is well known, the propagation of wave beams is well described using a parabolic equation. This equation is obtained from the Helmholtz equation for some arbitrary monochromatic component of the electric field

$$\frac{\partial^2 E}{\partial z^2} + \frac{\partial^2 E}{\partial x^2} + \frac{\partial^2 E}{\partial y^2} + k^2 n^2 (x, y, z)\, E = 0 \tag{32.1}$$

(where n is the refractive index of the medium and k is the wave number in vacuum) by introducing a rapidly oscillating phase factor of type exp(iknz) where the z axis is taken to be the beam propagation direction. Specifically, an ampltidue Ψ(x,y,z) slowly varying in space is introduced, given by the relations

$$E = n_0^{-1/2} \Psi \exp\left(ik \int n_0(z)\, dz\right), \quad n_0 \equiv (0, 0, z). \tag{32.2}$$

Next, discarding in Equation (32.1) the terms containing $\partial^2\Psi/\partial z^2$, dn_0/dz, and d^2n_o/dz^2, we can rewrite it in the form

$$i\lambda\frac{\partial\Psi}{\partial\xi} = -\frac{\lambda^2}{2}\left(\frac{\partial^2\Psi}{\partial x^2}+\frac{\partial^2\Psi}{\partial y^2}\right)+\frac{1}{2}\left[n_0^2(z)-n^2(x,y,z)\right]\Psi, \tag{32.3}$$

where the variable ξ is defined as

$$\xi = \int^z \frac{dz}{n_0(z)}, \tag{32.4}$$

and $\lambda=\Lambda/2\pi=k^{-1}$. Equation (32.3) is the analog of the Schrodinger quantum equation that describes two-dimensional motion of a particle of a unit mass in a potential $U(r)=(n^2_o-n^2)/2$, with the role of Planck's constant assumed by the wavelength Λ.

An ideal waveguide corresponds to a situation, when the potential U(r) is quadratic in the transverse coordinates x and y. The beam wave can then be focused into a point. Moreover, the beam does not spread out without limit, but its width either oscillates or remains constant. Production of such waveguides in real life is a rather difficult task, for one must be able to manufacture an optic fiber in which the refractive index decreases continuously like $n(r)=(n^2_o-\Omega^2r^2)^{1/2}$ as the distance from the axis increases.

We now recall that the influence of static gravitational fields on an electromagnetic field can be described in the framework of the Maxwell equations, by introducing effective dielectric and magnetic permeabilities which can be expressed in terms of the metric tensor g_{ik} as follows [34]:

$$\varepsilon_{ik} = \mu_{ik} = -g_{00}^{-1/2}[\det\|g_{ik}\|]^{1/2}\, g^{ik} \tag{32.5}$$

In weak gravitational fields

$$g_{00} \approx 1+2\varphi/c^2,\ g_{ik} \approx -\delta_{ik}(1-2\varphi/c^2) \tag{32.6}$$

$\varphi(r)$ being the Newton gravitational potential. Thus, the effective potential of the Schrodinger-like equation (32.3) equals

$$U(r) = 2[\varphi(x,y,z)-\varphi(0,0,z)]/c^2 \tag{32.7}$$

Now imagine the existence of "filamentary" mass distributions with constant density ρ inside the filaments. The potential (32.7) inside a filament is then obviously a quadratic function of the transverse coordinate $r=(x^2+y^2)^{1/2}$

$$U(r) = (1/2)\Omega^2r^2,\ \Omega^2 = 4\pi\rho G/c^2 \tag{32.8}$$

But then such a filament is precisely the ideal waveguide. Consequently, if radiation from a remote source travels to the observer inside such a filament it will, by virtue of the waveguide effect, not attenuate like $1/R^2$. Moreover, the source brightness will turn out to be much stronger than the brightness of analogous objects located at the same distance (i.e., for the same red shift). In other words, the apparent energy released by the source (say, a quazar) will be anomalously large. This property can be used either to explain the anomalous energy released by remote sources, or to observe similar "filaments."

These results were obtained earlier in References [31-33], but without the extra factor 2 in expression (32.7) for the effective potential, the contribution of the space components of the metric tensor to the effective refractive index was not taken into account (the same difference takes place in the Newton and Einstein formulas for the light deflection by the Sun).

We now proceed to quantitative estimates. A spherical wave from a source $E=(1/R)\exp(ikR)$ can be represented in the paraxial approximation in the form

$$E(z,r) = \frac{1}{z}\exp\left[ikz + \frac{ikr^2}{2z} - \frac{r^2}{2z^2}\right], \tag{32.9}$$

if the following expansion is used:

$$R = (z^2 + r^2)^{1/2} \approx z\,(1 + r^2/2z^2), \qquad r \ll z.$$

(Since in real cases n_0 differs extremely little from unity, we shall set hereafter $n_0=1$ in Equations (32.2) and (32.3), i.e., make the variables z and ξ identical.)

Assume that the starting point of a filament of length L is at a distance H from a source shifted away by a distance α from the filament axis in the x direction. The amplitude of the field Ψ [see Equation (32.2)] entering the waveguide is then

$$\Psi_{\text{in}} = \frac{1}{H}\exp\left\{\frac{(ikH-1)}{2H^2}\left[(x-a)^2 + y^2\right]\right\} \tag{32.10}$$

(so far in this case $R=(H^2+y^2+(x-\alpha)^2)^{1/2}$). The field at the exit from the filament is calculated from the equation

$$\Psi_f(x,y,H+L) = \int dx_1\,dy_1 G(x,y,H+L,x_1,y_1,H)\,\Psi_{\text{in}}(x_1,y_1,H), \tag{32.11}$$

where the funtion G is the propagator (Green's function) of Equation (32.3). For the potential (32.8), this equation is in fact the Schrodinger equation for a harmonic oscillator, so that its propagator is well known:

$$G(x,y,H+L,x_1,y_1,H) = \frac{\Omega}{2\pi i\lambda\sin(\Omega\lambda)} \times$$
$$\times \exp\left\{\frac{i\Omega}{2\lambda\sin(\Omega L)}\left[\cos\Omega L(x^2+y^2+x_1^2+y_1^2) - 2(xx_1+yy_1)\right]\right\}. \tag{32.12}$$

The integral (32.11) is then Gaussian and can be evaluated exactly

$$\Psi_f = \Omega H\,[\Omega H^2\cos(\Omega L) + (H+i\lambda)\sin(\Omega L)]^{-1}\exp\{-{}^1/_2\,[1 - ikH -$$
$$- ik\Omega H^2\,\mathrm{ctg}\,(\Omega L)]^{-1}\,[(x^2+y^2)\,[(\Omega Hk)^2 -$$
$$- \Omega k\,(i+kH)\,\mathrm{ctg}\,(\Omega L)] + 2xa\Omega k\sin^{-1}(\Omega L)\,(1+kH) -$$
$$- a^2\Omega k\,(i+kH)\,\mathrm{ctg}\,(\Omega L)]\}. \tag{32.13}$$

Note the parameter H drops out of the denominator of the pre-exponential factor if the length L satisfies the condition

$$\mathrm{tg}\,(\Omega L) = -\Omega H. \tag{32.14}$$

Assume that $\Omega H \ll 1$. Then $\mathrm{tg}(\Omega L)\approx 0$ and the minimum possible value L π/Ω. (The solution L=0 is untenable, since $\mathrm{tg}(\Omega L)$, while small, must be negative according to (32.14). Thus we have $\cos(\Omega L)$ -1, so that $\sin(\Omega L)$ $-\mathrm{tg}(\Omega L)=\Omega H$. In such a case, Equation (32.13) assumes the much simpler form

$$\Psi_f = \frac{1}{i\lambda}\exp\left\{-\frac{(1+i\lambda/H)}{2\lambda^2}\left[(x+a)^2+y^2\right]\right\}. \tag{32.15}$$

Consequently radiation emerging from a point with coordinates $(\alpha,0,0)$ is focused in a small vicinity (radius on the order of the wavelength) of a point with coordinates $(-\alpha,0,H+L)$. This means that rays from an extended source are focused inside the waveguide in such a way that at a distance L satisfying the condition (32.14) an inverted (since the signs of the coefficient α in (32.10) and (32.15) are opposite) formed image of the source, having the very same dimensions as the source.

Thus, the waveguide "draws," so to speak, the source closer to the observer: if the true distance to the source is R, its image brightness will correspond to that of a similar source but at the closer distance

$$R_{eff} = R - H - L.$$

If we do not assume that $\Omega H \ll 1$, we obtain in lieu of (32.15) the equation (in which we neglect the term $i\lambda/H$ compared with unity)

$$\Psi_f = \frac{(1+(\Omega H)^2)^{1/2}}{i\lambda}\exp\left\{-\frac{1+(\Omega H)^2}{2\lambda^2}\left[y^2+\left(x+\frac{a}{(1+(\Omega H)^2)^{1/2}}\right)^2\right]\right\}, \quad (32.16)$$

which states that in the general case the size of the image is decreased by $(1+(\Omega H)^2)^{1/2}$ times. The amplitude at each point increases here by the same factor, so that the brightness is $(R/R_{eff})^2$ times larger, as before.

In the opposite limiting case $\Omega H \gg 1$ we have $\mathrm{tg}\,\Omega L \to \infty$, so that $L \to \pi/2\Omega$. We see that the shortest focal length of the waveguide is

$$L_{foc} = \chi\pi/\Omega, \quad (32.17)$$

where x can range from 1/2 to 1. Hereafter we assume in the estimates that x=1, so that

$$L_f = (\pi c^2/2\rho G)^{1/2}. \quad (32.18)$$

Since the gravitational constant and the density of matter are small, the focal length is tremendous, for a density $\rho=10^{-29}$ g/cm^3, i.e., larger by an order of magnitude than the average density of the visible matter in the universe, we get

$$L_f \approx 1.5\cdot 10^{25}\ cm \approx 5\cdot 10^{24}\ Mps, \quad (32.19)$$

which is an order of magnitude larger than the visible part of the universe. Consequently, any hope of feasibility of the waveguide effect can be realistic if the density of matter inside the filaments is at least two orders of magnitude larger than the critical density 10^{-29} g/cm^3. But this leads to complications, since the absorption or scattering of light by the matter inside the filament increases with increase of density. Clearly, the focusing effect takes place so long as the following inequality is met

$$L_f \lesssim l_a = m_p/\rho\sigma \quad (32.20)$$

where m_p is the proton mass and σ is the cross section for absorption or scattering by hydrogen atoms (in a cold medium) or by electrons (in a hot fully ionized plasma). As a result, we obtain for the waveguide effect a certain critical density

$$\rho_* = 2Gm_p^2/\pi c^2\sigma^2. \quad (32.21)$$

If this cross section is assumed equal to the Thomson cross section (i.e., a hot plasma is considered):

$$\sigma = \sigma_T = \frac{8\pi}{3}\left(\frac{e^2}{m_e c^2}\right)^2 \approx 6.4\cdot 10^{-25}\ cm^2, \quad (32.22)$$

we obtain the critical density and the corresponding focusing length

$$\rho_T = 3\cdot 10^{-28}\ g/cm^3,\ L_{fT} = 10^4\ Mps \quad (32.23)$$

The critical density, however, can be increased by considering a cold gas. The scattering cross section is then strongly dependent on the frequency

$$\sigma = \sigma_T \, (\omega/\omega_0)^4, \tag{32.24}$$

where ω_0 is the atomic frequency; $\omega_0 = \mathrm{Ry}/\hbar \approx 2 \cdot 10^{16}$ sec^{-1}. In the radio band the scattering cross section can then be so small that practically no real restrictions are imposed on the density by absorption, and the focusing length, i.e., the minimum filament length, is determined only by the possible density of matter inside the filament. For densities $\rho \approx 10^{-24}$g/cm^3 (i.e., one proton per cubic centimeter), which can hardly be exceeded on a galactic scale, we obtain L≈100 Mpc, which is of the order of the cell dimensions in the large-scale structure of the universe. Therefore, the "filaments" can be gigantic strongly elongated galaxy clusters. A waveguide effect can arise if the sources are lcoated behind such a cluster on its axis, the observations are carried out in the radio band, and the source is invisible in the optical band.

We must also discuss conditions under which the gravitational deflection of the rays will not be "smeared out" by refraction due to the gradient of the bulk refractive index of the medium on account of the decrease of the density with increase of distance from the fiber axis (the density will not be strictly constant under real conditions). Assuming that the usual dielectric constant of a cold gas in the radio band (i.e., at $\omega < \omega_0$) is given by

$$\varepsilon = 1 + \frac{4\pi\rho e^2}{m_e m_p \omega_0^2}, \tag{32.25}$$

we obtain the value of the "waveguide potential" connected with the usual dielectric constant, on the periphery of the filament

$$U_{\Delta\rho} = 2\pi\Delta\rho e^2 / m_e m_p \omega_0^2, \tag{32.26}$$

where $\Delta\rho$ is the change of density. Comparing this equation with (32.8) we arrive at the upper bound of the filament thickness:

$$r > \left(2 \frac{\Delta\rho}{\rho} \frac{1}{G m_p m_e}\right)^{1/2} \frac{ec}{\omega_0}. \tag{32.27}$$

for $\Delta\rho/\rho \approx 1$ we then have $r > 10^{14}$ cm. Thus, there are in fact no real restrictions on the "filament" thickness, if it is recognized that the "filament" is a galaxy cluster.

The waveguide effect can take place not only for a cylindrical geometry of the masses, as considered above (axisymmetric waveguide) but also for flat geometry (planar waveguide), in the presence of a constant-density layer having a dimension in one direction (say, x) much smaller than the dimensions in the other two. The "waveguide potential" will then have the form

$$U(x) = 4\pi G \rho x^2 / c^2, \tag{32.28}$$

so that, accurate to a factor $(2)^{1/2}$, the focusing length remains the same as above. The role of such planar waveguides might be assumed by increased-density regions separating the cells in the large-scale structure of the Universe. The observed effect could be the following: the brightness of quasars whose radiation passes through the cell boundaries (along the boundaries) should be higher than the brightness of other quasars having the same red shift.

So far we have considered a situation in which the radiation from a point source entered a waveguide after traveling some distance in free space. In principle one can also conceive of another possibility, wherein a wide source is located directly at the entrance into the waveguide. Examples can be the known jets of matter ejected from galactic cores or possibly in nova outbursts.

Again approximating the entering wave beam by a Gaussian exponential, it is easy to show (see, e.g., [1]) that the beam width (which also determines, by virtue of the normalization condition, its intensity on the axis) will oscillate at double the "frequency" 2Ω in the direction of the propagation axis z. Consequently, the intensity on the axis will also oscillate about a certain mean value, but will not have a $1/R^2$ decrease. This will itself manifest again as an anomalously high brightness of the corresponding source.

There are no oscillations at all when the width of the initial beam σ, defined by the initial function

$$\Psi_{in} = A \exp\left[-(x^2 + y^2)/4\sigma^2\right], \tag{32.29}$$

is matched to the waveguide parameters in the following manner [1]:

$$\sigma^2 = \sigma_*^2 = \lambda/2\Omega = \lambda c\,(8\pi\rho G)^{-1/2} \tag{32.30}$$

(with a coefficient x=1 in (32.17)). The width is then quite acceptable even at a density $\rho=10^{-29}$ g/cm^3, i.e., at $L_{foc}\simeq 10^{29}$cm: for $\lambda=10^{-5}$ cm (red light) we have σ $5\cdot 10^{11}$ cm, which is only one order of magnitude larger than the sun's radius. It is important that in the considered case, there are no density-related restrictions on the "filament" length. The reason is that in the preceeding case the initial beam entering the waveguide was very wide, and a long filament was needed to significantly narrow it down. In the present case, however, if the width of the initial beam is already comparable with the matched width σ_* (32.30), the beam propagates in the ideal case without any change of width, whereas in the remaining cases its width will oscillate. Fortunately, as we have seen, the matched width is precisely of the order of the star diameter, and it depends very weakly on the "filament" density: $\sigma_* \propto \rho^{-1/4}$. Thus, if some jet of matter is emitted from a source such as an asymmetrically exploding star and is directed along the line of sight (in which case we do not in fact see the real jet), the probability of the waveguide effect becomes realistic enough. The most substantial restriction in this case is the requirement that the density be constant over a scale σ_*. It follows from (32.27) that at $r\approx\sigma_*$ the density variations should be smaller than

$$\frac{\Delta\rho}{\rho} < \frac{\lambda c m_e m_p}{e^2\lambda_0^2}\left(\frac{G}{32\pi\rho}\right)^{1/2}, \qquad \lambda_0 = \frac{c}{\omega_0} = \frac{c\hbar}{\mathrm{Ry}} \approx 1{,}5\cdot 10^{-6}\ \text{см}. \tag{32.31}$$

This condition can hardly be met in the optical band: for $\lambda=10^{-5}$ cm and $\rho=10^{-28}$ g/cm^3 we obtain the restriction $\Delta\rho/\rho<2\cdot 10^{-6}$. The restrictions are lifted starting with meter band of the radio waves (and higher). Consequently, even in the case presently considered the waveguide effect can appear only in the radio band, under the condition that the "filament" consists of cold (non-ionized) matter. If, however, the jet is an ionized gas, it is necessary to replace λ_o in (32.31) by wavelength λ of the radiation itself, and this condition is met in neither the radio nor the optical band.

Nevertheless, the jet cannot be short: since the brightness gain is purely geometric $(R/R_{eff})^2$, the jet length must be comparable with the distance to the source. It is not excluded, however, that the matter inside the waveguide can be in a nonequilibrium state and be capable of amplifying the radiation (a cosmic maser). Its length can then be substantially less.

The waveguide effect will not vanish if the axis of the "filament" or "string" is bent smoothly enough in space. Then, just as in the case of a gravitational lens, "twins" can be observed if part of the radiation comes to the observer directly from the source, and another part is captured by the bent waveguide. The "virtual" image can then turn out brighter than the "real" one.

Finally, we note that a waveguide effect can be observed, not only for electromagnetic radiation, but also for streams of neutrinos or other particles propagating inside the "filaments," since it is caused by the same mechanism – gravitational attraction. In addition, sufficiently strong clusters of elongated massive neutrinos could by themselves make up the most ideal waveguide, for in this case the gravitational effect would manifest itself in pure form, without factors that complicate the observation and are connected with absorption or refraction of electromagnetic waves in ordinary matter. In fact, the gravitational potential for the sphere with the constant matter density as well as for the cylinder with the constant matter density depends quadratically on the distance from the center of the sphere in the first case and on the distance from the cylinder axis in the second case. Therefore, any particle must perform a vibrational motion in this potential. Physically, the effect of focusing the photons in the cylindrical gravitational waveguide may be understood in the simplest way as follows. A photon with energy $\hbar E$ possesses the gravitational mass E/c^2. Thus it is attracted to the waveguide axis and vibrates in the perpendicular direction while propagating along the waveguide axis. Because of this, the photon beam does not go out of the waveguide. This description is the qualitative one, but it corresponds to the strict equations for the electromagnetic field in the presence of the gravitational potential. These qualitative arguments may be also applied for all the other massless fields. Since the quanta of all fields possess energy, the same waveguide effect must exist for them in the presence of the gravitational potential. This fact may have an essential meaning for neutrinos and gravitational waves mentioned above.

Let us make some additional remarks about the observation of the possible effects connected with the hypothesis of the existence of the gravitational waveguides in the Universe. So, if there is such a long distanced source (for example, a quasar) and there is a matter producing a waveguide between the source and the observer, then from the direction to the quasar, besides the anomally large photon fluxes, the intensive fluxes of neutrinos and intensive pulses of gravitational waves may come. This circumstance may change in a positive direction the evaluations of the maximal intensities of the gravitational waves for gravitational antennas, if the antenna is oriented in a proper direction. The presence of the waveguide may be found independently using completely different measurements in the optical or radio-frequency bands. Analogously in the case of the existence of the gravitational waveguides, the neutrinos detected on the earth could possess some angular anisotropy connected with the direction of the waveguide axis (cylindrical or planar) to the observer.

If one tries to explain the huge production of energy by quasars as the apparent effect due to the existence of the waveguide between the quasar (with a moderate energy production) and the observer on the earth, then some difficulties arise. Indeed, about one thousand quasars are known. Then the same thousand "filaments" with the axis oriented strictly to the earth must exist, if the waveguides have cylindrical geometry. Though the probability of the existence of such waveguides could be nonzero, nonetheless it seems to be very small.

However, the problem can be softened in the case of planar gravitational waveguides if one assumes that both the sources and the observer are placed in the plane of the waveguide. This explanation would give the extra arguments, pro and contra, different scenarios of the Universe evolution. In particular if the quasars and the observer are located on the surface of a "bubble," this surface could serve the planar waveguide. Anomaly large brightness of the quasars could testify indirectly to the scenario of the Universe evolution according to which all the quasars are placed on the bubble surface.[1]

Another remark on the consequences of the gravitational waveguide hypothesis is related to the possibility of existing the observable objects on the sky sphere, which could possess anomalously large (in comparison with their neighbors) angular motion velocities. Such a phenomenon could mean that one observes not the object itself, but its image transmitted through the moving gravitational waveguide. The waveguide itself could change its form. In this case, the image of the object could move with essentially different angular velocity than that of the observable neighbor objects whose light reaches the Earth directly (not through the waveguide). The discovery of long distanced objects with anomalous velocity could support the hypothesis of the gravitational waveguide effect. At last, the evolution of the waveguide (its destruction or change of the axis direction from the orientation to the Earth) could produce the effect of the disappearance (in the opposite case of the appearance) of the observed object. In this connection it seems interesting to analyze the results of the astronomical observations for a long period. The appearance or disappearance of some observable objects could not necessarily mean their real birth or death. There also exists the possibility to imitate their appearance or disappearance due to the appropriate change of the waveguide through which the object is observed. Let us note that the example of the gravitational lens effect (experimentally confirmed recently) gives hope to the possibility of experimentally discovering the gravitational waveguide by some methods, since the arguments in favor of the existence of gravitational lenses in the Universe are similar to the arguments in favor of the existence of gravitational waveguides. If the gravitational waveguide hypotheses is confirmed, it will essentially change our notion on the structure and distribution of the matter in the Universe.

Let us note that neutrino astronomy is only now at the beginning of its evolution. But in its further development, due to very small absorption, the neutrino could give essentially more informaiton about the gravitational "neutrinoguides" than one can receive from the photons observations.

33. CONCLUSION

In the previous Sections we have considered the possibilities to apply the correlated and coherent states in some physical systems: resonant circuit, Josephson junction, string, and discussed the possible role of the gravitational waveguide in the Universe. Let us note that there are some aspects which have not been discussed in this paper and they deserve additional consideration. So, in the case of Josephson junction it is interesting to study the possible effects connected with excitation of the junction, for example, by the electromagnetic or sound radiation which model the time-dependence of the capacitance and inductance of the equivalent quantum resonant circuit.

The interesting effect is the analog of the Casimir effect for the quantum resonant circuit and Josephson junction. The analogy is related to the fact that the energy of the

---

[1] The connection of the anomal brightness of the quasars with the hypothesis that they are placed on the surface of the "bubble" and with the hypothesis of the existence of the planar waveguide effect has been pointed out to the authors by Professor L. Michel. The authors are grateful for a fruitful discussion.

ground state $E_0=\hbar\omega/2$ depends on the inductance and capacitance through the frequency. The capacitance of the quantum resonant circuit, if it is considered in the frame of the model of two small plates, is inverse proportional to the distance of the ground state of the resonant circuit and Josephson junction decreases with the distance between the plates as the root squared from the distance. This corresponds to the attraction, the attraction force being inverse proportional to the root squared from the distance. This dependence on the distance differs from the dependence in the usual Casimir effect which leads to the Vander-Vaals forces. In the case when both effects are present, i.e., the plates serve as the capacitance plates in the quantum resonant circuit, the superposition of two effects takes place. But these effects are different ones: the first – Casimir effect – is due to the dependence of the ground state energy on the effective capacitance of the quantum resonant circuit. The second effect is, as well as the first one, a pure quantum effect and tends to zero when the Planck constant tends to zero. The inductance also tends to decrease, i.e., the quantum nature of the oscillation in the circuit and Josephson junction provides the squeezing effect (i.e. the linear sizes tend to decrease) of the inductance. The dependence of the force causing this additional quantum squeezing is proportional to the inductance in the power 3/2. The parametric excitation of the Josephson junction from the ground state is equivalent to the nonstationary Casimir effect.

Another effect to be studied is the effect of excitation of the connected Josephson junctions with the help of the parametric action on one of these junctions. So, acting on the junction by the laser light or by the sound wave, or changing the parameters of one junction by another, we excite the other junction connected with the first one and they both can be excited into the correlated state. This phenomenon is an analog of the nonstationary Casimir effect. Nevertheless, for the usual Casimir effect, the phenomenon can be easily understood. Since skwid, which is the set of the connected Josephson junctions, is described at small values of the variables by the quadratic Hamiltonians, all the properties of the quadratic quantum systems [1] may be demonstrated for these devices. The set of the connected resonant circuits in the continuous limit provides the line to which it is also described by the quadratic quantum Hamiltonian.

Let us note that in recent publications [35-40] such quadratic systems as a charge moving in the electromagnetic field [35-37], the damped oscillator [36], vibrational principle for the equations of motion [38], and the paraxial optical beams [39,40] have been considered. In studies [40-43] the model of the quantum resonant circuit was applied to describe the Josephson junction. In studies [45-51] different models of the quantum oscillator strings and chains have been considered.

Accepted for publication in December 1988

REFERENCES

1. Dodonov, V.V, Man'ko, V.I. "Invariants and correlated states of nonstationary quantum systems" Proc. Lebedev Phys. Inst., Moscow: Nauka, 1987, vol. 183, pp. 71-181.
2. Dodonov, V.V., Man'ko, V.I., Man'ko, O.V. "Quantum nonstationary oscillator" Proc. Lebedev Phys. Inst., Moscow: Nauka, 1989, vol. 191, pp. 185-224.
3. Leznov, A.N., Man'ko, V.I.,Savel'yev, M.V. "Soliton solutions to non-linear equations and group representation theory" Proc. Lebedev Phys. Inst., Moscow: Nauka, 1986, vol. 165, pp. 65-207.

4. Glauber, R.J. "Coherent and incoherent states of the radiation field" *Phys. Rev.*, 1963, vol. 131, no. 6, pp. 2766-2788.
5. Schrodinger, E. "Zum Heisenbergschen Unscharfeprinzip" *Ber. Kgl. Akad. Wiss.*, Berlin, 1930, S. 296-303.
6. Robertson, H.P. "A general formulation of the uncertainty principle and its classical interpretation" *Phys Rev. A*, 1930, vol. 35, no. 5, pp. 667.
7. Dodonov, V.V., Kurmyshev, E.V., Man'ko, V.I. "Correlated coherent states" Proc. Lebedev Phys. Inst., Moscow: Nauka, 1986, vol. 176, pp. 128-150.
8. Stoler, D. "Equivalence classes of minimum uncertainty packets" *Phys. Rev. D.*, 1970, vol. 1, no. 12, pp. 3217-3219.
9. Dodonov, V.V., Man'ko, V.I. "Universal invariants of quantum systems and generalized uncertainty relation" Proc. Second Intern. Seminar on Group Theoretical Methods in Physics, Zvenigorod (Russia), 1982, Moscow: Nauka, 1983, vol. 2, pp. 11-34.
10. Caldirola, P. "Forze non concervative nella meccanica quantistica" *Nuovo cim.*, 1941, vol. 18, no. 9, pp. 393-400.
11. Kanai, E. "On the quantization of the dissipative systems" *Progr. Theor. Phys.*, 1948, vol. 3, no. 4, pp. 440-441.
12. Dodonov, V.V., Man'ko, V.I. "Wigner functions of a damped quantum oscillator" Proc. Second Intern. Seminar on Group Theoretical Methods in Physics, Zvenigorod (Russia), 1982, Moscow: Nauka, 1983, vol. 2, pp. 109-122.
13. Dodonov, V.V., Man'ko, O.V. "Relaxation of quantum particle in magnetic field" *Teor. i Mat. Fizika*, 1985, vol. 65, no. 1, pp. 93-107.
14. Dodonov, V.V., Man'ko, O.V. "Quantum damped oscillator in a magnetic field" *Physica A*, 1985, vol. 130, no. 1/2, pp. 353-366.
15. Malkin, I.A., Man'ko, V.I. "Dinamicheskie simmetrii i kogerentnye sostoyaniya kvantovykh sistem" [Dynamical symmetries and coherent states of quantum systems]Moscow: Nauka, 1979, p. 320.
16. Dodonov, V.V., Man'ko, V.I. "Linear adiabatic invariants and Berry phase" Proc. Third Workshop on quantum optics, Dubna (Russia), 1988, Singapore: World Sci. Publ., 1989.
17. Malkin, I.A., Man'ko, V.I., Trifonov, D.A. "Linear adiabatic invariants and coherent states" *J. Math. Phys.*, 1973, vol. 14, no. 5, pp. 576-582.
18. Berry, M.V. "Topological phases for quantum system" *J. Phys. A.*, 1985, vol. 18, no. 1, pp. 15-27.
19. Engeneer, M.H., Ghosh, G. "Berry's phase as the asymptotic limit of an exact evolution: An example" *J. Phys. A*, 1988, vol. 21, no. 2, pp. L95-L98.
20. Dodonov, V.V., Kurmyshev, E.V., Man'ko, V.I. "Generalized uncertainty relation and correlated coherent states" *Phys. Lett. A*, 1980, vol. 79, no. 2/3, pp. 150-152.
21. Kurlsrud, R.M. "Adiabatic invariants of the harmonic oscillator" *Phys. Rev.*, 1957, vol. 106, pp. 205-207.
22. Ghosh, G., Dutta-Roy, B. "The Berry phase and Hannay angle" *Phys. Rev. D*, 1988, vol. 37,no. 6, pp. 1709-1711.
23. Feynman, R., Hibbs, A. "Kvantovaya mechanika i integraly po traektoriyam" [Quantum mechanics and path integrals] Moscow: Mir, 1968, p. 382.
24. Yurke, B. "Squeezed-state generation using a Josephson parametric amplifier" *J. Opt. Soc. Amer. B*, 1987, vol. 4, no. 10, pp. 1551-1557.
25. Josephson, B.D. "Supercurrent through barriers" *Adv. in Phys.*, 1965, vol. 14, no. 56, pp. 419-451.
26. Likharev, K.K. "Vvedenie v dinamiku dzhozefsonovskikh perekhodov" [Introduction in dynamics of Josephson junctions] Moscow: Nauka, 1985, p. 320.

27. Golub A.A., Grimal'skiy, O.V. "Quantum energy levels of two contact interferometer of direct current" *Sov. Phys. JETP*, vol. 94, no. 2, pp. 733-743.
28. Lempitskiy, S.V. "Infrared radiation influence on critical current of Josephson tunnel junction" *Sov. Phys. JETP*, 1988, vol. 94, no 11, pp. 331-341.
29. Henly, E.M., Thirring, W. "Elementary quantum field theory," New York, etc.: McGraw-Hill Company, Inc., 1962.
30. Kittel, C. "Introduction to solid state physics" New York: John Wiley & Sons, Inc.; London: Chapman & Hall, Ltd., 1956.
31. Dodonov, V.V., Man'ko, V.I. "Gravitational waveguide" Preprint, Lebedev Phys. Inst., no. 255, Moscow, 1988, p. 13; *J. of Soviet Laser Research*, Plenum Publ. Co., 1989, vol. 10, no. 3, pp. 240-247.
32. Man'ko, V.I. "Invariants and coherent states in fiber optics" In: *Lie Methods in optics*, eds. J.S. Mondragon, K.B. Wolf, N.Y., etc. Springer Verlag, 1986, pp. 193-207. (Lect. Notes Phys.; no. 250).
33. Dodonov, V.V., Man'ko, V.I. "Gravitational waveguide and coherent states of radiation" Abstract of 7-th conference on gravitation, Erevan: University Press, 1988, pp. 478-479.
34. Landau, L.D., Lifshits, E.M. "Teoriya polya" [Field theory], Moscow: Nauka, 1973, p. 329.
35. Marmo, G., Rubano, C. "Alternative Lagrangians for a charged particle in a magnetic field" *Phys. Lett. A*, 1987, vol. 119, pp. 321-324.
36. Yeon, K.H., Um, C.I., George, T.F., "Coherent states for the damped harmonic oscillator" *Phys. Rev. A*, 1987, vol. 36, pp. 5287-5291.
37. Serimaa, O.T. "Gauge independent Wigner functions: General formulation" *Phys. Rev. A*, 1986, vol. 33, pp. 2913-2927.
38. Osborn, T.A., Molzahn, F.H. "Structural connections between two semiclassical approximations: WKB and Wigner-Kirkwood approximation" *Phys. Rev. A*, 1986, vol. 34, pp. 1696-1707.
39. Dodonov, V.V., Man'ko, O.V. "Universal invariants of paraxial optical beams" Proc. Third Intern. Seminar on Group Theoretical Methods in Physics, Yurmala (Russia), 1985, Moscow: Nauka, 1986, vol. 2, pp. 432-441.
40. Dodonov, V.V., Man'ko, O.V. "Universal invariants of paraxial optical beams" "Komp'yuternaya optika" [Computer optics] eds. A. Prokhorov and E. Velikhov. Moscow: Intern. Center of Scientific and Technological Information of Socialist Countries, 1987, issue 1, Physical foundations, pp. 84-90.
41. Dodonov, V.V., Man'ko, V.I., Man'ko, O.V. "Correlated states in quantum electronics (resonant circuit)" Preprint, Lebedev Phys. Inst., no. 89, Moscow, 1988, p. 16; *Journal of Soviet Laser Research*, Plenum Publ. Co., 1990, vol. 10, no. 5, pp 413-420.
42. Dodonov, V.V., Man'ko, O.V. "Correlated states in quantum electronics (oscillatory circuit)" Interaction of Electromagnetic Field with Condensed Matter (Directions in Condensed Matter Physics - vol. 7), Eds. N.N. Bogolubov, A.S. Shumovsky and V.I. Yukalov, Singapore: World Sci. Publ., 1990, pp. 310-325.
43. Dodonov, V.V., Man'ko, V.I., Man'ko, O.V. "Correlated states and quantum noises of resonant circuit" Abstracts of Third Conference on Quantum Metrology and Fundamental Physical Constants, Leningrad, 1988, Leningrad: Gosstandart press., 1988, p. 217.
44. Man'ko, O.V. "Correlated states of quantum chain", *Group Theoretical Methods in Physics*, Proc. of XVIII Intern. Colloquium, Moscow, 1990, eds. V.V. Dodonov, V.I. Man'ko, N.Y., etc.: Springer-Verlag, 1991, pp. 461-468 (Lect. Notes in Phys., no. 382).
45. Man'ko, O.V. "Coherent states of the quantum parametric damped chain" in: "*Quantum field theory, quantum mechanics and quantum optics. Pt. 1 Symmetries*

and algebraic structures in physics. Proc. of the XVIII Inter. Colloquium on Group Theoretical Methods in Physics, Moscow 1990", eds. V.V. Dodonov, V.I. Man'ko, N.Y., Nova Science Publ., 1991, pp. 237-243 (Proc. Lebedev Phys. Inst., vol. 187).

46. Dodonov, V.V., George, T.F., Man'ko, O.V., Um, C.I., Yeon, K.H., "Propagators for quantum oscillator chains" In:"Proc. of Inter. Workshop on Squeezed and Correlated States, Moscow, 1990" *J. of Soviet Laser Research,* Plenum Publ. Co., 1991, vol. 12, no. 5, pp. 385-395.
47. Dodonov, V.V., Man'ko, O.V., Man'ko, V.I. "Nonstationary parametric chain of oscillators" Proc. Lebedev Phys. Inst., Moscow: Nauka, 1992, vol. 208, pp. 174-206.
48. Dodonov, V.V., Man'ko, O.V., Man'ko, V.I. "Correlated states of quantum oscillator and quantum chain with δ-kicked frequency" Preprint Lebedeb Phys. Inst., Moscow, 1992, no. 139, p. 34 *J. of Soviet Laser Research,* Plenum Publ. Co., 1992, vol. 13, no. 3.
49. Dodonov, V.V., George, T.F., Man'ko, O.V., Um, C.I., Yeon, K.H. "Exact solutions for a mode of the electromagnetic field in resonant with time-dependent characteristics of the internal medium" in: Proc. of Intern. Workshop "Squeezing groups and quantum mechanics" Baku, 1991 "*J. of Soviet Laser Research,* Plenum Publ. Co., 1992, vol. 13, no. 4.
50. Dodonov, V.V., Man'ko, O.V., Man'ko, V.I. "Squeezing in quantum parametric chain" *Nuovo cim. A,* 1991.
51. Man'ko, O.V. "Quantum parametric chain in Wigner representation" in: "Abstracts of Second Intern. Wigner Symposium, Goslar, 1991" Arnold Sommerfeld Inst., Technical University of Clausthal and Department of Mathematics, Florida Atlantic University, p. 65, report 17P. 33.

SUBJECT INDEX